Human Impacts on Amazonia

Ethnobotanist Dr. Darrell Addison Posey (1947–2001) with two Kayapo friends on a forest expedition in Kayapo, Gorotire, Amazon–Brazil, to learn about the uses of medicinal plants in Amazonia. Dr. Posey was a champion for the intellectual property rights of indigenous peoples. His work was instrumental in acknowledging a series of rights for traditional peoples that were included in the 1992 biodiversity convention at the Rio Earth Summit. (Photo: Mark Edwards/Still Pictures)

Human Impacts on Amazonia

*The Role of Traditional Ecological Knowledge
in Conservation and Development*

Edited by
Darrell Addison Posey and Michael J. Balick

COLUMBIA UNIVERSITY PRESS   NEW YORK

COLUMBIA UNIVERSITY PRESS
Publishers Since 1893
New York   Chichester, West Sussex

Copyright © 2006 Columbia University Press
All rights Reserved

Library of Congress Cataloging-in-Publication Data

Human impacts on Amazonia : the role of traditional ecological knowledge
   in conservation and development / edited by Darrell Addison Posey and
   Michael J. Balick.
          p.   cm. — (Biology and resource management series)
      Includes bibliographical references and index.
      ISBN 0-231-10588-6 (alk. paper) — ISBN 0-231-10589-4 (pbk. : alk. paper)
         1. Indigenous peoples — Ecology — Amazon River Region.
   2. Traditional ecological knowledge — Amazon River Region.   3. Indians of
   South America — Ethnobotany — Amazon River Region.   4. Nature — Effect of
   human beings on Amazon River Region.   5. Soil degradation — Amazon River
   Region.   6. Environmental degradation — Amazon River Region.   7. Amazon
   River Region — Environmental conditions.   I. Posey, Darrell Addison, 1947–2001.
   II. Balick, Michael J., 1952–   III. Biology and resource management in the
   tropics series.

GF532.A4H85   2006
304.2'09811 — dc22   2005053931

∞

Columbia University Press books are printed on
permanent and durable acid-free paper.
Printed in the United States of America

c   10   9   8   7   6   5   4   3   2   1

p   10   9   8   7   6   5   4   3   2   1

Contents

Preface

Amazonia is usually thought of as a vast wilderness—a huge "empty" region, "full" of pristine nature. But for the contributors to this volume, the Amazon Basin is indelibly sculpted by human impacts on its history, landscapes, and ecosystems. And for the indigenous peoples of Amazonia, the forests, savannas, hills, and streams are their homes, gardens, backyards, hunting reserves, and spiritual retreats.

Charles Clement (chapter 3) describes a considerable variety of pre-Columbian "domesticated landscapes" that are defined using genetic, botanical, and ethnobotanical methods. He also proposes that through indigenous coevolution with plants, Amazonia became one of the major world centers for crop domestication.

But domestication cannot be used as the only yardstick for human manipulation of plants. Jan Salick (chapter 12) analyses the conundrum some native Amazonians still face when deciding whether to collect or cultivate plants such as ipecac (*Carapichea ipecacuanha* (Brot.) L. Andersson). In some cases it has been preferable to improve useful plants in the setting of a domesticated landscape rather than to bring them into human care–dependent "full" domestication. The importance of these nondomesticated (or semidomesticated) species has been underestimated and underesteemed by scientists.

The history of Amazonia for most people begins with the "discovery" of the New World in 1492. Tragically, after European contact and colonialism Amazonia was marked forever by a long sequence of human and ecological disasters. John Hemming (chapter 1) traces some of Amazonia's colonial history and its terrible impact on native Amazonians, and points out that postcolonial development continues to wreak havoc upon the subregions and their peoples. A major excrescence of colonization was the introduction of diseases by missionaries, migrants, and slaves and their masters. Nancy Leys Stepan (chapter 2) reminds us that the nature of the tropics in Amazonia has been to some degree defined to the outside world by the early visitors who constructed its identity—in terms both positive and negative.

Disease is not the only foe of modern Amazonia and Amazonians. Elizabeth Allen (chapter 4) details one of many environmental disasters that have left garish wounds on the area—the devastating forest fires of 1998 in the Brazilian state of Roraima. James Ratter, Felipe Ribeiro, and Samuel Bridgewater (chapter 5) discuss how *cerrado* vegetation is under extraordinary threat of destruction. And Christopher Barrow (chapter 6) points out how the luxurious wetlands and riverine ecosystems of Amazonia are disappearing under the unseeing eyes of the rest of the world.

Peter Furley (chapter 7) and Stephen Nortcliff (chapter 8) warn us that the soils that literally underlie the existence of Amazonia are being eroded and lost at unimaginable rates. Development projects mushroom throughout the region, with developers giving little thought to the degrading impact on the soil. If developers care so little about such a fundamental element of Amazonia, it should come as no surprise that they show little or no sensitivity toward native Amazonians and biodiversity conservation.

Such insensitivity makes no economic sense according to Philip Fearnside (chapter 9), who argues that standing forests (and the diversity of life-forms associated with them) make far more economic sense than do charred, chainsawed forests. Fearnside proposes a development paradigm based on the benefits accrued from providing environmental services for forests and forest products.

"Benefits" usually connotes economic gain, rather than social or ecological progress. Since colonization, Amazonia has been viewed as a source of raw materials and natural products. The proliferation of searches in tropical ecosystems for commercially valuable genetic and biochemical resources, especially for the pharmaceutical, biotechnological, and agricultural industries, has made "bioprospecting" a household word. Frequently cited figures illustrate the enormous market for natural-product-based pharmaceuticals (US$43 billion per year) and seeds derived from traditional crop varieties (US$50 billion per year); the calculations for other natural compounds (those useful for hair and body products, oils, essences, dyes, colorings, etc.) are equally impressive.

Interest in traditional knowledge about the uses of flora and fauna has also dramatically increased because some companies predict that research and development costs could be cut by as much as 40 percent if traditional knowledge is used to help guide research and development efforts. Such a savings could be remarkable for global pharmaceutical companies that usually invest US$200 million or more in research and development for a single new drug. Commercial production of medicinal plants, whether for research or as raw materials, provides additional sources of income to those involved. But the issue is more complicated than that. Claudio Pinheiro (chapter 13) illustrates the problem, citing the large-scale production of the medicinal forest plant *jaborandi* (*Pilocarpus microphyllus*), which has benefited the pharmaceutical company but has left the local communities impoverished.

William Milliken (chapter 15) alerts us that outsiders have been given too much free rein over the resources of native Amazonians. Instead, outsiders should be a source of guidance for local peoples' efforts. He explains that one indigenous group, the Yanomami, needs advice and support for their attempts to sell their own herbal remedies and develop their own "science parks" for ecological and ethnoecological study.

But entering the market economy, particularly outside of one's immediate region, is not so simple. Alcida Ramos (chapter 16) cautions that most indigenous communities have very little or no experience with the financial and legal relationships necessary to exploit market possibilities. And Stephen Nugent (chapter 17) sounds a powerful warning message: the notion of indigenous knowledge as a material "resource" that can simply be seized for efficient exploitation is profoundly flawed. Local knowledge is embedded in a web of social, cultural, and spiritual values that can be weakened or destroyed by market forces.

Thus successful projects require policy and legal supports that guarantee respect for local values. Michael Richards (chapter 11) warns that attempts by donors and funders to promote market-based forestry tend to undermine indigenous ("gift economy") institutions by allowing the term "forestry" to refer only to planted trees. That ignores indigenous strategies for forest utilization and conservation, which are embedded in more complicated integrated resource-management systems that value forests for more than their wood and timber products.

Perhaps the best way to use traditional knowledge is to let the local people themselves guide and control its use and application. Christine Padoch and Miguel Pinedo-Vásquez (chapter 10), for example, detail how the "invisible practices" of villagers in the Napo-Amazon floodplain are effective in developing sustainable commercial timber extraction. Mark Harris (chapter 14) describes how the practices of peasant riverine communities in the Lower Amazon are an efficient means to conserve biodiversity. But Harris also concludes that "The extent to which 'traditional ecological knowledge' may have a role in the future of the floodplains depends almost entirely on the ability of its inhabitants to secure legal ownership of land and water areas

in their vicinity." Ramos and Milliken emphasize that land, water, and resource rights are not the only legal protections required. Native Amazonians
also must have protection of their intellectual property, genetic resources,
and traditional knowledge.

Why have such basic rights and guarantees of benefits been so slow in
coming? Although the world now recognizes the economic potentials for
traditional resources, Amazonian governments continue to foster policies
that destroy tropical ecosystems and the native Amazonians (e.g., *índios,
campesinos, caboclos, peoes, colonos,* and *caicaras*) who live in, manage, and
conserve them. Since the beginning of colonial domination, indigenous and
traditional peoples have been treated as "backward and primitive," and as
such are viewed as barriers to development and modernization. Those attitudes fuel and justify policies that strip native Amazonians of their lands,
territories, and resources—all in the name of development, progress, and
conservation. Scientists, too, continue to believe that traditional knowledge
is mere folklore; this in turn leads to the replacement of local-knowledge
specialists and leaders with alien scientific and technical "experts."

Amazonian countries find it difficult to modernize and reverse social,
economic, and ideological problems rooted in five hundred years of their
own political tradition. In part this is because recognition of the importance
of traditional resources and the knowledge of native Amazonian communities obliterates the hegemonic ideologies used to expropriate the land and
resources of native Amazonians. Anthony Hall (chapter 20) believes that
Latin American countries need to look upon mobilized local populations
as valuable social capital that will provide the underpinnings for sustainable
development. Indeed, according to Clóvis Cavalcanti (chapter 19), local
concepts and systems of exchange may provide the fuel for the development
of quite radical ethnoeconomic models of human coexistence with nature.

There is agreement among authors of this volume that, without the collaboration of scientists, policymakers, conservationists, industrialists, and
native Amazonians, there is little hope for conservation or environmental
and economic sustainability. As Joanna Overing (chapter 18) puts it, "we
should learn from indigenous peoples the advantages of dialogical and
shared knowledge practices." To some degree, the future of Amazonia and
its peoples might be positively influenced by those who find inspiration in
this book, written by individuals who have a strong personal commitment
to the region. In his introductory chapter, Michael Balick calls for everyone
interested in the future of Amazonia to do whatever is necessary (whether
political, institutional, or personal) to place more responsibility for research,
conservation, and other activities back into the hands of local peoples.

Miguel Hilario-Manenima, a Shipibo-Conibo participant at the Oxford
Amazon Conference,[1] summarized the main message from the contributors
to this volume in this manner:

> Amazonia has been a place where indigenous people have lived,
> fished, and hunted since time immemorial. [Any] disruption of this
> physical and spiritual interconnectedness leaves native Amazonians
> malnourished and sick. As the environment is plundered, contami-
> nated, and destroyed—so is the loss of our knowledge of plants, herbs,
> trees, and conservation techniques. When the plants are gone, our
> knowledge will be gone; when our knowledge is gone, our soul will
> be gone; when our soul is gone, then we as peoples will cease to exist.
> A failure to understand this indigenous equation will cause the failure
> of all humanity.

Darrell Addison Posey
Oxford University

---

1. The first Oxford Amazon Conference was held at Linacre College from June 5
through 6, 1998. The conference was organized for the University of Oxford Centre for
Brazilian Studies (established in October 1997 and directed by Professor Leslie Bethell)
by Dr. Darrell Addison Posey, Director, Traditional Resource Rights Programme, Oxford
Centre for the Environment, Ethics, and Society, Mansfield College. The theme of
the conference was "Human Impacts on the Environments of Brazilian Amazonia:
Does Traditional Ecological Knowledge Play a Role in the Future of the Region?" This
volume comprises contributed papers from that conference.
     The two-and-one-half-day conference generated considerable debate, and in addition
to delivering papers, participants made a number of recommendations (appendix). This
volume follows the organization of the conference papers.

Coeditor's Note

Darrell Addison Posey was one of those fortunate people who wove his passion for life, fascination with knowledge, and quest for truth into a distinguished and productive academic career. I first met Darrell one evening while standing on a Brazilian customs line that did not seem to move; both of us were headed to our respective field sites. Darrell was bringing the latest technology into the region. Fortunately, he was adequately prepared to respond to any questions that might be raised over the contents of his suitcase. After both of us made the green light, we went our separate ways: he to an indigenous village to study ethnobiology and I to a *caboclo* village to study the interaction between palms and people. Over the years, we corresponded frequently, and during my time at Oxford when I was a Visiting Fellow at Green College, we had many fascinating discussions. While at Oxford, my family and I would take breaks from the wonderful bibliographic collections available to us and wander over to Darrell's cottage for a barbecue and relaxing swim.

As all of those who knew Darrell realized, he was the sort of person who put himself on the line for his beliefs. Some of his work in the human rights arena brought him more notice and controversy than he desired, but his search for truth, justice, and humanity was respected even by his critics. I was delighted when he invited me to participate in his first Amazon symposium, held at Linacre College, Oxford, in June of 1998. My role was to listen to the presentations and discussions and to craft summary comments at the end of the meeting. When Darrell began to think about where this collection of papers should be published, I invited him to contact Colum-

bia University Press, which publishes the "Biology and Resource Management Series" that I coedit with Kent Redford, Chuck Peters, and Anthony
Anderson. Sadly, Darrell was fighting a terrible illness at that time and this
manuscript did not make it to the Press in final form. Following Darrell's
untimely passing in March 2001, and after discussions with the series editors
and with Darrell's associates at Oxford, I agreed to take on the coeditorship of
this volume. Under any circumstances the task of assembling many individual papers is a difficult one; this time it was made more complicated by the
state of Darrell's possessions. Some of the manuscripts were missing; others
were only fragments or rudimentary drafts. Additionally some symposium
participants were no longer able to contribute to this endeavor. Nevertheless,
the contributors to this volume—to whom I am extremely grateful—felt that
it was important to complete the project that Darrell had begun. In the end,
decisions about what to include and how to organize the book were based on
helpful comments from several reviewers. However, if mistakes were made,
through error or omission, the fault exists with the coeditor.

Once when we were swimming in Darrell's serene Oxford pond, his haven from an often difficult world, he told me how he had learned of the
passing of his mentor, Beptopoop. Without any word from the Kayapó village where Beptopoop lived, Darrell sensed a farewell from his friend when,
as he put it, Beptopoop's spirit flew over the pond. "I felt he was there. I felt
he was with me, and I knew he had come to impart his last teachings and
say goodbye," Darrell noted. "Of course that was very important to me," he
continued, "and during these past few years I have begun to understand this
next level of his knowledge." Although Darrell's illness and passing were
filled with suffering, he was surrounded by friends and others who admired
him as a human being, an activist, a creator of ideas, and a fine researcher,
and for his passion for all he believed in. It would not surprise me if, before
his spirit ascended to the next place, Darrell took a journey over his pond
and his beloved Oxford, and perhaps around the Kayapó village at Gorotire,
one last time. That would certainly have been his style.

                                                      Michael J. Balick
                                        The New York Botanical Garden

Acknowledgments

We are grateful to the Oxford Centre for Brazilian Studies, the British Council in Brasília and Recife, and the U.K. Department for International Development for funding this conference. Michael Balick's collaboration on this project was funded by The New York Botanical Garden through the Philecology Trust, The Edward John Noble Foundation, and The MetLife Foundation through Dr. Balick's appointment as a MetLife Fellow. We are most grateful to Elizabeth Pecchia for her care in typing many versions of the manuscript; to Willa Capraro for organizing the files received from the Posey archives and for helping to craft the material into a cohesive volume; to Robin Smith at Columbia University Press for advice in getting the project back on track; to Leslie Bethell and Kristina Plenderleith for assistance in securing the original manuscript and to the latter for generously agreeing to proofread the final manuscript; to Darrell's brother, Richard A. Posey, for his help in seeing the project to its completion; and finally, to Darrell's friends and colleagues, who showed their devotion to him through the completion of this work. *Obrigado* and *gracias* to you all.

Human Impacts on Amazonia

Thoughts on the Future of Amazonia

The Region, Residents, Researchers, and Realities

Michael J. Balick

Being a futurist is a tricky business; as someone once said, "The easiest way to make God laugh is to tell God of your future plans." To ponder the future it is necessary to consider the past and the present. Perhaps as recently as forty or fifty years ago, Amazonia was considered a limitless treasure with extraordinary resources. The major questions asked at that time focused on identifying Amazonia's biological diversity and determining which of its natural resources had the most value for extraction — usually the answer was timber. Today Amazonia is in a critical phase, one in which the region is perceived to be an endangered and limited resource. As we see in this volume, and as is usually the case in discussions of biological resources, the watchword now is "sustainability." One important contemporary question is, "What can be taken out while leaving the whole intact?" If there are no interventions, and by that I mean significant actions that lead to more effective global conservation, Amazonia may be propelled into another phase, this time characterized as disastrous. Some suggest that this phase may come as soon as the year 2050 and will be one in which Amazonia is viewed as a remnant resource or even an exhausted resource. By then the question will be, "How can *Homo sapiens* survive as a species?"

An understanding of the present and future phases requires different research needs. Without effective conservation programs in Amazonia, there is great cause for alarm. Collectively, we should admit that many of our conservation efforts, both regional and global, have been less than effective. All too often our efforts have been scattered and uncoordinated, championed by individuals rather than institutions, and of short duration due to funding constraints.

Overall, a new paradigm is required for undertaking the conservation enterprise in the next millennium. Indeed, I fear that Haiti — a beautiful country with extraordinary people — may portend the fate of the Amazon Valley in the year 2050. Over the last few decades, most of the tropical forest cover of Haiti has been destroyed. While conservation activities focus on the few forest fragments that remain, many rural people in Haiti struggle to survive by cultivating the land and collecting every scrap of wood from the forests that they can find to fuel their cooking fires. Effective regional conservation requires new strategies, fresh ideas, and many billions of dollars for funding on-the-ground activities.

In other areas of the world, the deforestation doomsday scenario will come before 2050. During the next several decades, major areas of tropical forest in Asia, the Pacific, and the Americas will disappear as human intervention takes its toll. Some areas, of course, will benefit from reforestation by humans. However Amazonia will face growing pressure. As global ecosystems lose their ability to sustain human life, the potential for violence will surely grow: issues of land tenure and human rights will become increasingly significant.

This volume deals principally with the influence of Western culture on Amazonia and its people. But a deep look inside Western culture reveals that elements of indigenous beliefs are beginning to penetrate our own societies. For example, Amazonian shamanism is for sale to those who feel they need it. In the United Kingdom and the United States there are weekend courses on shamanistic thinking, belief, and practices. A sacred shamanic plant (*Banisteriopsis caapi*) has even been used in a service in the Church of England (*The Guardian*, June 6, 1999). At the end of the second millennium, spirituality appears to be a growth industry for some of those who live in Western cultures, and Amazonia and its resources play an important role in this trend.

University curricula require new perspectives, specifically to enhance the interplay between the social and natural sciences. Such thinking is an effective way of addressing critical problems of great human concern. Increased opportunities for short-term training, both *in situ* and *ex situ*, will be necessary as well.

Scientific investigation is often condemned to recurring research cycles that are reminiscent of the semiclinical definition of insanity: repetition of the same activity over and over with the expectation of a different outcome each time. It is time to construct a detailed record of information about past research projects, and use this database to help break such wasteful cycles of effort.

There are many criticisms of so-called big science, specifically of the resources megaprojects consume. Perhaps, however, such large endeavors have value as a model for strengthening Amazonian science. Networks of researchers, linked projects, and standardized data-collection methodologies all could help attract greater support for the work discussed at this conference. The G7 and GEF programs could be a start in this direction. Perhaps a "Union of Amazonian Researchers" could be formed. First on the agenda, from the perspective of this botanist, might be to find out why it is easier to get a logging permit in some areas of Amazonia than to get a permit to carry out nondestructive biological inventories and conservation biology research.

In general, governmental and international agencies fund the ethnosciences in an ad hoc fashion, and the resources allocated are too few to address long-term critical research questions. The increasing number of conferences devoted to aspects of Amazonia and other tropical forest regions threatens to drown us all in a sea of scientific and activist meetings. What is the long-term value of so many meetings? Let's say that an uninformed estimate of the actual cost of these meetings is around US$50 million per year. If each organizer were to agree to a 15 percent value-added tax on each meeting — to be allocated to on-the-ground work — additional millions of dollars might become available in support of unfunded or underfunded projects. Fantasy, perhaps.

In the area of intellectual property, it is clear from this and other volumes that we are dealing with a moving target. Accepted practice today becomes a target for criticism tomorrow, as indeed the train has moved on to the next station. One ongoing debate involves the ownership of information gathered by ethnoscientists. Is such information the property of the scientist, the community from which it is gathered, the local government, the national government, or the group that supported the research? More often than not, intellectual property is discussed in terms of the scientist-investigator, local and national governments, and industrial sponsors, but what about the community from which the information is obtained? Intellectual property issues encompass resource-management information, mythology, cosmological beliefs, and shamanism. As mentioned earlier, shamanism is a growth industry; the myths and cosmological beliefs of other cultures are being taught to eager Western students who pay their course fees with little thought for the source — the providers — of the information. Thus, another question that must be addressed is the best way to collect, store, and teach information developed by indigenous peoples while compensating those who have shared their knowledge.

Several contributors to this volume touch on the marketing aspects of Amazonian products. My own observations on this have shown that, far too often, companies that claim to be doing something for the environment actually donate very little of their profits to effective programs that yield results. Some of our ethnobotany students have contacted companies that claim to donate a significant percentage of corporate profits to environmental causes. In actuality, as the students found out, several of those companies gave either nothing or a few hundred dollars per year; the strategy in those cases was to use an accounting trick to "reduce" profits. Integrity is needed here.

Contributions to this volume also underscore the importance of guaranteeing that Amazonian peoples have greater input into Amazonian issues, policies, and decision making. There is a need to increase efforts to train and support greater numbers of Amazonian peoples to ensure that the responsibilities for research, conservation, and other related endeavors are returned to local people. Local people can have a greater impact on local issues than can outsiders.

It is also time to reflect on the personal commitments that each scientist can make toward enhancing the research climate in Amazonia, improving human rights, helping to resolve issues of land tenure, creating bioconservation parks, and participating in other nontraditional academic activities. If each of us makes one such contribution through our research endeavors, then the overall picture would begin to improve, at least to some small degree. Working together, learning the languages of each other's disciplines, cultivating academic tolerance, and moving forward as a unified and powerful group would be one way to help ensure that Amazonia remains a viable ecosystem throughout the millennium.

1 Romance and Reality

The First European Vision of Brazilian Indians

John Hemming

The Romance

The first encounter between the Portuguese and the indigenous peoples of Brazil occurred on April 22, 1500, at Monte Pascoal in Porto Seguro, some 800 kilometers north of Rio de Janeiro. As we shall see, it was a meeting that was to have an extraordinary impact on European intellectuals of the time.

A fleet bound for the Cape of Good Hope and India was blown westward; sailors following land birds brought the fleet to the coast of Brazil. Luckily for us, that fleet contained a brilliant observer, Pero Vaz de Caminha. He described the excitement of the new discovery and its exotic people to the king of Portugal, in a long letter that was carried back to Europe in a ship loaded with Brazilian produce. The letter was so vivid and accurate that it has been hailed as Brazil's first anthropological source.

Vaz de Caminha (1500) told how the European discoverers celebrated the first mass on Brazilian soil. Their admiral, Pero Álvares Cabral, ordered a cross to be made. The Indians avidly watched the Europeans cutting the wood. Vaz de Caminha reported: "Many of them came there to be with the carpenters. I believe that they did this more to see the iron tools with which they were working than to see the cross itself. For they have nothing made of iron. They cut their wood and boards with wedge-shaped stones fixed into pieces of wood and firmly tied between two sticks." During the ensuing mass, the Indians carefully imitated everything done by the strangers who had appeared in their midst. "And at the elevation of the host, when we knelt, they also placed themselves with hands uplifted, so quietly that I assure Your Majesty that they gave us much edification."[1]

That story contains two elements that were to become constants in the colonization of Brazil. One was the natives' obsession with metal cutting tools. As Vaz de Caminha (1500) explained, the Indians could fell trees only laboriously, using stone axes. To them, the strength and sharpness of metal were miraculous. To this day, metal axes, machetes, and knives remain the tools with which isolated tribes are seduced, leading to their eventual contact. Initially, these are the only possessions of our society that newly contacted people really covet. I myself have been present when contact was first made with four indigenous groups in different parts of Brazil, and I can vouch for the power of attraction of metal blades.

The other constant was the Indians' apparent willingness to accept Christianity. Vaz de Caminha (1500) wrote to his king: "It appears that they do not have nor understand any faith. May it please God to bring them to a knowledge of our holy religion. For truly these people are good and of pure simplicity. Any belief we wish to give them may easily be stamped upon them, for the Lord has given them fine bodies and handsome faces like good men."

Thus were the ideas of the handsome noble savage and the blank slate introduced. When the first Jesuits reached Brazil fifty years later, their pious leader Manoel da Nóbrega was thrilled by the natives' apparent lack of religion. "A few letters will suffice here, for it is all a blank page. All we need do is to inscribe on it at will the necessary virtues, to be zealous in ensuring that the Creator is known to these creatures of his" (Nóbrega 1931, letter of Aug 19, 1549).

The first missionaries were unaware of the Indians' intricate spiritual world and rich mythology. The apparent innocence and simplicity of the native population suggested to the newcomers that here was a blank page upon which they might inscribe their own religion. The most striking manifestation of the Brazilians' simplicity was their nakedness. Two handsome warriors came aboard Cabral's ship and they were "naked and without any covering: they pay no more attention to concealing or exposing their private parts than to showing their faces. In this respect they are very innocent" (Vaz de Caminha 1500).

---

1. Unless otherwise noted, English translations in the text have been rendered by the author of this paper; selected examples of English versions of the original works have been included in the bibliography for the convenience of the reader.

Portuguese sailors, coming from the morally circumscribed world of fifteenth-century Iberia, were dazzled by the sight of unclothed women. Vaz de Caminha (1500) described them as being "just as naked as the men, and most pleasing to the eye ... [but] exposed with such innocence that there was no shame there." One girl in particular stood out. Vaz de Caminha wrote to the king that "she was so well built and so well curved and her privy part (what a one she had!) was so gracious that many women of our country, on seeing such charms, would be ashamed that theirs were not like hers."

The first observer noted that the Brazilians had no domestic animals or any cereals other than manioc. "Despite this, they are stronger and better fed than we are with all the wheat and vegetables we eat.... They are well cared for and very clean" (Vaz de Caminha 1500). Medieval Europeans washed themselves only rarely, so the natives' cleanliness made a big impression. The French priest Yves d'Évreux (1615) remarked that "the reward always given to purity is integrity accompanied by a good smell. They are very careful to keep their bodies free from any filth. They bathe their entire bodies very often ... rubbing all parts to remove dirt. The women never fail to comb themselves frequently." Another French pastor, Jean de Léry ([1578] 1980), admitted flogging women who did not wear clothing. "But it was never within our power to make them dress. As an excuse for always remaining naked, they cited their custom of ... diving into streams with their entire bodies like ducks, more than a dozen times a day. They said that it would be too much effort to undress that often! Isn't that a beautiful and cogent reason?"

Early observers viewed not only the Indians as pure and innocent, but their government and society as well. Amerigo Vespucci, the boastful Florentine who visited Brazil in 1502 as a passenger in a Portuguese flotilla, made such an impression with his description of these wonderful people that two continents were named after him. Vespucci lived with the Brazilian Indians for twenty-seven days. From that experience he concluded that "they do not recognize the immortality of the soul. They have no private property, because everything is common. They have no boundaries of kingdoms or provinces, and no king! They obey nobody, and each man is lord unto himself. There is no justice and no gratitude, which to them is unnecessary because it is not part of their code.... They are a very prolific people, but do not designate heirs because they hold no property" (Vespucci [1503] 1974:285, trans. Morison).

Vespucci—whose letter was rapidly translated and disseminated throughout Europe—was putting forth some very subversive ideas. The Portuguese chronicler Pero de Magalhães [de] Gandavo noticed that the Tupi language (like modern Japanese) did not use the letters f, l, or r. By coincidence, those were the first letters in the Portuguese words for faith, law, and king. So Gandavo (1576) concluded that, though the Indians had "nem fei, nem lei, nem rei," yet they thrived without these three pillars of European social order.

We now know that these observations were flawed. It was incorrect to say that the Indians had no faith, since they believed in an elaborate world of spirits, legend, and shamanism. Nor were they devoid of laws. Even today, native villages appear very tranquil, but they are highly conservative and regimented. Their conformity functioned without any need for codified law or a legal profession.

It was, however, reasonably correct to say that Brazilians had no kings. Their tribes had chiefs, but these were only primus inter pares. Each family was a self-sufficient entity and each head of household ranked roughly equally in tribal society. Michel de Montaigne (1580) asked a Tupinamba chief what advantages or powers he gained from leading five thousand men. The Indian answered simply, "To march first into battle."

Chroniclers also contrasted Indian generosity with European avarice. Gandavo (1576), in an important passage, described a society based on hunting and communism: "Each man is able to provide for himself, without expecting any legacy in order to be rich, other than the growth that nature bestows on all creatures.... They have no private property and do not try to acquire it as other men do. They thus live free from greed and inordinate desire for riches that are so prevalent among other nations.... All Indians live without owning property or tilled fields, which would be a source of worry. They have no class distinctions, or notions of dignity and ceremonial. And they do not need them. For all are equal in every respect, and so in harmony with their environment that they all live justly and in conformity with the laws of nature."

All this fitted with European fantasies of the noble savage. The simple and generous Brazilians were portrayed as living in the midst of an opulent nature. Amerigo Vespucci ([1503] 1974) was impressed by the intricacy of native architecture and the abundance of foods. He marveled at the diet of fruits, herbs, game, and fish, and "great quantities of shellfish—crabs, oysters, lobsters, crayfish and many other things that the sea produces." He portrayed Brazil as a delightful place whose evergreen trees yielded "the sweetest aromatic perfumes and ... an infinite variety of fruit." Vespucci told his readers about the flowers, birds, and exotic animals of the new world that was to be named after him. He concluded, "I fancied myself to be near the terrestrial paradise."

There was a long tradition of these romantic visions. They were a return to the golden age of Ovid's *Metamorphoses*, a book much read at that time.

Writing of his thirteenth-century travels, Marco Polo praised primitive man. In *The Divine Comedy* (1307–1321), Dante's Master Brunetto Latini said that Africans were innocent, naked, and not greedy for gold. And the Belgian John Mandeville (14th c.) described savages as "good people, full of all virtues and free from all vice and sin."

In the year 1500, when Cabral's sailors first saw Brazil, the Italian Pietro Martire d'Anghiera ([1530] 1969) wrote that among peoples of the new world "the land belongs to all, just as the sun and water do. 'Mine and thine,' the seeds of all evils, do not exist for these people.... They live in a golden age. They do not surround their properties with ditches, walls, or hedges, but live in open gardens. They have no laws or books or judges, but naturally follow goodness."

Such visions of indigenous peoples may have been exaggerated and inaccurate, but they contained a powerful political message. In 1508, Erasmus of Rotterdam was in London, staying with his friend Thomas More. While there, he wrote the satire *In Praise of Folly* (1509); its heroine, Folly, came from the Fortunate Islands, where happy people lived in a state of nature and the land yielded abundant food. Thomas More's *Utopia* (1516) had a leading character who was described as being one of the people left at Cabo Frio by the expedition on which Vespucci was a passenger. More declared that in utopia man should live by nature and his own instincts, and he praised the Indians' disdain for material possessions. *Utopia* had a great influence when it was first published and translated into many languages—and again when retranslated by Jean-Jacques Rousseau in the eighteenth century.

It was the French who most avidly developed the theme of the noble savage. French sailors from Normandy traded actively with Brazilian Indians, and they brought some of their Tupinamba friends back to France. The French established colonies at Rio de Janeiro from 1555 to 1560 and at Saint-Louis (São Luis) in Maranhão from 1612 to 1615. Both colonies were rapidly extinguished by the Portuguese, but each produced two splendid chroniclers; those early writers greatly added to ethnographic knowledge of indigenous peoples and partly added to romantic misconceptions about them. The four were Jean de Léry and André Thevet at Rio de Janeiro, and Claude d'Abbeville and Yves d'Évreux at Maranhão.

François Rabelais knew Thevet and other Frenchmen who had visited Brazil, and he drew on Erasmus and More in the first part of *Pantagruel* (1532). The poet Pierre de Ronsard (1550) idealized the innocence of the Tupinamba Indians, who were subject to no one, "but live as they please and answer to no one: they themselves are their law, their senate, and their king." He begged that these noble people be left alone: "Live, you fortunate people, without distress or cares. Live joyously. I wish that I could live like you."

The political theorist who was most closely identified with Brazilian Indians was Michel de Montaigne. He had studied in Bordeaux under the Portuguese André de Gouveia, a teacher who inspired his students to find the meaning of man in nature. Montaigne drew on the works of both Léry and Thevet; he owned a collection of Brazilian artifacts, and he interviewed three Tupinamba whom he met in Rouen in 1562. In his famous essay *Des Cannibales* (1580), Montaigne concluded that Indian virtues led to subversive revolutionary doctrines. He imagined a conversation between some Tupinamba and the French King Charles IX. When asked what they thought about France, the Indians replied that they had noticed some Frenchmen gorged with all manner of possessions while there were emaciated beggars at their gates. They found it strange that "those needy opposites should suffer such injustices, and that they did not take the others by the throat or set fire to their houses." The contrast with the Indians' communal and communistic society was obvious. Montaigne was particularly struck by the modesty of Brazilian chiefs who had no special powers, no aura of majesty, no luxurious establishments. He copied Léry's description of a fierce intertribal battle, but he claimed that it was fought from altruistic love of honor rather than for revenge. He also followed Ronsard, Gandavo, and others in noting that Indians flourished without an established church, monarchy, or written law.

The initial fascination with Brazilian Indians was soon eclipsed by the sensational conquests of the more sophisticated Maya, Aztec, and Inca empires. By the mid-seventeenth century the Portuguese had driven all other European powers out of what is now Brazil, and they were so paranoid about the possibility that others might learn about the country that they suppressed published reports about it. Curiosity about Indian societies had also ceased. However lack of fresh ethnographic information did not inhibit the political theorists. They continued to draw on the accurate but romanticized accounts of Vespucci, Thevet (1575), and Léry, and on the conclusions that Montaigne and others had derived from them. In Shakespeare's *The Tempest*, Gonzalo expounded the virtues of primitive society—although he was ridiculed by Sebastian and Antonio for it. The Dutch political theorist Hugo Grotius (1625) cited the simple communal life of American Indians as a basic social model. Francis Bacon (1626) and Tommaso Campanella (1623) owed much to Thomas More's vision of utopia. And three of the great theorists of the seventeenth century (Pufendorf in Germany, John Locke in England, and Baruch de Spinoza in Holland), wrote about the simplicity and nobility of indigenous Americans.

In the eighteenth century it was again French philosophers who used Brazilians as a model. Charles Louis de Secondat Montesquieu (1748) stressed the freedom and equality of Indians, whose lack of possessions removed any motive for robbery or exploitation. Candide, the hero of Voltaire's eponymous book (1759), traveled across Brazil from the Jesuit theocracy in Paraguay to an idyllic El Dorado in the Guianas. This was a wonderfully rich land, but devoid of priests and monks. A kindly old man told Candide that his people did not pray because they had nothing to ask of God. "He has given us all we need, and we thank him ceaselessly—[but we do so without a priesthood], for we are all priests."

Jean-Jacques Rousseau believed strongly in the ideal of the noble savage and the natural goodness of man. His novel *Émile* starts with the declaration that "all is goodness when it emerges from the hands of the Creator; all degenerates in the hands of man" (Rousseau [1762a] 1992, trans. Foxley). Rousseau derived from Jean de Léry the notion of Indian babies running free without swaddling clothes, and he urged mothers to breast-feed their infants; he even rewarded those who did so with hat ribbons that he made himself.

Rousseau seemed unaware of one aspect of tribal societies that would have appealed to him: indigenous villages are true democracies, miniature city-states in which every family head has a say in deliberations held in the men's hut. But in *Du Contrat Social*, Rousseau (1762b) propounded the ideas of natural nobility; equality; lack of laws, rulers, or clergy; and indifference to possessions—all concepts that originated in the earlier reports about the natives of Brazil.

These ideas reappeared in Denis Diderot's famous entry on "Sauvages" in his *Encyclopédie* (1751–1765). He admired the Indians living unconquered in their forests—for their love of liberty, their courage in the face of pain and death, and their naked innocence. He concluded that these virtues made them more open-minded and better able to hear the voice of reason.

All this had a direct influence on the independence of the United States, the Rights of Man, and the French Revolution. More recently it influenced Communism: both Marx and Lenin read More's *Utopia* and praised the indigenous Americans' indifference to gold.

The Reality

So much for the romance. As we know, back in Brazil the reality was far removed from the beguiling theories of European philosophers. The Indians were not the paragons of the noble-savage concept—they had the usual human qualities and failings, and some of their practices and attitudes fell short of the Christian ideal. But they suffered appallingly from the European invasion.

Relations between Indians and Europeans started amicably enough. A chief told Claude d'Abbeville (1614), "In the beginning, the Portuguese did nothing but trade with us, without wishing to live here in any other way. At that time they freely slept with our daughters, which our women considered a great honor."

The honeymoon period between Europeans and Brazilians soon degenerated into colonization, missionizing, oppression, and enslavement. The Portuguese started to plant permanent colonies in Brazil from the 1530s onward. They discovered that the region near Bahia and Pernambuco was ideal for growing sugar; but that very lucrative crop was labor-intensive. The Indians, who had been so generous in giving brazilwood logs and food to the first explorers, were not prepared to toil in the cane fields and sugar mills. To them, the notion of one man working for another was abhorrent; they had no desire for accumulated wealth or surpluses, and agriculture was women's work and unworthy of male hunters and warriors. So the only way in which the Portuguese could get labor for their sugar was to enslave the Indians—and when that failed, to import black slaves from Africa.

The church, and the Jesuits in particular, condemned slavery—of indigenous people with copper-colored skin, that is, but not of blacks. However, missionaries connived in making their nominally "free" Indians perform forced labor for the colonists. In return, the Indians received ludicrous payment: lengths of cotton cloth that they did not need and that they had made in the first place.

The Portuguese crown also compromised and vacillated over the crucial issue of slavery. While condemning it on moral grounds, Portuguese law permitted a wide range of exceptions—in order to keep the colonists happy and the sugar (and later gold) flowing to the mother country. As a face-saving euphemism, it was declared that Indians captured in intertribal wars could be "ransomed" from certain execution by their captors—just as Christians captured by the Moors were ransomed. Anyone saved from death in that way was forced to work as a slave for the rest of his life. Abuses of such a weird system were of course rampant. As a Tupinamba explained, "after exhausting the slaves ransomed from prisoners of war, they wanted to have our own children" (Abbeville 1614).

The euphoria of the first missionaries also degenerated into disillusionment. The earliest Jesuits were wrong to think that indigenous peoples were a blank slate on which they could inscribe Christian belief. The Indians were intrigued by the strangers' new religion. But those who appeared to convert to it wanted to add to their own culture, not to abandon it. They had no intention of giving up cherished customs such as festivals with drinking and dancing, belief in legends and the spirit world, shamanism, polygamy for important warriors, or even (in a few tribes) ritual cannibalism.

One reason why Indians turned away from Christianity was its failure to prevent the destruction of their people by imported diseases. Preconquest peoples of Brazil and Amazonia were magnificently healthy. The Jesuit leader Manoel da Nóbrega (1931) wrote, "I never heard it said that anyone here died of fever, but only of old age" (letter from Porto Seguro, 1550). Indigenous peoples were active, well fed, and extremely fit. But they were fatally vulnerable to imported diseases against which they had no inherited immunity. Smallpox, measles, cholera, and pulmonary diseases such as tuberculosis and influenza quickly killed tough and healthy natives. The tragedy was exacerbated by the missionaries' total lack of medical knowledge about the nature of the diseases or their treatment. All that the Jesuits could do was watch helplessly while their congregations died en masse.

There are many harrowing accounts of the early epidemics. One Jesuit wrote, "Another illness engulfed them, far worse than any other. This was a form of smallpox or pox so loathsome and evil-smelling that none could stand the great stench that emerged from them. For this reason, many died untended" (Leite 1956–1960:55; letter from Antonio Blasques to Diego Miron, Bahia, May 31, 1564). Another reported that in the twenty-five years after 1550 some 60,000 Indians were baptized near Bahia. But in that same period they were reduced to one village with only 300 men—a depopulation of 98 percent! José de Anchieta (1933) wrote in despair that "the number of people who died here in Bahia in the past twenty years seems unbelievable." Leonardo do Vale sailed along the colonized coast of Brazil and saw piteous destruction and piles of corpses. "Where previously villages had five hundred fighting men, there would now be fewer than twenty" (Leite 1956–1960:12; letter from Leonardo do Vale to Gonçalo Vaz de Melo, Bahia, May 12, 1563). And the epidemics struck uncontacted tribes. "The Indians say that this is nothing in comparison with the mortality raging through the forests" (ibid.).

That annihilation by disease caused the world's worst demographic catastrophe. It is the greatest of all constants in Brazilian indigenous history. In 1500 there were 3.5 million or more indigenous people in Brazil. There are now 350,000, a mere 0.6 percent of Brazil's total population. Disease mortality is the reason why South America is essentially a European continent rather than a native-American one.

Different tribes reacted in different ways to these terrible cultural shocks. Some fought back, courageously and tenaciously, during decades of struggle. But tribes rarely presented a unified resistance, and some made the classic error of trying to enlist foreign colonists into their intertribal feuds. So they were picked off one by one, and eventually crushed by the invaders' superior firepower.

Other tribes tried to escape by migrating away from the colonization frontier. On one occasion in the 1580s, 84 Tupinamba villages containing 60,000 people decided to move inland from the coast of northern Bahia. They told a Jesuit that they were fleeing in terror: "We must go! We must go before the Portuguese arrive! These Portuguese will not leave us in peace.... They will surely enslave us ourselves and our wives and children" (Anchieta 1933: 374–375).

A few despairing tribes tried to establish their own versions of Christianity. Throughout Brazilian history there have been messianic movements in which preachers led their flocks to a promised land, a "land without evils" far beyond the frontier of colonization. Adherents hoped there would be an upheaval that would make Indians dominant over whites. But such a reversal of fortunes never came: utopia remained only a romantic notion of European philosophers. So it was hardly surprising that in April 2000 Indians from all parts of Brazil protested about the celebration of Cabral's landing at Monte Pascoal five centuries earlier. They saw the Portuguese as their eventual nemesis. That was the reality neither Vespucci nor the other early observers could foresee.

References

Abbeville, Claude d'. 1614. *Histoire de la Mission des Pères Capucins en l'Isle de Maragnan et Terres Circonfines*. Paris: Imprimerie François Huby.

Abbeville, Claude d'. 1963. *Histoire de la Mission des Pères Capucins en l'Isle de Maragnan et Terres Circonfines*. Ed. A. Métraux and J. Lafaye. Vienna: Akademische Druck u. Verlaganstalt.

Anchieta, José de, SJ. 1933. *Cartas, Informações, Fragmentos Históricos e Sermões*. Ed. A. de Alcântara Machado. Rio de Janeiro: Biblioteca Nacional.

Anghiera, Pietro Martire d'. [1530] 1969. *De Orbe Novo*. In S. Buarque de Holanda, ed., *Visão do Paraíso*, p. 180. Brasiliana 333. São Paulo: Editora Nacional.

Bacon, Francis, Baron Verulam. 1626. *The New Atlantis*. London.

Campanella, Tommaso. 1623. *Civitis Solis*. Naples.

Campanella, Tommaso. 1981. *The City of the Sun*. Trans. D. J. Donna. Berkeley: University of California Press.

Diderot, Denis. 1751–1765. "Sauvages." In *Encyclopédie*. Paris: Briasson.

Diderot, Denis. 1967. *Denis Diderot's The Encyclopedia*. Trans. and ed. J. Gendzier. New York: Harper & Row.

Erasmus, Desiderius. 1509. *Erasmi Rotterdami Encomium Moriae (Stultitiae Laus)*. Basle.

Erasmus, Desiderius. Rpt., 1972. *In Praise of Folly* (1509). Trans. F. Eichenberg. New York: Aquarius Press.

Évreux, Yves d'. 1615. *Suite de l'Histoire des Choses Plus Memorables Advenues en Maragnon, ès Années 1613 and 1614*. Paris: L'Imprimerie de François Huby.

Évreux, Yves d'. 1864. *Voyage dans le Nord du Brésil*. Ed. F. Denis. Leipzig and Paris: Librairie A. Franck.

Gandavo, Pero de Magalhães [de]. 1576. *Tratado da Terra do Brasil: Historia da Provincia Santa Cruz*. Lisbon.

Gandavo, Pero de Magalhães [de]. 1922. *The Histories of Brazil*. Trans. J. B. Stetson. New York: The Cortes Society.

Grotius, Hugo. 1625. *De Jure Belli ac Pacis*. Paris.

Grotius, Hugo. 1922. *The Law of War and Peace*. Trans. W. S. M. Knight. London: Sweet and Maxwell.

Hemming, John. 1995. *Red Gold, The Conquest of the Brazilian Indians*. Rev. ed. London: Papermac.

Leite, Serafim Soares, ed. 1956–1960. *Monumenta Brasiliae*. Vol. 4. Rome.

Léry, Jean de. [1578] 1980. *Histoire d'un Voyage Fait en la Terre du Brésil*. La Rochelle: Antoine Chuppin. In. J-C. Morisot, ed., *Le Voyage au Brésil de Jean de Léry*. Geneva: Librairie Droz.

Mandeville, Sir John [Jean de Bourgogne of Liège?]. 14th c. Untitled manuscript. The travels of Sir John Mandeville. British Library.

Mandeville, Sir John. 1973. *Voyages and Travels*. Trans. M. C. Seymour. London and New York: Oxford University Press.

Montaigne, Michel de. 1580. "Des cannibales." In *Essais*, ch. 31. Bordeaux: Simon Millanges.

Montaigne, Michel de. 1991. "The Cannibals." In M. A. Screch, trans., *The Essays of Michel de Montaigne*, ch. 31. London: Allen Lane.

Montesquieu, Charles Louis de Secondat, Baron de. 1748. *L'Esprit des Lois*. Paris.

Montesquieu, Charles Louis de Secondat, Baron de. 1977. *The Spirit of Laws*. Trans. D. W. Carrithers. Berkeley: University of California Press.

More, Sir Thomas. 1516. *Utopia*. London.

More, Sir Thomas. 1995. *Utopia*. Trans. and ed. G. M. Logan, R. M. Adams, and C. H. Miller. Cambridge: Cambridge University Press.

Nóbrega, Manoel da, SJ. 1931. *Cartas do Brasil, 1549–1560*. Ed. E. S. Soares Leite, A. do Vale Cabral, and R. Garcia. Rio de Janeiro: Imprensa Nacional.

Rabelais, François. 1532. *Pantagruel*. Ste. Lucie: Lyon.

Rabelais, François. 1990. *Gargantua and Pantagruel*. Trans. Sir T. Urquhart and R. Motteux. Chicago and London: Encyclopedia Britannica.

Ronsard, Pierre de. 1550. *Ode Contre Fortune*. Paris: Chez la Veuve Maurice.

Ronsard, Pierre de. 1997. *Odes*. Trans. and ed. G. Castor and T. Cave. Manchester: Manchester University Press.

Rousseau, Jean-Jacques. [1762a] 1992. *Émile, ou de l'Éducation*. Paris. Trans. B. Foxley and ed. P. D. Jimack as *Emile*. London: Everyman's Library.

Rousseau, Jean-Jacques. 1762b. *Du Contrat Social*. Geneva.

Rousseau, Jean-Jacques. 1993. *The Social Contract*. Trans. G. Cole. London: Dent.

Thevet, André, OFM. 1575. *La Cosmographie Universelle*. Paris: Chez Pierre l'Huillier.

Vaz de Caminha, Pero. 1500. Letter to King Manoel, Porto Seguro, 1 May 1500. In J. Cortesão, ed., *A Carta de Pero Vaz de Caminha*.

Vaz de Caminha, Pero. 1937. *The Voyages of Pedro Álvares Cabral to Brazil and India*. Trans. W. Brooks Greenlee. London: Hakluyt Society, 2nd series, 81:3–33.

Vespucci, Amerigo. [1503] 1974. *Mundus Novus* (letter to Lorenzo di Pier Francesco de' Medici). Lisbon. Trans. S. E. Morison as *The European Discovery of America: The Southern Voyages, 1492–1616*. New York: Oxford University Press.

Voltaire (François Marie Arouet). 1759. *Candide, ou l'Optimisme*. Paris.

Voltaire (François Marie Arouet). 1990. *Candide, or Optimism*. Trans. R. M. Adams. New York and London: Norton.

2   Constructing Tropical Nature

Nancy Leys Stepan

The process of deconstructing the myths and errors that have plagued interpretations of, and policies toward, Amazonia—perhaps the most exemplary site of the tropical—is now proceeding apace.

I think it was David Arnold, the specialist on India, who was the first to use the term "tropicality" in our contemporary sense, to indicate the constructed and discursive character of the tropics (Arnold 2000). By tropicality, Arnold means to indicate that the tropical in the Western tradition is more than a purely empirical or geographic term, but represents a fundamentally European, outsider, and imaginative view of large parts of the social and natural world. He points to the way in which the tropical has been positioned as an area of pure nature, as opposed to one of culture and history—as an area that contrasts with the temperate world, which it both opposes and helps constitute. Arnold's main interest lies in the emergence of tropical medicine (Arnold 1993). Other historians have implicated other disciplines in the construction of the tropics; David L. Livingstone, for example, has examined the long influence of "moral climatology" in geography, and the connections between the idea of climatic tropicality and a racialized anthropology (Livingstone 1994). A special issue of the *Singapore Journal of Tropical Geography* (2002), called "Constructing the Tropics," has articles on aspects of natural history, biogeography, landscape painting, land surveying, and landscape aesthetics design. My recent book *Picturing Tropical Nature*

(Stepan 2001) looks at tropical representations of place, race, and diseases, with a special emphasis on the visual images that, I argue, are extremely important in creating popular understanding of, and therefore the discursive production of, the tropics.

These recent studies of tropicality are related to, though somewhat different from, the extremely important books by Gerbi (1973), Holanda (1987), Glacken (1967), Brading (1991), Pagden (1993), and others on the extraordinary sixteenth- to eighteenth-century literary and scientific discussions of New World nature. It is also worth remembering that in the 1920s the Brazilian sociologist Gilberto Freyre developed his own concept of tropicality (*tropicalidade*), starting with his classic works of historical and national interpretation (e.g., *Masters and Slaves*, 1946). By the 1950s Freyre had established a permanent seminar in Recife, at which the social and natural sciences were used to investigate the impact of the distinctive environment and peoples in the tropics on the emergence of a distinctive Luso-civilization in the tropics—a *New World in the Tropics*, as one of his books called it (Freyre 1963). The science of the tropics, or "tropicology," that Freyre promoted as a way to systematize knowledge about the tropics, embraced nearly every discipline of knowledge—tropical nutrition and diet, tropical medicine and disease, tropical fishery, tropical ecology, tropical aesthetics, and tropical architecture (Congresso Brasileiro de Tropicologia 1987).

Freyre's tropicalism lacked, of course, the currently fashionable postmodern, "deconstructive," ironic approach to nature, in which the very possibility of an independent nature is treated with skepticism. Freyre was serious about it—about reinventing his own idea of tropical realities. In drawing attention to tropicality, Freyre tried on the one hand to give recognition to the special regional, geographical, climatic, and human dimensions of tropical regions without, on the other hand, falling into the trap of the tropical determinism characteristic of Western thought, a determinism that had resulted in depicting the tropical world as one of natural biodiversity but of cultural and civilizational poverty. In emphasizing the importance of the cultural dimensions in his concept of tropicality, Freyre aimed to rescue his country from charges of tropical backwardness and decay. Whether Freyre was able to achieve much with his version of tropicology is another matter. But the fact that he tried suggests the enduring power of the concept.

The more recent reconfigurations of the debate on tropicality, with their emphasis on nineteenth- and twentieth-century science in constructions of tropical nature, reflect, I think, the surge of new interest in tropical environmentalism, and the desire to understand the historical roots and complex strands of ideas that have produced a persistent imaginative geography of the tropics whose chief feature is its geographical determinism.

My comments in this article are intended to highlight some general aspects of the history of the concept of tropicality. They are divided into three parts. First, I characterize the style or manner by which tropicality has been constituted historically and conceptually. Second, I discuss briefly the idea of the "anti-tropical," using this term to refer to efforts to challenge the meanings attached to the tropical. And third, I reflect on the current status of tropicality, when many new narratives, some Edenic and some not so Edenic, are being put forward in ecology, anthropology, medicine, and the social sciences. Does the concept of "the tropical" still have any value in this postmodern age, or should we conclude, with Lévi-Strauss (1973), that as a concept the tropical is out of date and should be abandoned?

The Humboldtian Tropics

The idea that there is something distinctive and usually negative about hot places is, as we know, extremely old, going back to classical times; but it acquired its modern features and its intellectual force in the late-eighteenth and nineteenth centuries, the era of colonial expansion and the utilitarian exploitation of the natural world. I stress here the modernity of the concept of tropicality, because of its enduring influence on our modern scientific disciplines, science of course being the most authoritative of languages for describing nature.

This modern version of tropicality is marked by many continuities in its main themes, images, and preoccupations. It is disconcerting to see the similarities between the ideas of the early-nineteenth-century explorer and traveling naturalist Alexander von Humboldt (discussed below) and those, for example, of the twentieth-century French geographer Pierre Gourou, whose 1947 work *The Tropical World* Arnold has analyzed recently (Arnold 2000).

I have chosen Humboldt here to examine the style in which tropicality was constructed because his life's work was critical to the scientific separation of the tropical world from the temperate and to establishing what was characteristic about the tropical. Humboldt was one of the last great polymaths of the nineteenth century; revered by historians of science as the founder of modern physical geography, he has more recently been attacked by postmodernists as *the* imperialist of knowledge, whose very gaze—the knowing I/eye—supposedly appropriated tropical nature wholesale for consumption and exploitation (Pratt 1992). Many of Humboldt's ideas were old and familiar; but to them he added new ones, especially a language of aesthetics, an emphasis on scientific precision, and above all an identification of the tropics with the New World. Humboldt's five-year journey between 1799 and 1804 in the Americas was recounted in his *Personal Narrative of Travels*, which was read by nearly every European traveler to the American tropics that followed him (Humboldt 1966). His views were influential long after his pre-evolutionary biology and eighteenth-century-enlightenment optimism had been replaced by evolutionary biology and late-nineteenth-century social pessimism.

So what was his view of the tropics? First, it was global in scope; it embraced the *cosmos* (the title of his encyclopedic account of the natural world) (Humboldt 1848–1858). Humboldt suggested there existed a "tropical world," made up of characteristic ensembles of animals and plants, human beings, and later, diseases. Tropical biodiversity, which early on was recognized as extraordinary, nonetheless was viewed at the same time as a kind of homogeneity, as the tropical jungle came to stand as a trope for all of tropical nature.

Second, Humboldt believed that the tropics should be understood as an aesthetic as well as a scientific space. Humboldt (1849) suggested that the natural world should be approached as though it were a set of aspects or views (I refer here, for example, to his *Views of Nature*). This was a way of seeing and representing the natural world through the eyes of a spectator, from a distance or a height, as though nature were best appreciated as a fine painting. For Humboldt, nature in the tropics was sublime because it could evoke in the alert viewer a sense of awe in the presence of the vast and the mysterious. The apparent monotony of the dark forests, with their undifferentiated mass of vegetation and their slow, broad rivers, had to be *given* interest by the human mind; without this mental and aesthetic involvement, Humboldt believed, much of nature was *not* of interest and could not compensate for the hardships experienced in traveling through it. To this aesthetic appreciation of tropical nature, Humboldt also brought a huge number of the best precision instruments his personal wealth could buy; he collected a vast array of data, putting it into his published works and his visual diagrams.

Third, Humboldt viewed the tropics as fundamentally vegetative in character. Humboldt's seemingly arbitrary selection of plants over animals to represent the special character of the tropics was connected to his concern to infuse the empirical study of nature with human emotions, thereby preventing the rupture between the objective and the subjective approach to the natural world that he believed modern science threatened. In his view, the greatest emotional impression nature could make on the human senses was communicated by the mass of vegetation, rather than by fleeting animals. This vegetative mass was greater in the tropics than in the temperate world because the heat and the humidity of the tropical climate produced an intensity and fecundity of nature not found in cooler climates. Humboldt emphasized especially the greater vigor of organic life in the tropics, compared with that in the temperate world (Humboldt 1966).

These views were conveyed through Humboldt's use of innovative pictures and diagrams, such as his isothermal lines, which presented the globe encircled by climatic bands of "regular sinuosity" (to use Dettelbach's deft phrase [1993]). Within these bands, different places were connected in a common tropicality that set limits on the kinds of plants and animals (and peoples) that could exist within them. Humboldt's maps helped draw attention away from the debate about the merits of Old World nature versus New World nature which had preoccupied geographers for much of the eighteenth century and toward a new image of the globe divided into climatic bands. In the process of geographic reconceptualization, the northern United States was reconfigured as fundamentally temperate and potentially capable of producing civilization; in contrast, the southern American continent was pictured as a space whose lowland heat and humidity, so favorable to the production of nature, would pose a permanent challenge to the emergence of high civilization.

This latter idea was central to the dystopian elements that were joined to the utopian in Humboldtian physical geography. Humboldt's geography, like that of the enlightenment philosophers, was very much a political geography. To many thinkers at the time, political liberty and civilization were closely tied to temperate climates, with the tropical areas of the earth being viewed in terms largely outside human history. Humboldt, of course, traveled through many parts of the settled Spanish colonies in the Americas, and wrote on mining and trade; nonetheless, his general theme was the unsuitability of the hottest parts of the New World for sustaining high culture, which he believed was found only in the cold high Andes and the high plateaus of Mexico. He saw the indigenous populations of the Amazon Valley as making up only isolated tribes; the very fertility of the soil itself, Humboldt argued, prevented the development of the intellectual faculties and therefore of civilization. This fact—that apparently large areas of the southern Americas were destined to be places of nature, not of culture—did not cause Humboldt despair. Human beings in the tropics, he remarked, got used to the idea of a world "that supports only plants and animals; where the savage has never uttered either a shout of joy or the plaintive accents of sorrow" (Humboldt 1966:14–15).

Humboldt's approach to tropical places implied that their vistas were important mainly to European viewers. Indeed, according to Humboldt, the great advantage of tropical over temperate vistas was their potential to influence the European imagination. The European visitor, unlike the native inhabitant who simply happened to live in the tropics, possessed, said Humboldt, the language, arts, and sciences needed to appropriate tropical nature, to transform it into a higher aesthetic, and through imagination and painting, "create within himself a world as free and imperishable as the world in which it emanates" (Humboldt 1849:231). Humboldt therefore urged Europeans and North Americans, scientists and artists, to follow him to the tropics, to see tropical nature for themselves, and to communicate its physiognomy in pictures and words.

Popularizing Tropical Nature: The Amazonian Contribution

And follow him they did: artists such as Johann Moritz Rugendas, Timothy Peale, and Frederic Edwin Church (Church retraced parts of Humboldt's original route and climbed, as Humboldt had, the mountain peak of Chimborazo, and then painted it); and naturalists such as Karl Friedrich Philip von Martius, Johann Baptist von Spix, Richard Spruce, Alfred Russel Wallace, Henry Walter Bates, and Louis Agassiz (even William James went to the Amazon to see if tropical nature collecting was for him; he found it was not) (Stepan 2001:91).

Several points can be made about the mid-to-late-nineteenth-century development of tropicality. First, we note the way Amazonia comes into clearer focus as the primary exemplification of the tropical. Travelers had of course been there before; but Amazonia, in its popular representations, is largely a nineteenth-century phenomenon. By mid-century, many of the most important tropicalists, Humboldt's heirs in this regard, had been in the Amazon for years — Spruce for fourteen, Bates for eleven, Wallace for four. In Europe there was a growing demand for books on natural history, popular geography, ethnography, adventure, and travel. Such books were in fact second only to novels in popularity in the nineteenth century, and the difference between scientific and fictional genres was not as great as one might think. Many natural history books were illustrated, which of course added to their appeal. If we analyze the way they were illustrated, we can see how they contributed much of the misinformation or exaggerations by which the tropical world was shown. For instance, a picture would show a number of animals in a tropical ensemble, in a way they could never be seen together at the same time or place in reality. By selection, there was a perhaps natural preference for representing species that were bizarre, deadly, and exotic, at the expense of more ordinary and mundane ones.

Most of the ideas in Humboldt's work were repeated in the work of the nineteenth-century naturalists, even though the new evolutionary outlook, following the publication of Darwin's *On the Origin of Species* in 1859, was alien to the Humboldtian world view (Darwin 1964). The naturalists, for example, repeated Humboldt's views about the superfertility of the tropics and its connection to the low capacities of tropical peoples. Natives were found to be remarkably uncurious (especially about visiting Europeans), and generally "indolent." As Hecht and Cockburn (1989) have shown, almost to a person the European naturalists failed to understand indigenous harvesting techniques and their relation to the ecology of the tropical forest. Instead European visitors saw a lack, an absence of cultivation and culture owing to the hot climate, and especially the lassitude of the native and mixed populations. "Where is the population to come from to develop the resources of this fine country?" inquired the British naturalist Henry Walter Bates, referring to Amazonia. "They might plant orchards or the choicest fruit trees ... grow Indian corn, rear cattle or hogs, as intelligent settlers from Europe would certainly do, instead of relying on the produce of their small plantations, and living on a meagre diet of fish and farinha" (Bates 1892:117, 139). The image of the carefully manufactured landscape of Europe, the tidy farm, remained the ideal in the European imagination; and it was to the European that the task of overcoming the tropics, and producing its development, was assigned. Given the devastating effects that cattle ranching has had on the vulnerable soils in the Amazon in the twentieth century, the comments of these nineteenth-century naturalists make ironic, indeed painful, reading today; but they were common. The Swiss-born and naturalized American scientist Louis Agassiz, for example, dreamed of a European population of millions in the Amazon, and took pride in the fact that he had helped persuade the Brazilian emperor, Dom Pedro II, to open up the Amazonian waterways to European trade and communication (Stepan 2001:92).

But as the Amazon was opened up to commerce, trade, and exploitation, its image darkened; the tropical sublime was replaced by an image of tropical degeneration. Many factors contributed to this shift in representations at the end of the nineteenth century, paramount among them being perceptions of disease. The diseased tropical Amazon is not a theme in Bates's or Wallace's writings, even though both suffered from malaria, and it was disease, as well as loneliness, that eventually sent Wallace back to London after four years of nature collecting. It was not really until the early-twentieth-century rubber boom-and-bust, when a series of medical investigations carried out by Brazilian physicians revealed the truly appalling rates of malaria mortality, that the Amazon came to be viewed as a morbid site of devastating disease. By then, the new tropical medicine, based on parasitology and the vector theory of disease transmission, had emerged to emphasize the idea that tropical spaces were spaces of pathology (Stepan 2003).

The new picture of tropical disease merged with, and was reinforced by, the growing racism in scientific thought. Humboldt's universalism and enlightenment optimism was replaced with polygenic views of racial difference. Tropical hybridization was singled out for study and censure, as a dangerous process hastening the degeneration that the tropical climate produced. But though the fear that Europeans would be unable to acclimatize in the tropics gave rise to a huge literature, on the whole it was believed that it was *only* through European rationality that the fruits of the tropical world could be reaped.

Anti-Tropicalism

One of the insights the concept of tropicality provides is that, being a human invention, it takes cultural work to maintain; it is also open to challenge and change. I present here just a few examples of what I have called elsewhere "anti-tropicalism," by which I mean an alternative or revisionist view of tropical spaces that challenges the conventional stereotypes of the tropical world.

The nineteenth-century naturalist Alfred Russel Wallace, for example, was a maverick and social radical who is especially interesting, historically, for opposing the notion of tropicality that was standard in his day. He developed a conservationist and environmentalist outlook, expressed especially in his ironically titled book *The Wonderful Century* (1898), an outlook that he acquired during years of fieldwork in the Amazon and in the tropical environments of the Malay Archipelago. In such places he observed the devastating results that European export monoculture could have on tropical environments, and especially the human harm it caused (such as famines). Richard Grove (1995), in his book *Green Imperialism*, points out that many of the first scientists to sound environmentalist warnings had worked in tropical colonies, where serving for many years, they could see at first hand how European systems of cultivation and settlement had quite the opposite impact than had been hoped for. Imperialism thus set in motion the movement

for conservationism and a kind of environmentalism (thus reinforcing perhaps the long-standing image of the tropics as a place to be thought of largely
in terms of nature, not culture).

One of the most interesting shifts in tropical perceptions, though, involves those people—artists, writers, scientists—living in tropical countries
who responded to European representations of the tropics with counterrep-
resentations and images of their own. It is an indication of the power of the
European idea that the tropics stood for nature and race degeneration that so
many of the significant works of interpretation produced by Latin Americans
rely on these same terms. For example, in his famous work, *Rebellion in the
Backlands*, the writer Euclides da Cunha (1944) gives a brilliant account of
the armed revolt of poor Brazilians in the northeast. His book opens with
two long chapters on the land and on man, locating the rebellion in relation
to the environment/climate and race, the two most important biological parameters of tropicality in European thought. When Euclides da Cunha went
to the Amazon in 1905 to report on the rubber boom, his account conjured
up for the reader an image of nature at its very birth; Genesis, he said, was
newly writing itself there (Cunha 1976). It was as though geography, geology, natural history, and ethnology provided both the burdensome, or negative, narratives of Brazil *and* the inescapable terms by which the country
could even be imagined or described at all.

This is especially evident, I think, in the works produced by the scientific,
artistic, and literary modernists of the 1920s and 1930s (my focus is on Brazil
because this is the area I know best). Many of these works were preoccupied
with the problem of civilization in the tropics—whether such a thing could
exist, given the climatic and other conditions that had long been taken to
be an impediment to culture, or what forms it might take. In a familiar process of cultural appropriation, intellectuals turned to Europe for new ideas
in art, architecture, science, and literature, which they proposed to remake
in indigenous terms. In the 1922 Modern Art Week, held in São Paulo to
celebrate one hundred years of independence, European stereotypes were
attacked in order to declare a new artistic independence.

The tone was set by the "Pau-Brazil Poetry" Manifesto written by the poet
Oswald de Andrade, in which he asserted the identity between the tropical and the modern (Ades 1989:310–311). He called for a Brazilian poetry
for export, like brazilwood itself, a poetry that would be like the country:
barbaric and modern, skeptical and naive, of the jungle and of the school.
The other Andrade, Mário, who like many other artists and writers sought to
claim an authenticity and originality in homegrown Brazilian cultural forms
and popular traditions, wanted to overcome the dread in the face of nature,

a dread that permeated Brazilians' attitudes toward their environment and culture. Writing to the painter Tarsila do Amaral, in Paris, he said, "Tarsila! Tarsila, go back within yourself.... Leave Paris! Tarsila! Come to the virgin forest." (M. de Andrade, quoted in Herkenhoff 1995:243); and come back to Brazil she did, to make her mark as a modernist painter whose pictures comment wittily on, and reformulate, many of the stereotypes of tropicalism as nature.

Mário de Andrade made his own journey of discovery to Brazil in 1924 (Andrade 1993), during which he traveled through the Brazilian Amazon, camera in hand, westward all the way to Iquitos, Peru. Like the scientists who had explored the Amazon a few years before him, Mário de Andrade did not look for, nor did he find, a tropical paradise. But neither did he find the disagreeable portrait of racial degeneration in the Amazon that Agassiz had found fifty years before. He found poverty and disease, but also the bedrock of Brazil's real national creativity, in the ordinary people he encountered. *Macunaíma*, the title of his literary masterpiece and the name of its fictional hero, is a Brazilian Everyman, a black man who could transform himself into white, a person who belongs to the Amazon and to the city, a universal hero of modern times (Andrade 1984). The story is a wonderful parody of the "Brazil as natural history" genre, with the author drawing on academic and scientific studies of ethnology, linguistics, and natural history in order to stuff his text with lists of rivers, animals, plants, and Indian vocabularies. Brazil, like the novel's hero, says Andrade, is an Indian primitive; a cunning trickster; a liar; a sensual charmer; a multiple, unstable being (a being "without any character," as the novel's subtitle puts it) who is also a new universal of modernity. Brazil cannot choose between these different selves, but can only accept their contradictions; this is what it means to be modern. To be tropical is to be modern.

My last example of many efforts to reimagine the meaning of tropicality is the work of the brilliant modernist landscapist Roberto Burle Marx, whose public garden of Flamengo is such a distinctive feature of modern Rio de Janeiro. Burle Marx's chosen medium of expression, the designed landscape, engaged him directly with the tangible, material stuff of nature that had dominated foreign interpretations of the country. Using tropical plants as signature elements, Burle Marx constructed austere, abstract landscape gardens that made no effort to recreate the feeling of an immersion in a tropical Eden; in their pared-down abstractions, they could not be further, visually, from the expected Western picture of tropical nature as an overgrown and entangled jungle. They are thus both a celebration of the tropical and its reconfiguration (Stepan 1991:220–239).

In making these gardens, which have often puzzled garden historians and critics, Burle Marx relied on geometrical patterns in order to give, he said, "a full place to the existence of man" (Leenhardt 1994:55). The garden was his means of making the environment adequate to the demands of civilization and human creativity. In European history, tropical nature had been largely represented as forest, or jungle, something generally disliked because it was disorganized, overexuberant, unbalanced, and excessive, especially in relation to human occupation and the development of civilization. "It seems that all people encounter in our nature is designated jungle, and because it is jungle, it is of no use," he commented (Burle Marx, quoted in Cals 1995:87). Burle Marx's solution to the problematic of tropicality was not to Europeanize the garden (as the elite historically had done), but to question this view of tropical nature itself by placing tropical nature in the artificial setting of the human-made landscape, which he then manipulated to suggest alternative possibilities.

He combined plants with human artifacts, such as stone pillars; he massed together many examples of a single plant species, creating wide swaths of color in gardens, as though challenging Humboldt's idea that the tropics lacked social plants (i.e., large tracts of the same plant) and therefore social life. He mixed and matched plants from Amazonia with those of the arid *sertão* (scrubby backlands in the northeast) and with plants from the temperate regions farther south. He also used tropical plants from India. They all, he seems to imply, belong here, in the so-called tropics. He experimented with creating ecological niches to accommodate plant combinations not seen in nature. He was no purist; indeed, we could say that in so combining plants, and plants with artifacts (such as his strange sculpture posts made of palm tree stumps, or metal, or stone), he indicates the real heterogeneity of human and natural elements that in fact make up the tropical world, while commenting as well on the processes of exchange and acclimatization of plants that have characterized the modern age.

Another feature of Burle Marx's gardens that relates to his concern to reconfigure the meaning of the tropical is their urbanism; his well-known willingness to mould the landscape to achieve particular effects—to actually move mountains, or mountains of earth—can be interpreted as part of a system of urban reference, rather than merely a willful demonstration of the power of his art to conquer nature. Rio's own famous mountains, he seems to be saying, are artifacts too, creations of human civilization in the tropics—they have been bored through to make tunnels for transport, cleared of vegetation, and more recently, replanted with trees. Nature, his gardens suggest, is always culture before it is nature. And this is as true of the famous "tropical nature" as it is of any other kind.

Tropical Nature as Garden

Yet in all these reworkings of the tropical, we see that scientists and artists found it difficult to escape from the very terms they were interrogating. This leads me to ask whether the concept of the tropical is one that we *can* refashion in order to express our understanding of the complex social and natural world we find in places situated near the equator.

I might in this regard start with something Claude Lévi-Strauss said. In his famous *Tristes Tropiques*, Lévi-Strauss (1973) confessed that an unanticipated invitation to go to Brazil in the 1930s had immediately conjured up for him a mental picture of a world utterly different from the one he knew. "I imagined exotic countries to be the exact opposite of ours," he wrote, "and the term 'antipodes' had a richer and more naive significance for me than merely its literal meaning. I would have been most surprised if anyone could have told me that an animal or a vegetable species could have the same appearance on both sides of the globe. I expected each animal, tree or blade of grass to be radically different, and its tropical nature to be glaringly obvious at a glance.... Looked at from the outside," he added, "tropical nature seemed to be of a quite different order from the kind of nature we are familiar with; it displayed a higher degree of presence and permanence. As in the Douanier Rousseau paintings of exotic landscapes, living entities attained the dignity of objects" (1973:55). But after a few months of living in the tropics, Lévi-Strauss changed his mind. He concluded that the tropics, as a concept, was simply out of date (1973:106).

Must we agree with Lévi-Strauss? Where does the concept of the tropical stand today? Certainly the stories we tell about the tropical regions are different from those of the past. Instead of the jungle, we now speak of the rain forest; instead of imagining Amazonia as pristine, unspoiled, virgin forest, we speak of how the Amazon region has been shaped, and in this sense produced, by Amerindians over the course of centuries (the preconquest population being estimated today as much larger than previously thought, and archeologically more sophisticated). Instead of superfecundity, we worry about the fragility of the forests. The detailed steps leading, conceptually, from the old stories to these new ones, have barely been outlined historically, let alone told; but conceptually and representationally, our view of Amazonia is undergoing radical revision.

But of course, we should not expect there to be easy consensus on how to narrate the tropics. Just because we now call tropical nature a rain forest instead of a jungle does not mean that we are now able to see it, and represent it, correctly, once and for all, as we could not do in the past. For *my* rain forest may not be the same as *your* rain forest, or an *Amerindian's* rain forest. There are different visions of Eden, which often conflict with each other. To some environmentalists, tropical rain forests are the last remaining forms of wilderness, exemplars of pure nature that must be left alone at all costs. To others, tropical nature signifies biodiversity, or a genetic library, a rich and endangered source of genetic information, which must be actively harvested, extracted, or "read" before it disappears. To some, sustainable development is compatible with preserving the rain forests; to others, it is not. To the remaining Amerindian populations, the Amazonian rain forest is not a source of world commodities or an aesthetic landscape in the European sense, but rather a local, lived-in space, and perhaps, a land title. To call something a rain forest is not, therefore, to settle what its significance is, either scientifically or emotionally.

Even the proper place of human beings in Amazonia is disputed. Historically, tropical nature was imagined by means of an erasure of human history and culture. Today, of course, so-called development in the Amazon is occurring at a staggering rate; the vast geographical terrain is filling up with new populations, much as the European naturalists in the nineteenth century hoped it would. New cities, roads, schools, and hospital clinics are sprouting up, along with mining, cattle ranching, and other agribusinesses that turn the natural resources of the world into commodities. All of these activities are destructive of the Amazon region as it once was; they are aspects of the culture-nature exchange that are as old as human beings, though their pace is unprecedented, and their consequences are of a different kind than those experienced before (because the land beneath the tropical canopy is relatively infertile, once the cover provided by the trees is gone the soil cannot support agriculture for long, as deforested temperate land can). But the human inhabitants of the tropical rain forests (estimated at 50 million people across the globe) have their claim to a space in the world, and it is only by seeing Amazonia as a human-nature hybrid that we can devise adequate policies toward its tropical environment.

Nature in the tropics, then, like other kinds of nature, is a heterogeneous thing, a mix of the natural and the artificial, the human and the nonhuman, the organic and the nonorganic; it is both a physical space of living and nonliving things and a human invention, as the landscape gardener Roberto Burle Marx grasped. Indeed, I would like to propose, as a kind of conclusion to this overview of the concept of tropicality, that in looking to the future, and in considering whether the concept of the tropical itself has a future, thinking in terms of a tropical *garden* may be a useful way of rethinking the human relationship to the natural world in a period when concepts of untouched "wilderness" and "pure nature" have lost scientific credibility or possibility. A garden might be thought of as both an expression of, and a metaphor for, the culture-nature nexus. As a human invention it is an artifice; but as a human construction it is also made out of the material, malleable stuff of the world. It is therefore unpredictable and not always easy to control (which is why a gardener's work is never done). A garden provides aesthetic pleasure, and an escape from the built environment of the city (something that becomes increasingly important as the city environment becomes everyone's). But as a human invention and aesthetic escape, to be successful as well as environmentally sustainable, a garden has to respect the ecological interdependence of species, both vegetable and animal, and allow untidiness to be part of the garden's achievement. This is an idea that Peter Coates (1998:177), in *Nature: Western Attitudes since Ancient Times*, has also discussed, namely the view of some environmentalists that the garden, rather than wilderness, is the appropriate metaphor to use as we think about how to repair our relation with nature.

Of course, as tropical nature becomes an endangered species, it may well be that a garden (or a theme park, or a picture, or a representation) is all that we will have left; that tropical nature will exist only as a Disneyland imitation, or as a greenhouse rain forest built for amusement in a Las Vegas hotel, which will even put on a show of a tropical downpour for its restaurant guests, while managing to keep the diners dry.

References

Ades, D. 1989. *Art in Latin America: The Modern Era, 1820–1980*. New Haven: Yale University Press.

Andrade, M. de. 1984. *Macunaíma: A Hero Without Any Character*. New York: Random House.

Andrade, M. de. 1993. *Fotógrafo e Turista Aprendiz*. São Paulo: Instituto de Estudos Brasileiros.

Arnold, D. 1993. *Colonizing the Body: State Medicine and Epidemic Disease in Nineteenth-Century India*. Berkeley: University of California Press.

Arnold, D. 2000. Illusory riches: Representations of the tropical world, 1840–1950. *Singapore Journal of Tropical Geography* 21(1):6–18.

Bates, H. D. 1892. *The Naturalist on the River Amazons*. With a memoir of the author by E. Clodd. London: J. Murray.

Brading, D. 1991. *The First American: The Spanish Monarchy, Creole Patriots, and the Liberal State, 1492–1867*. Cambridge: Cambridge University Press.

Cals, S. 1995. *Roberto Burle Marx: Uma Fotobiografia*. Rio de Janeiro: Gráfica Editora Hamburg.

Coates, P. 1998. *Nature: Western Attitudes since Ancient Times*. Cambridge: Polity Press.

Congresso Brasileiro de Tropicologia. 1987. *Ciência para os Trópicos*. Recife: Fundaj, Editora Massangana.

Cunha, E. da. 1944. *Rebellion in the Backlands*. Trans. S. Putnam. Chicago: Chicago University Press. (Orig. pub. 1902 as *Os Sertões*.)

Cunha, E. da. 1976. *Um Paraiso Perdido: Renunião dos Ensaios Amazonicos*. Petrópolis: Editora Vozes.

Darwin, C. 1964. *On the Origin of Species*. Facsimile of the first (1859) edition. Cambridge, Massachusetts: Harvard University Press.

Dettelbach, M. 1993. Humboldtian science. In N. Jardine, J. A. Secord, and E. C. Spary, eds., *Cultures of Natural History*, pp. 287–304. Cambridge: Cambridge University Press.

Freyre, G. 1946. *Masters and Slaves: A Study in the Development of Brazilian Civilization*. New York: Knopf.

Freyre, G. 1963. *New World in the Tropics: The Culture of Modern Brazil*. New York: Vintage Books.

Gerbi, A. 1973. *The Dispute of the New World: The History of a Polemic, 1750–1900*. Trans. J. Moyle. Pittsburgh: University of Pittsburgh Press.

Glacken, C. 1967. *Traces on the Rhodian Shore. Nature and Culture in Western Thought from Ancient Times to the End of the Eighteenth Century*. Berkeley: University of California Press.

Grove, R. 1995. *Green Imperialism: Colonial Expansion, Tropical Island Edens and the Origins of Environmentalism, 1600–1860*. Cambridge: Cambridge University Press.

Hecht, S. and A. Cockburn. 1989. *The Fate of the Forest: Developers, Destroyers, and Defenders of the Amazon*. London: Verso.

Herkenhoff, P. 1995. The jungle in Brazilian modern design. *Journal of Decorative and Propaganda Arts* 21:239–259.

Holanda, S. B. de. 1987. *Visión del Paraíso: Motivos Edénicos en el Descubrimiento y Colonización del Brasil*. Caracas, Venezuela: Biblioteca Ayacucho.

Humboldt, A. von. 1848–1858. *Cosmos: A Sketch of a Physical Description of the Universe*. Trans. E. C. Otté. 5 vols. London: H. G. Bohn.

Humboldt, A. von. 1849. *Views of Nature; or, Contemplations on the Sublime Phenomena of Creation*. Trans. E. C. Otté and H. G. Bohn. London: H. G. Bohn.

Humboldt, A. von. 1966. *Personal Narrative of Travels to the Equinoctial Regions of the New Continent, During the Years 1799–1804*. Trans. H. M. Williams. 7 vols in 6. Vol. 3:14–15. London: Longman (facsimile of 1818–1829 ed.); New York: AMS Press.

Leenhardt, J. 1994. Questions à Roberto Burle Marx. In J. Leenhardt, ed., *Dans les Jardins de Roberto Burle Marx*, pp. 53–75. Crestet: Actes Sud.

Lévi-Strauss, C. 1973. *Tristes Tropiques*. Trans. J. Weightman and D. Weightman. London: Cape. (Orig. pub. 1955.)

Livingstone, D. L. 1994. Climate's moral economy: Science, race, and place in post-Darwinian British and American geography. In N. Smith and A. Dolewska, eds., *Geography and Empire*, pp. 132–154. Oxford: Blackwell.

Pagden, A. 1993. *European Encounters with the New World: From Renaissance to Romanticism*. New Haven: Yale University Press.

Pratt, M. L. 1992. *Imperial Eyes: Travel Writing and Transculturation*. London: Routledge.

Singapore Journal of Tropical Geography. 2002. Special issue, "Constructing the Tropics," 21(10).

Stepan, N. L. 2001. *Picturing Tropical Nature*. London and Ithaca: Reaktion Books and Cornell University Press.

Stepan, N. L. 2003. The only serious terror in these regions: Malaria control in the Brazilian Amazon. In D. Armus, ed., *Disease in the History of Modern Latin America: From Malaria to AIDS*, pp. 25–50. Durham, North Carolina: Duke University Press.

Wallace, A. R. 1898. *The Wonderful Century: Its Successes and Failures*. New York: Dodd, Mead.

3 Demand for Two Classes of Traditional Agroecological Knowledge in Modern Amazonia

Charles R. Clement

There were probably 5–7 million people living in Amazonia at the time of European contact; it has been estimated that population densities in some areas were in excess of 25 individuals/km^2, although because of apparently low carrying capacity on the *terra firme* (dry uplands) interfluves, it was probably lower than one individual/km^2 in the region as a whole (Denevan 1992a). In order to feed and clothe that population, native Amazonian peoples domesticated many of the region's landscapes and at least 86 native plant species to varying degrees (Clement 1999). These two types of domestication (landscape and plant) are here considered to be separate, though interacting, classes of traditional agroecological knowledge. Both have roles to play in the search for sustainable agricultural development, and by extension sustainable development in general, in Amazonia and elsewhere in the humid tropics. The research and development (R&D) effort devoted to each class is quite different, however, with landscape domestication receiving the lion's share of R&D funding while indigenous Amazonian crop domestication languishes.

This paper reviews the concepts of landscape and crop domestication, and the sociopolitical, economic, and biological "costs of Amazonia"—to borrow from the better known phrase "costs of Brazil"—are examined to help us understand why the modern market economy penalizes the use of knowledge about domesticated landscapes, but would accept knowledge about domesticated crops if it were developed. Finally this paper suggests that traditional knowledge about landscapes and crops can contribute much more to sustainable agricultural development in modern Amazonia if R&D priorities are changed.

Landscape Domestication

Landscape domestication is a conscious process by which human manipulation results in changes in a landscape's ecology and in the demograph-
ics of its plant and animal populations, thereby creating a landscape that
is more productive and congenial for humans (Harris 1989). The intensity
of manipulation may vary widely, from promotion through management to
cultivation (table 3.1).

Pristine environments were probably rare in Amazonia at contact (Denevan 1992b; Smith 1995; Balée 1998). Promoted landscapes are manipulated
to favor desirable individual plants or plant populations (Groube 1989) and
are likely to remain promoted long after human interference ceases. Accidental promotion may also have been important, for example when fires

TABLE 3.1 Types of Landscape Domestication and Human Interventions, and
Permanence of Landscape in the Environment

Landscape[1]	Human intervention[2]	Permanence[3]
Pristine	No human intervention	Indefinite
Promoted	Minimal forest clearance, expansion of forest fringes	Long-term
Managed	Partial forest clearance, expansion of forest fringes, transplanting of desirable individual plants or planting of individual seeds, amendments to enhance plant growth, reduction of competition	Mid- to long-term
Cultivated	Forest felling and burning, localized or extensive tillage, seedbed preparation, transplanting seedlings and planting seeds, weeding, pruning, fertilizing, mulching, watering	Short-term
Swidden-fallow	Cultivated, then managed	Mid- to long-term

[1] The types of landscapes are listed along the continuum from pristine to cultivated. The list
does not include environmental modification for human habitations (e.g., the habitation
mounds on Marajó Island) (Roosevelt 1991).
[2] Incomplete listing of the human interventions necessary to create these landscapes.
[3] Permanence in the environment after human abandonment.
Source: After Clement 1999.

escaped from control in swidden clearings and favored the growth of a useful species such as the *babaçu* palm (*Attalea speciosa*). Managed landscapes are intensively manipulated to increase the abundance and diversity of food and other useful plant species (Alcorn 1989; Anderson and Posey 1989; Groube 1989), and the effects tend to remain long after humans have abandoned the area. Cultivated landscapes are created by the complete transformation of the biotic landscape to favor the growth of a few selected food plants and other useful species (Harlan 1992). The biotic components of this artificial landscape do not survive long after human abandonment because the changes that favor the growth of the human-selected species also favor the growth of weeds and the invasion of other secondary forest species. However a cultivated landscape may take a long time to return to a natural state. The abiotic transformations practiced in this landscape (e.g., the agricultural earthworks in various parts of lowland northern South America) often survive for long periods. (See Clement 1999 for further discussion of landscape domestication.)

The swidden-fallow sequence is the combination of a cultivated landscape (the swidden), which yields well for a few years but becomes progressively more difficult to weed and tend as soil fertility decreases, followed by the management of useful weeds and volunteer or transplanted shrubs and trees at progressively lower intensities until a managed secondary forest (the fallow) results (Denevan and Padoch 1987). The managed fallow may remain long after humans have abandoned it. It is the most common indigenous practice in use today, but may not have had the same importance before contact (Roosevelt 1989).

Balée (1989), Frickel (1978), Posey (1985), and others have identified various types of formerly domesticated landscapes in Amazonia by their higher-than-expected numbers of useful plant species. Some early promoted and managed landscapes that can still be seen today fall under Balée's (1989) definition of anthropogenic forest types, including palm, bamboo, and liana forests, and forest islands. Other previously managed or cultivated landscapes are recognizable four hundred years after contact because of their exceptionally high densities of incipient-domesticated and semidomesticated species, such as *bacuri* (*Platonia insignis*), Brazil nut (*Bertholletia excelsa*), *piquiá* (*Caryocar villosum*), *taperebá* (*Spondias mombin*), and *tucumã* (*Astrocaryum tucuma*) (Frikel 1978).

Patches of *terra preta do índio* (Anthrosols) also indicate early landscape domestication; these are usually located on bluffs overlooking rivers or previous river channels, and are occasionally found on the interfluvial plateaus (Smith 1980). Terra preta do índio is thought to have originated from kitchen middens and other areas of debris accumulation (Sombroek 1966; Smith 1980; Herrera et al. 1988). Sombroek (1966) also mentions *terra mulatta*, which has many chemical and physical characteristics of terra preta do índio but lacks the ceramic or other easily visible artifacts that typify the latter. The extent and importance of terra preta do índio and terra mulatta in the Amazon basin are not yet clear, but they are the subjects of a new international R&D effort (Lehmann et al. 2004). Several of the abandoned terra preta do índio sites that I have examined contain numerous incipient-domesticated, semidomesticated, and domesticated species; it is unlikely, however, that the sites contain exclusively pre-Columbian vegetation because their rich soils are also favored by modern inhabitants. Nonetheless they are clear examples of landscapes that were domesticated at contact.

The traditional ecological knowledge that created, managed, and maintained these domesticated landscapes may have much to offer modern attempts to develop sustainable agricultural systems in Amazonia (Moran 1993). Specific indigenous practices are often used, either without change or with modification, in agroforestry and forest management R&D today (Posey 1985) and have been widely studied over the last few decades (e.g., Denevan and Padoch 1987; Posey and Balée 1989; Plotkin and Famolare 1992; Altieri 1995).

Crop Domestication

Plant domestication is a coevolutionary process by which human selection on the phenotypes of promoted, managed, or cultivated plant populations results in changes in the population's phenotypes and genotypes that make them more useful to humans and better adapted to human intervention in the landscape. As with landscape domestication, there are degrees of crop domestication that reflect the intensity of selection and the time during which selection was practiced (table 3.2).

The majority of species in any biome are wild, for only about 3,000 species have been domesticated to any degree worldwide (Harlan 1992). Incidentally coevolved populations volunteer and adapt to human-disturbed environments without human selection, though they may undergo genetic change in the process. Weeds are incidentally coevolved populations and can enter the domestication process if humans start to select for their useful traits, and manage or cultivate them (Harlan 1992). Incipient-domesticated populations are modified by human selection and adaptation to landscape manipulation, but the average phenotype is still within the range of variation in the wild population for the trait(s) subject to selection. Semidomesticated

TABLE 3.2 Degrees of Crop Modification and Effects of Selection on Variation and Ecological Adaptation

| Population type[1] | Variation[2] | | Ecological adaption[3] |
	Phenotype	Genotype	
Wild	= wild type	= wild type	environments where naturally selected
Incidentally coevolved	? different	? different	domesticated landscapes
Incipient-domesticated	< wild type	< wild type	promoted and wild landscapes
Semidomesticated	> wild type	<< wild type	managed, promoted, and wild landscapes
Domesticated	>> wild type	<<< wild type	cultivated and extensively managed landscapes
Landrace domesticated	= semidomesticated or domesticated	= semidomesticated or domesticated	managed and cultivated landscapes

1. Degrees of crop modification shown along the crop domestication continuum from wild to domesticated.
2. Effects of selection on phenotypic and genotypic variation.
3. Consequences of phenotypic and genotypic variation on the ecological adaptation of the population.
Source: After Clement 1999.

populations are significantly modified by human selection and adaptation to landscape manipulation, with average phenotypes diverging from the variation found in the wild population for the trait(s) subject to selection. The population retains sufficient ecological adaptability to survive in the wild if human intervention ceases, but the phenotypic variation selected for by humans gradually disappears in the natural environment, although that may take dozens of generations. Domesticated populations may be phenotypically similar to semidomesticates or even more variable, but their ecological adaptation is reduced to being able to survive only in human-created environments, specifically in cultivated or intensively managed landscapes (Harlan 1992). If human intervention ceases, the population dies out in short order. In clonally propagated crops, a single genotype may be the domesticate, but in most cases is lost soon after it is abandoned. A landrace is a domesticated or semidomesticated population selected in a cultivated landscape within a restricted geographical region by an individual ethnic group (see Clement 1999 for more details). It is important to recognize that crop domestication interacts closely with landscape domestication, although the reverse is not always the case.

At least 19 domesticates, 29 semidomesticates, and 38 incipient-domesticates native to Amazonia can still be found among traditional (indigenous and peasant) and modern human populations (Clement 1999). This number of domesticates represents roughly 60 percent of the crops probably present in Amazonia at contact and roughly 30 percent of all crops domesticated in the Americas (León [1992] lists 257 crops in the Americas at contact). That makes Amazonia the third most important area of plant domestication in the Americas, after Mesoamerica and the Andes. Additionally, there were at least another 55 crops in Amazonia brought in from what is today northeastern Brazil, northern lowland South America, the Caribbean, Mesoamerica, and even the low-elevation Andes (Patiño 1963, 1964; León 1992; Clement 1999). In a biome dominated by trees, it is not surprising that 71 (80 percent) of the various types of Amazonian domesticates are woody perennials, although that may be an artifact of differential genetic erosion that was a consequence of human population loss after contact.

When human populations in Amerindian villages and urban centers were devastated by disease, slavery, war, and missionization during the centuries after contact, a large number of domesticated crop populations, which depended upon human intervention, became extinct, leading to the erosion of crop genetic diversity. The loss of villages probably equaled or exceeded the decline in population, estimated at 90–95 percent in the 100–150 years postcontact (Denevan 1992a). Clement (1999) hypothesized that 90 percent of the crop genetic diversity in Amazonia at contact was lost over the next centuries. The rate of loss depended upon a crop population's life history, physical stature, ecosystem of origin, degree of domestication, and the domesticated landscape to which it was adapted and in which it occurred when abandoned.

The traditional ecological knowledge that created, managed, and maintained these domesticated plant populations is similar to what plant breeders do today. Traditional human populations still conserve more crop genetic resources than do institutions (Oldfield and Alcorn 1987) and continue to create new variation. The crops themselves are widely used in Amazonia and some have become important worldwide (Clement 1989) — for example, manioc (*Manihot esculenta*), pineapple (*Ananas comosus*), and rubber

(*Hevea brasiliensis*). Because they are already adapted to the Amazonian environment and traditional agroecosystems, they may have much to offer modern attempts to develop sustainable agricultural systems in Amazonia (Clement 1992, 1997). This crop heritage has been partially inventoried and evaluated (e.g., Prance and Kallunki 1984; Prance and Balick 1990; Hernández Bermejo and León 1992; Clement 1999), but is not being as intensively developed as its potential merits (Clement 1992).

Using Traditional Knowledge

Information on the types of domesticated landscapes and the variety of domesticated indigenous crops has practical as well as theoretical value. It helps set priorities for genetic resources prospecting. It identifies species that have a high likelihood of being easily adapted to modern agroforestry and horticultural practices, and that represent potential market opportunities for a region with too few current options. And it provides examples of how to modify or transform the landscape to increase its economic value without completely destroying biodiversity and thus foreclosing future options.

Vigorous efforts to utilize and conserve these genetic resources and the ecosystems and agroecosystems where they still occur are vital tasks if modern Amazonian nations hope to emulate their Amerindian predecessors, who developed apparently sustainable agriculture in ecosystems that challenge and often defeat the agricultural practices imported from temperate zones today (Clement 1997). While that is easily said and generally agreed to, there are major difficulties in getting there from here, not the least of which are the undeveloped state of Amazonia and, paradoxically, its wealth of biodiversity. Traditional knowledge may hold answers to the second impediment, but only enlightened political decisions can change the course of Amazonian economic development. Consequently, the limitations to development are worth reviewing because they affect demand for traditional knowledge.

The Costs of Amazonia

Clement (1997) recently examined the biologic and socioeconomic limitations to new crop development in Amazonia. In fact, these limitations apply to sustainable development in general. Here I will lump these limitations under the title "the costs of Amazonia" (which has a direct relation to the costs of under- and distorted development in Brazil as a whole and is referred to by the economic community and popular media as "the costs of Brazil") and expand the analysis somewhat. Although these costs apply to other Amazonian countries as well, the main costs of Amazonia on agricultural development in Brazil are:

• *Small and poor markets.* The human population of Brazilian Amazonia is about 17 million and growing rapidly. Nonetheless this is only 10 percent of Brazil's population, living in 50 percent of the national territory. As in the rest of Brazil, the great majority of the Amazonian population is poor (e.g., in Manaus 5 percent of the population own 80 percent of the wealth, and in the hinterlands these proportions are more skewed still), surviving on one or two minimum salaries (in June 1998 the minimum salary was R$130 = US$113). For the producer it means that once subsistence is accounted for, excess production must be exported from the production area or from Amazonia as a whole. Exports are immediately limited by other costs of Amazonia.

• *Distance to markets.* Although Amazonia's population is largely urban today, it is concentrated in two major and a few minor urban centers (Becker 1995). Most of Amazonia is a long way from these centers; hence it is expensive to get produce to market. Brazil's main markets (e.g., São Paulo, Rio de Janeiro, Belo Horizonte) are especially far from the interior of Amazonia. The barrier of distance is compounded by transportation difficulties.

• *Deficient transportation infrastructure.* Before the military regime (1964–1985), transportation in Amazonia was essentially fluvial, and the political-economic system of the region provided at least partial support for the necessary infrastructure. During the military regime that infrastructure was essentially abandoned in favor of highways, including the famous Transamazon system, although these were never made truly perennial owing to lack of financial commitment to pave them. To facilitate their construction, the highways followed the interfluves; these were generally areas with the lowest human carrying capacity due to poor soils, and they did not connect well to the fluvial network. Since the end of the military regime, policy has shifted from government to government, with the current government supporting a major fluvial infrastructure project (the Madeira River Hydrovia) and several highways. Nevertheless, the transportation infrastructure in Amazonia remains deficient, making it expensive to transport goods within Amazonia and to areas outside the region.

• *Deficient storage and transshipping infrastructure.* The same policy inconsistencies applied to transportation infrastructure were applied to storage and transshipping infrastructure during and after the military regime. The need for action was recognized, but no coherent policy was developed; that resulted in absurd investments, such as a rice drying, storage, and transshipment facility that was built in far western Amazonia where no farmers were or are interested in rice. The Madeira River Hydrovia is well supported by storage and transshipping facilities at the terminals, but the new highway policy is not. These inconsistent policies increase costs for most users.

• *Deficient commercial networks.* When the military regime abandoned fluvial transport, the commercial network and the infrastructure on which it depended degraded. Highways have not been effective substitutes for rivers, and few alternative commercial networks have evolved to take production from the hinterlands to the urban centers and to the rest of Brazil. The resultant inefficiency further raises costs.

• *Distorted tax structure.* The military regime's decision to integrate Amazonia into Brazil included the creation of tax incentives to attract southern Brazilian businessmen to the region. Incentives included the Manaus Free Zone and cattle-ranching subsidies (including tax breaks). Large southern businesses and multinational corporations engaging in inappropriate development projects often pay lower taxes than do local businesses that produce, process, and commercialize Amazonian products. Consequently, the tax structure has effectively guided the Amazonian economy away from indigenous products. Although much needs to be done, some of the distortions are being addressed (e.g., tax breaks for some Amazonian products), some have been removed (e.g., new cattle-ranching incentives), and some have been scheduled for removal (e.g., the Manaus Free Zone).

• *Deficient research, development, and extension.* Although the economic potential of native Amazonian products has been recognized since contact (Hecht and Cockburn 1990), local R&D has always been poorly funded and staffed, resulting in little benefit to Amazonia (Smith et al. 1998). The little that has been done in Amazonia has attracted attention elsewhere, however, leading to the continued exodus of Amazonian products to areas where non-Amazonians make a tidy profit; rubber is the most famous example, and *guaraná* (*Paullinia cupana*) and *pupunha* (*Bactris gasipaes*) are recent and current examples, respectively. The majority of state governments in Amazonia do not consider R&D a priority; only Pará and Amazonas have created their constitutionally mandated Research Support Funds. Extension is in no better shape (Smith et al. 1998). Consequently, there is little available information in Amazonia about indigenous crops, and what there is does not go to the right places (e.g., the extension service and local businesses).

• *Poor soils.* Most (95 percent) Amazonian soils are nutrient poor and acidic. As a result, they are inappropriate for conventional agriculture without high fertilizer inputs. Even with high inputs, conventional agriculture may not be sustainable (NRC-CSAEHT 1993).

 • *Abundant rainfall.* Most of the soils in Amazonia are nutrient poor because they have been thoroughly weathered by millions of years of abundant rainfall (NRC-CSAEHT 1993). Although the abundance of rain favors crop growth, it also favors weed growth, and because wet soils compact more easily, it inhibits mechanization as well.

 • *High temperatures.* The mean annual temperature in Amazonia is about 26°C; like rainfall, that provides favorable conditions for growth of tropical and subtropical plants, and also for weeds and pests of all kinds. The high temperatures combined with abundant rainfall cause rapid degradation of biomass (NRC-CSAEHT 1993), resulting in insignificant humification (humus is more efficient at retaining nutrients in the soil than are minerals).

 • *Biodiversity.* Yes, biodiversity! The same biodiversity that contains so many potential economic options also contains a wonderful diversity of herbivorous animals and insects, pathogenic fungi, bacteria, viruses, and weeds. As soon as humans put too many individuals of the same crop together in a small space, one or more components of Amazonia's rich biodiversity appears to use the new resource (cf. Altieri 1995).

Many of these costs of Amazonia were also faced by precontact populations, who devised ways of overcoming or avoiding them. Many of the economic limitations never arose because the economy was local; that explains why the early chronicles contain frequent mention of the abundance of food all along the main river (Carvajal 1894). Only high-value products, including crop germplasm (León 1987), were traded beyond the local market (Bruhns 1994). This is the reason so many non-Amazonian crops were found in Amazonia at contact and the reason so many crops have left Amazonia, both before and since contact. Perhaps the main thing we can learn from precontact populations, however, is how they avoided the biological limitations of the costs of Amazonia.

Demand for Traditional Ecological Knowledge

All of the domesticated landscapes identified in this paper were highly heterogeneous, either in terms of species (especially promoted and managed landscapes) or in terms of genetic diversity (all landscapes). Even the species monocultures of manioc used in cultivated landscapes were composed of numerous cultivars (Chernela 1987). Thus diversity—in landscapes and in crops—was key to human adaptation to Amazonia, and it is precisely that diversity that has eroded most seriously since contact (Clement 1999). Nevertheless the identification of diversity as a key, perhaps *the* key, solution to avoiding biological limitations in turn identifies priorities for research, development, and extension (R&D&E).

The recognition of the importance of diversity is not new, of course. Brazil's EMBRAPA (Empresa Brasileira de Pesquisa Agropecuária) system clearly recognized the importance of diverse agroecosystems in the early 1990s and transformed each of its Amazonian research institutes into agroforestry R&D centers that were designed to study and develop heterogeneous production systems (domesticated landscapes). Although significant advances are being made today, it has also become clear that only low-diversity agroforestry systems are economically attractive (Smith et al. 1998), partly because of the current economic model of free-trade corporate capitalism (Daly 1993; Prugh 1995) and partly because of the political-economic costs of Amazonia. Diverse agroecosystems tend to be both more labor intensive and more knowledge intensive than monocultures (Altieri 1995), and hence more expensive. Additionally, diverse agroecosystems tend not to allow a scale of production that is economically efficient, and therefore lucrative, in the current economic model (Duncan J. Macqueen, DFID, pers. comm., 1999). Consequently, their produce is at a disadvantage in the market compared with similar products imported from conventional agroecosystems in other parts of Brazil and the world. Homma (1990) identified the same limitations on extractive resources and economies, which are essentially based on the same traditional agroecological knowledge as agroforestry systems.

Thus R&D alone will not be sufficient to develop highly diverse production systems based on traditional agroecological knowledge. At a minimum, other costs of Amazonia must also be modified to allow these systems to prosper, although changes in the current economic model are probably even more important (e.g., Prugh 1995). Reducing these costs is urgent, not only to alleviate rural poverty through increased productivity of marketable crops, but also to conserve biodiversity and crop genetic resources in the more complex agroecosystems pioneered by indigenous peoples and maintained by traditional populations in Amazonia. Conservation will only occur, however, if it is economically interesting for producers; that is, if there is demand from the market.

Diverse systems are only one part of the equation. An ample diversity of crops containing significant genetic diversity—while producing a relatively uniform product for market or agroindustry—is also essential. There is currently strong demand worldwide for new crops for food and industry (e.g.,

Janick 1996; Smartt and Haq 1997) and that is less constrained by the economic model than is traditional agroecological knowledge. This is because a
new product tends to have a high unit value when first introduced (because
it is rare, exotic, Amazonian, etc.), so that demand is high enough to pay
for overcoming the costs of Amazonia and Brazil. If demand continues to
increase, however, the sequence outlined by Homma (1990) is activated and
the new crop will be taken out of Amazonia to an area with lower production costs, ultimately outcompeting Amazonian production—rubber is the
classic example.

Unfortunately, no modern Amazonian nation has an effective program
to develop new crops, although all recognize its importance and have some
kind of program in place. Brazil is putting a considerable amount of money
into that effort through such programs as Programa de Apoio ao Desenvolvimento Científico e Tecnológico (PADCT III), Projetos de Pesquisa
Dirigida do Programa Piloto para a Preservação das Florestas Tropicais do
Brasil (PPD/PPG7), Programa de Desenvolvimento de Tecnologias Agrícolas para o Brasil (PRODETAB), and Programa Nacional de Biodiversidade
(PRONABIO). Although PADCT III encourages partnerships between businesses and R&D institutions, none of these programs has a clear plan of
action to establish priorities and direct R&D at the market. One plan might
be to require all applied crop R&D to base research priorities on a production-to-commercialization system (p-cs), something that is done elsewhere
in the world and that offers the promise of providing useable results quickly,
thus capturing political support as well.

A p-cs-based approach would make all R&D planning and execution market-aware by forcing researchers to think about the product from germplasm
to market, wherever that market might be. The first step in researching any
new or conventional crop would be to develop a p-cs model to organize everything that is known about the crop and its current markets (table 3.3). If
there is no market for a given crop, an existing market for a similar crop can
be used to develop a preliminary model. To be useful, the model must also
include costs of production, transformation, and commercialization. With
a model like this, both knowledge gaps and excessively expensive stages in
the p-cs can be identified and R&D can then be directed to fill gaps and
reduce costs. A purely market-driven model will not recognize diversity as a
necessary element, as was shown in the agroforestry systems work mentioned
above. This p-cs proposal will be difficult to implement, however, because
the R&D&E institutions and staff must be reeducated to work with it. The
EMBRAPA system has recognized that something like that is required, although it has not yet been able to implement it fully.

TABLE 3.3 Main Components of a Production-to-Commercialization System Model to Guide Research and Development with New Crops/Products in Amazonia

Production	Transformation	Commercialization
Germplasm prospecting	Postharvest physiology	Market identification
Germplasm enhancement	Postharvest handling	Competing products
Propagation	Preprocessing locally	Market profiles
Plantation design and density	Product development	Market agents
Fertilization	Processing locally/ centrally	Storage and transshipping
Plantation management	By-product processing	Transportation
Harvesting	Packaging	Consumer profiles

Note: Each component must be evaluated in terms of available knowledge and costs, so that Research, Development, and Extension makes the new crop/product more competitive in local, regional, national, and international markets. Although the components are listed in three columns, there are important interactions among them.
Source: Clement et al. 2000.

Another way to develop an abundance of highly diverse new crops would be to create a new-crops program in Amazonia. That would need only a little more money than is currently available from the programs mentioned above. Brazil once had such a program, called Pioneer Crops, but it was eliminated in 1990 during one of Brazil's recurrent economic crises (Rizzini and Mors 1995). A key component of the program was a coordinating committee that helped institutions and research groups identify priorities (in terms of both species and R&D objectives within species), obtain available information before starting out, and develop solid projects that were market-aware. The Rede de Recursos Genéticos da Amazônia (GENAMAZ) initiative, based at the Superintendência para o Desenvolvimento da Amazônia (SUDAM) in Belém, Pará, was working toward that, but was hobbled by its parent ministry—the federal Planning Ministry—which funded GENAMAZ on a contingency basis! SUDAM was recently closed and its replacement, the Amazonian Development Agency, has not yet indicated the priority it will give GENAMAZ. At the regional level, the Procitrópicos and Tropigen networks are also working toward that, but are underfunded by their sponsoring governments.

A major question exists about how to convince Amazonian governments to invest in R&D&E when there are so many other segments of society clamoring for investment. A priori, an apparently good argument is that some R&D&E has produced good results. Two examples are pupunha, which is being widely planted for its heart-of-palm, and *cupuaçu* (*Theobroma grandiflorum*), which is being planted for both its aromatic flavorful pulp and its cacao-like seed. However, a close look at each example reveals that more pupunha is planted in other areas of Brazil than in Amazonia and that even cupuaçu is following cacao to Bahia. Thus investments in R&D&E alone are not enough—other costs of Amazonia must be addressed at the same time. That makes the task of convincing Amazonian state governments to invest in R&D&E very difficult.

Nonetheless investments in Amazonian crops and landscapes will provide returns, if not for Brazilian Amazonia, then at least for other parts of Amazonia and the world. Given the difficult political situation, however, international support to improve interinstitutional, inter-Amazonian, and international integration, and to plan and execute market-aware R&D&E, is essential in order to accelerate current activities before more of this Amazonian heritage is lost forever.

Conclusions

Both classes of traditional Amazonian agroecological knowledge (i.e., landscape and crop domestication) have potential to contribute to sustainable agricultural development in Amazonia. The modern market economy is actively seeking new crops and products, including some of the domesticated crops of Amazonia. This is a real demand from today's market, but little is being done to supply it. Global society needs sustainable agroecosystems—perhaps like the domesticated landscapes that once occurred throughout Amazonia—although the modern market economy doesn't recognize this yet. This is a real demand from a future market; even though this market doesn't yet exist, there is work being done to supply it.

If current market demand is used to set priorities, Brazil's agricultural R&D effort has its priorities backward. On the one hand, there is considerable effort to translate traditional Amazonian agroecological knowledge about landscape domestication into viable agroforestry systems and forest management practices, even though the market is unresponsive. On the other hand, there is little effort to develop new crops for the local, national, and international markets from the considerable crop heritage left by indigenous Amazonian societies and conserved by modern traditional peoples, even though the market is responsive to new crops for food and industry. Increasing R&D efforts to develop new crops without decreasing R&D efforts to develop sustainable agroecosystems in Amazonia would make Brazil's agricultural R&D effort in Amazonia both more market driven and more responsible to future generations of Amazonians.

Acknowledgments

I thank Drs. Darrell A. Posey and Leslie Bethell of the Centre for Brazilian Studies, Oxford University, Oxford, UK, for the invitation to present this paper; John R. Palmer and the Department for International Development (DFID/UK) for the support that made the presentation possible; and Darrell A. Posey and Duncan J. Macqueen (DFID/UK) for numerous useful criticisms and suggestions on the manuscript.

References

Alcorn, J. B. 1989. Process as resource: The traditional agricultural ideology of Bora and Huastec resource management and its implications for research. In D. A. Posey and W. Balée, eds., *Resource Management in Amazonia: Indigenous and Folk Strategies*, pp. 63–77. Advances in Economic Botany 7. Bronx: New York Botanical Garden.

Altieri, M. A. 1995. *Agroecology—the Science of Sustainable Agriculture*, 2nd ed. Boulder, Colorado: Westview Press.

Anderson, A. B. and D. A. Posey. 1989. Management of a tropical scrub savanna by the Gorotire Kayapó of Brazil. In D. A. Posey and W. Balée, eds., *Resource Management in Amazonia: Indigenous and Folk Strategies*, pp. 159–173. Advances in Economic Botany 7. Bronx: New York Botanical Garden.

Balée, W. 1989. The culture of Amazonian forests. In D. A. Posey and W. Balée, eds., *Resource Management in Amazonia: Indigenous and Folk Strategies*, pp. 1–21. Advances in Economic Botany 7. Bronx: New York Botanical Garden.

Balée, W. 1998. Operational equivalence of primary and secondary forests in Amazonia: A case study from eastern Bolivia. Talk presented at Human Impacts on Environments of Brazilian Amazonia: Does Traditional Ecological Knowledge Have a Role in the Future of the Region? Organized by Darrell A. Posey, The Centre for Brazilian Studies, Oxford University, Oxford, UK, June 5–6, 1998.

Becker, B. K. 1995. Undoing myths: The Amazon—an urbanized forest. In M. Clüsener-Godt and I. Sachs, eds., *Brazilian Perspectives on Sustainable Devel-*

opment of the Amazon Region, pp. 53–89. Man and the Biosphere 15. Paris: UNESCO and Carnforth, Parthenon Publications.

Bruhns, K. O. 1994. *Ancient South America*. Cambridge: Cambridge University Press.

Carvajal, G. de. 1894. *Descubrimiento del Río de las Amazonas*. Sevilla, Spain: Imprenta de E. Rasco.

Chernela, J. M. 1987. Os cultivares de mandioca na área do Uaupés (Tukâno). In B. G. Ribeiro, ed., *Suma Etnológica Brasileira. 1. Etnobiologia*, pp. 151–158. Petrópolis, RJ, Brasil: Financiadora de Estudos e Projetos.

Clement, C. R. 1989. A center of crop genetic diversity in western Amazonia. *BioScience* 39:624–631.

Clement, C. R. 1992. Los cultivos de la Amazonia y Orinoquia: Origen, decadencia y futuro. In J. E. Hernández Bermejo and J. León, eds., *Cultivos Marginados: Otra Perspectiva de 1492*, pp. 193–201. Colección FAO: Producción y protección vegetal, no. 26. Córdoba: Jardín Botánico de Córdoba and Rome, Food and Agriculture Organization (FAO).

Clement, C. R. 1997. Environmental impacts of, and biological and socio-economic limitations on new crop development in Brazilian Amazonia. In J. Smartt and N. Haq, eds., *Domestication, Production and Utilization of New Crops*, pp. 134–146. Southampton, England: International Centre for Underutilised Crops, University of Southampton.

Clement, C. R. 1999. 1492 and the loss of Amazonian crop genetic resources. I. The relation between domestication and human population decline. *Economic Botany* 53(2):188–202.

Clement, C. R., J. T. Farias Neto, J. E. Urano de Carvalho, A. G. Claret de Souza, T. M. Souza Gondim, J. F. Silva Lédo, and A. A. Müller. 2000. Fruteiras nativas da Amazônia: O longo caminho entre caracterização e utilização. In T. B. Cavalcanti and B. M. Teles Walter, eds, *Tópicos Atuais em Botânica*, pp. 253–257. Palestras Convidadas do 51° Congresso Nacional de Botânica, Soc. Botânica do Brasil, Brasília, DF, July 23-29, 2000.

Daly, H. E. 1993. The perils of free trade. *Scientific American* 269:50–57.

Denevan, W. M. 1992a. Native American populations in 1492: Recent research and a revised hemispheric estimate. In W. M. Denevan, ed., *The Native Population of the Americas in 1492*, pp. xvii–xxxvii. Madison: University of Wisconsin Press.

Denevan, W. M. 1992b. The pristine myth: The landscape of the Americas in 1492. *Annals of the Association of American Geographers* 82:369–385.

Denevan, W. M. and C. Padoch, eds. 1987. *Swidden-fallow Agroforestry in the Peruvian Amazon*. Advances in Economic Botany 5. Bronx: New York Botanical Garden.

Frickel, P. 1978. Áreas de arboricultura pre-agrícola na Amazônia: Notas preliminares. *Revista Antropológica* 21:45–52.

Groube, L. 1989. The taming of the rain forests: A model for Late Pleistocene forest exploitation in New Guinea. In D. R. Harris and G. C. Hillman, eds., *Foraging and Farming: The Evolution of Plant Exploitation*, pp. 292–304. London: Unwin Hyman.

Harlan, J. R. 1992. *Crops and Man*, 2nd ed. Madison, Wisconsin: American Society of Agronomy/Crop Science Society of America.

Harris, D. R. 1989. An evolutionary continuum of people-plant interaction. In D. R. Harris and G. C. Hillman, eds., *Foraging and Farming: The Evolution of Plant Exploitation*, pp. 11–26. London: Unwin Hyman.

Hecht, S. and A. Cockburn. 1990. *The Fate of the Forest: Developers, Destroyers and Defenders of the Amazon*. New York: Harper Perennial.

Hernández Bermejo, J. E. and J. León, eds. 1992. *Cultivos Marginados—Otra Perspectiva de 1492*. Colección FAO: Producción y protección vegetal, no. 26. Córdoba: Jardín Botánico de Córdoba and Rome, Food and Agriculture Organization (FAO).

Herrera de Turbay, L. F., S. Mora Camargo, and I. Cavelier de Ferrero. 1988. Araracuara: Selección y tecnologia en el primer milenio A.D. *Colombia Amazonica* 3:75–87.

Homma, A. K. O. 1990. *A dinâmica do extrativismo vegetal na Amazônia: Uma interpretação teórica*. Documentos 53. Belém, Pará: Centro de Pesquisas Agropecuárias do Trópico Umido (CPATU/EMBRAPA).

Janick, J., ed. 1996. *Progress in New Crops*. Alexandria, Virginia: American Society for Horticultural Sciences Press.

Lehmann, J., D. Kern, B. Glaser, and W. Woods, eds. 2004. *Amazonian Dark Earths—Origin, Properties, and Management*. Dordrecht: Kluwer Academic Publishers.

León, J. 1987. *Botánica de los Cultivos Tropicales*. San José, Costa Rica: Instituto Interamericano para la Cooperación Agrícola (IICA).

León, J. 1992. Los recursos fitogenéticos del Nuevo Mundo. In J. E. Hernández Bermejo and J. León, eds., *Cultivos Marginados: Otra Perspectiva de 1492*, pp. 3–22. Colección FAO: Producción y protección vegetal, no. 26. Córdoba: Jardín Botánico de Córdoba and Rome, Food and Agriculture Organization (FAO).

Moran, E. F. 1993. *Through Amazonian Eyes: The Human Ecology of Amazonian Populations*. Iowa City: University of Iowa Press.

NRC-CSAEHT. 1993. *Sustainable agriculture and the environment in the humid tropics*. Washington, DC: National Research Council, Committee on Sustainable Agriculture and the Environment in the Humid Tropics, National Academy Press.

Oldfield, M. L. and J. B. Alcorn. 1987. Conservation of traditional agroecosystems. *BioScience* 37:199–208.

Patiño, V. M. 1963. *Plantas Cultivadas y Animales Domésticos en América Equinoccial. Tomo I. Frutales*. Cali, Colombia: Imprenta Departamental.

Patiño, V. M. 1964. *Plantas Cultivadas y Animales Domésticos en América Equinoccial. Tomo II. Plantas alimenticias.* Cali, Colombia: Imprenta Departamental.

Plotkin, M. and L. Famolare, eds. 1992. *Sustainable Harvest and Marketing of Rain Forest Products.* Washington, DC: Island Press.

Posey, D. A. 1985. Indigenous management of tropical forest ecosystems: The case of the Kayapó Indians of the Brazilian Amazon. *Agroforestry Systems* 3:139–158.

Posey, D. A. and W. Balée, eds. 1989. *Resource Management in Amazonia: Indigenous and Folk Strategies.* Advances in Economic Botany 7. Bronx: New York Botanical Garden.

Prance, G. T. and M. J. Balick, eds. 1990. *New Directions in the Study of Plants and People.* Advances in Economic Botany 8. Bronx: New York Botanical Garden.

Prance, G. T. and J. A. Kallunki, eds. 1984. *Ethnobotany in the Neotropics.* Advances in Economic Botany 1. Bronx: New York Botanical Garden.

Prugh, T. 1995. *Natural Capital and Human Economic Survival.* Solomons, Maryland: International Society for Ecological Economics Press.

Rizzini, C. T. and W. B. Mors. 1995. *Botânica Econômica Brasileira,* 2nd ed. Rio de Janeiro: Âmbito Cultural.

Roosevelt, A. C. 1989. Resource management in Amazonia before the conquest: Beyond ethnographic projection. In D. A. Posey and W. Balée, eds., *Resource Management in Amazonia: Indigenous and Folk Strategies,* pp. 30–62. Advances in Economic Botany 7. Bronx: New York Botanical Garden.

Roosevelt, A. C. 1991. *Moundbuilders of the Amazon: Geophysical Archaeology on Marajó Island, Brazil.* San Diego: Academic Press.

Smartt, J. and N. Haq, eds. 1997. *Domestication, Production and Utilization of New Crops.* Southampton, England: International Centre for Underutilized Crops, University of Southampton.

Smith, N. J. H. 1980. Anthrosols and human carrying capacity in Amazonia. *Annals of the Association of American Geographers* 70:553–566.

Smith, N. J. H. 1995. Human-induced landscape changes in Amazonia and implications for development. In B. L. Turner II, A. Gómez Sal, F. González Bernáldez, and F. di Castri, eds., *Global Land Use Change—A Perspective from the Columbian Encounter,* pp. 221–251. Madrid: Consejo Superior de Investigaciones Científicas.

Smith, N., J. Dubois, D. Current, E. Lutz, and C. Clement. 1998. *Agroforestry Experiences in the Brazilian Amazon: Constraints and Opportunities.* Brasília, DF: Pilot Program to Conserve the Brazilian Rain Forest, Ministério do Meio Ambiente, Recursos Hídricos e Amazônia Legal and World Bank.

Sombroek, W. G. 1966. *Amazon Soils: A Reconnaissance of the Soils of the Brazilian Amazon Region.* Wageningen: Centre for Agricultural Publications and Documentation.

4 Fire in Roraima, 1998—Politics and Human Impact

What Role for Indigenous People in Brazilian Amazonia?

Elizabeth Allen

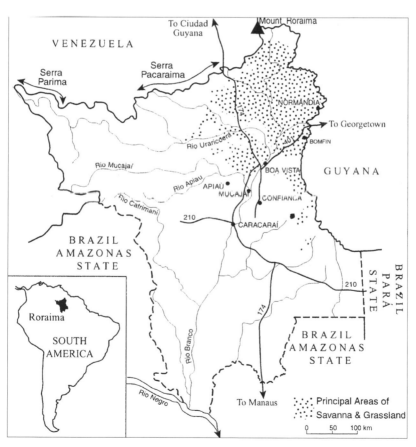

FIGURE 4.1 The state of Roraima, Brazil.

Some of the most dramatic and potentially most disastrous instances of human impact on Brazilian Amazonia in recent years were the fires that swept through the northernmost state of Roraima (figure 4.1) from January to April 1998. They presented a threat to the environment that seized the world's headlines and created a major embarrassment for the president of Brazil and his government at a critical stage in the campaign for presidential and congressional elections. More important, the fires, and their causes and the response to them, brought into sharp focus and to the front of public debate crucial issues about the nature of "human impact" throughout Amazonia, not least for its indigenous population.

The Roraima fires of 1998, and others that accompanied or soon followed them, are a reminder that human impact is, inescapably, a political issue. It involves human choices, priorities, and vested interests, and the political mechanisms both to express and to control them. In the case of Roraima, it involves politics at federal, state, and municipal levels. It also includes the politics of both international collaboration and resistance to it. This paper reflects on the fires of 1998 in a limited case study that raises important, still open, questions about human impact in Brazilian Amazonia, as well as in Amazonia more generally.

The State of Roraima

The Roraima conflagration of 1998, the largest fires by that date to affect Amazonia, burned for 63 days over a front of some 250 miles (400 kilometers), and had 2,000 focal points. It consumed vegetation in up to 25 percent of the area of the state, affected more than 39,000 people, and created damage estimated at R$15 million. It burned 10,000 to 15,000 square kilometers of intact forest (Nepstad et al. 1999; Woods Hole Research Center 2000). Like other crises in Brazil in recent years, the fires, initially limited in scope, threatened—for a mixture of reasons involving human error, confusion and unpreparedness—to turn into a national disaster (Allen 1994).

The fires occurred within a context of rapid and broad changes throughout Amazonia (Hemming, chapter 1 this volume), when in a period of about thirty years some 15 percent of the humid forest of Brazil and some 50 percent of the *cerrado*, or dry forest, were destroyed. The fires, in other words, have to be seen at one level as only another phase following the 1980 "decade of destruction," which means that by August 1998 about 532,000 square kilometers of Brazilian Amazonian rain forest had been cut down. Some 17,000 square kilometers of cover were lost in 1998 alone, a large part of which was due to the "devastating fires in Roraima in the early part of the year" (LARR-Brazil Report 1999:8).

Prior to those fires, the state of Roraima had not attracted the same level of national and international concern about tropical forest destruction as had some other parts of Amazonia. Destruction of the forests became more central to debate in Brazil after the return of civilian government in 1985; this was especially so in terms of indigenous land rights, particularly those of the Yanomami. The invasions of *garimpeiros* (wildcat gold prospectors) also threatened Yanomami rights, and even their survival. Otherwise, Roraima— Brazil's northernmost state, adjacent to Venezuela and Guyana—remained largely on the fringes of national consciousness, isolated by distance and poor communications (though less so than before the advent of air transport and the road-building programs of the 1970s).

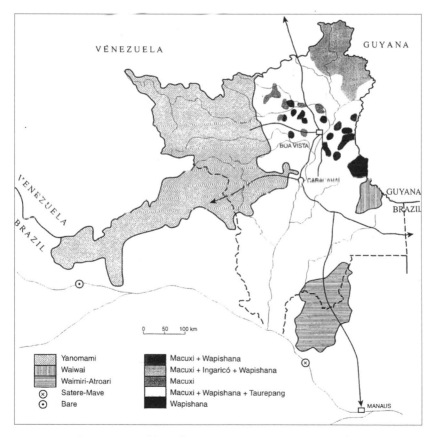

FIGURE 4.2 Areas occupied by indigenous groups in Roraima, Brazil, in 1998.

Roraima is a region of savanna, contact/transition forest, and rain forest. Traditionally it posed no serious issues of frontier dispute and fewer points of internal tension or violence than other parts of Amazonia, such as Pará or Acre. Most of the state is relatively low lying and drained by the Rio Branco and its tributaries. There are mountainous regions in the north and west of the state which rise to over 2,772 meters at Mount Roraima and are mainly covered by humid forest. Roraima has one national park, one national forest, and three ecological stations.

Roraima remained sparsely populated into the 1990s. It is about the size of Great Britain, with an area exceeding 230,000 square kilometers and a population under 250,000 (IBGE 1992). It became increasingly attractive to internal migration, in common with the rest of Amazonia. Recent research by the Brazilian Institute of Geography and Statistics (IBGE) suggests that Roraima is the Brazilian state that currently most attracts migrants, drawn by the availability of land and the development of new *municípios* (the smallest level of government in Brazil) (*Vejá* 1999:71).

The state has the third largest indigenous population in Brazil (about 39,000 people, or 14 percent of the country's total), and throughout Roraima the Yanomami, Macuxi, Wapixana, Ingarico, Tauripang, Yekuana, Waiwai, and Waimiri-Atroari make up substantial portions of the rural population. The number of Yanomami is estimated at 9,000–16,000 in Brazil, with an additional 12,000–14,000 in Venezuelan territory (Grimes 1992; CCPY 1995). Figure 4.2 shows the locations of indigenous areas in Roraima. In the municípios around the capital of Boa Vista, 17 percent of the population is Indian; in Normandie, 92.7 percent; and around Bomfim, near the Guyanese border, 23 percent (Allen 1989).

Roraima's savannas traditionally have been dominated by ranching, but over the last forty years they have seen the same processes of spontaneous and directed changes that have taken place elsewhere in Amazonia. In January 1978 just some 100 square kilometers were deforested, but by August 1996 the figure had grown to 5,361 square kilometers (IBAMA 1998b). Recent developments include new towns and farms; government settlement schemes; the development of large-scale commercial farming for rice, cattle, and horticulture; and steadily increasing mineral exploitation (Hemming 1990; Furley 1994; MacMillan 1995).

While the burning of biomass in the eastern Amazonian area and in the cerrados of central Brazil has been of great and increasing concern since the 1980s, the burning in the state of Roraima has been historically insignificant. Roraima contributes very little to the total burning that takes place in Brazil each year, and the wildfires of 1998 were a unique occurrence for this part of northern Brazil (Kirchhoff and Escada 1998:13).

In strategic terms, Roraima is also considered to be of prime importance to Brazil because of its location and its available and potential natural resources. Since 1985, those have been the fundamental reasons for military and economic development under the *Calha Norte Project* (Allen 1992a, 1992b, 1992c). Roraima now has road links from Boa Vista south to Manaus, north to Venezuela, and northeast to Guyana and the Caribbean coast, accelerating the process of both national and international integration.

Fire and Indigenous Practice

It was within this context that the 1998 fires in Roraima became, firstly, top international news, and secondly and somewhat belatedly, cause for national concern. Fire has always played a key role in clearance, agriculture, and settlement in Roraima, as in other similar environments. Unfortunately, the indigenous experience was either overlooked or deliberately ignored in Roraima, as in Amazonia more generally, even though in other parts of the humid tropics it is now being given high priority. For this reason it is worth stressing the indigenous practice in the use of fire in Roraima.

Natural fires caused by lightning, particularly in areas of deciduous vegetation such as the cerrado, have had a regenerative effect without which many seeds would not germinate, but it is hard to evaluate the influence of early natural fires. There is evidence that fires occurred in the humid forest six thousand two-hundred-sixty years ago, although it is not clear whether they were natural or man made (Jordan 1989; Schaefer 1994:14). Some research indicates that fire has been a moderate-level disturbance for tropical rain forests, as long as it has occurred repeatedly, infrequently, and at low intensity. Grassland and savanna areas with a lengthy dry season have burned most often, but the remains of charcoal, with or without human artifacts, suggest that within the last six thousand years there have been fires in areas of human settlement as well as in primary rain forest (Sanford et al. 1985).

Evidence at sites where recent wildfires have killed mature rainforest trees points to human origin, but only a slight alteration of climate in the northern Amazonian forests could generate wildfires that would burn well-developed *terra firme* (noninundated) and *igapó* (inundated) forests. Although there is indication that fire occurred in the forests well before the earliest traces of human presence—and destroyed large areas of forest, perhaps because of drier conditions—it is generally accepted that the existence of charcoal in mature forest soils is solely the result of ancient slash-and-burn agriculture (Sanford et al. 1985).

Indigenous groups in Amazonia have used fire for at least the past two thousand years as part of their adaptation to both rainforest and savanna environments—for hunting, as a process to encourage selected vegetation regeneration, and to prevent the invasion of savanna areas by forest species (San Jose and Farinas 1971; Posey and Balée 1989; Eden 1990; Schaefer 1994). Even in areas considered at first sight to be "untouched," soil studies in this region have revealed carbon evidence of fires dating back two hundred to three hundred years over wide areas of the humid forest (Jordan 1989:14).

Fire was, and still is, an important tool for indigenous peoples. They knew how to use it in ways that met not only their immediate demands but also modified their environment, and this view is one that is now occupying studies in a variety of disciplines, including ethnobotany, anthropology, archaeology, history, botany, and forestry. (For research on the indigenous peoples of the Americas, see, for example, Boyd 1999. For the history of indigenous use of forest fires in Mexico, see International Forest Fire News 1998a; Asbjømsen, Næss, and Torres 1999; Goldammer 1999b).

Examples of environmental modification by fire are there to be seen in the Amazon region: the Tupi, for example, used to practice coivara (a burning of land to destroy excess leaves and branches) in the lower Amazon before European contact in America (Kirchhoff and Escada 1998:4). Early European visitors to the Roraima region, such as Captain Lobo D'Almada in 1787, recorded the use of anthropogenic fires in this part of Brazil (Schaefer 1994).

Depending upon the particular environment within Amazonia, there are variations in the way fire is used by indigenous peoples, and these practices may hold some clues for the ways in which fire could be contained and controlled in the future. One of the fundamental differences between modern techniques of environmental development and the indigenous approach to the Amazonian environment is that the traditional methods utilize complex systems of agriculture and intensive soil management to overcome the "poor soils" and "fragility" of the Amazonian area. Indigenous knowledge systems are rich in management techniques for poor soils, including the use of fire to enrich production (e.g., Hecht and Posey 1990).

The basic reason for the indigenous use of fire is that in subsistence agricultural practice fire is an essential prerequisite to planting, and makes manual work much easier. It is particularly useful for increasing soil nutrients (particularly calcium, magnesium, and potassium) in the short term, as well as for clearing weeds. This traditional mode of fire clearance affects relatively small areas and allows cultivation for up to four years. It relies on a long-term rotational system, as it requires decades to replace the losses from the ecosystem (Jordan 1989; Goldammer 1990).

In the areas of the Amazon Basin that have blackwater rivers—such as the Rio Negro in northern Brazil, the Río Vaupés in southwestern Colombia, and the Venezuelan tributaries that flow into the Amazon River—forests have a relatively low biomass and good insolation. This promotes a good quality burn of cut vegetation, which in turn determines the yield of the subsequent crops. In some cases the biomass dries so well in some areas that after as few as 20 rainless days the burn may go beyond the cleared areas, setting hundreds of hectares on fire (Clark and Uhl 1987).

In these types of regions, indigenous peoples leave the cleared land to return to fallow in order to encourage regeneration before the area becomes too nutrient-poor to return to forest. Regeneration is further promoted by the planting of fruit trees; this in turn encourages the proliferation of birds and bats, whose pollination and seed-bearing droppings accelerate the speed of forest replacement. This kind of indigenous practice could be utilized to replace forest burned by uncontrolled fires, but the process is long and slow. Moran (1993), for example, argues that, even with these practices, regeneration of forest in these areas will take one hundred years. The slow process of regeneration is echoed in other types of Amazonian environments. It is claimed that in upland forests of the terra firme, for example, where weed invasion, rather than poor soils, forces the indigenous population to move to new sites, secondary succession of vegetation can take between ten and one hundred (or more) years (Sanchez et al. 1982; Uhl and Clark 1983; Fearnside 1992). That, too, suggests that the use of indigenous techniques and "natural" restoration of areas burned by fires, whether deliberately or accidentally, can be seen only as long-term solutions.

Research in both archaeology and ecology shows that fire has been used for a long time by indigenous peoples in Amazonia. This is seen in the strong correlation between the black-earth areas in Amazonia and the presence of liana forest, for example in the lower valleys of the Tocantins and Xingú rivers. Liana forest has an unusual concentration of plants that are useful to human populations and occur more often in areas having anthropogenic soil patches; that kind of vegetation is not the result of spontaneous fires: it only occurs in regions where the forest has been cut and allowed to dry. Those areas may have been occupied by pre-Columbian chiefdoms (Balée 1989; Moran 1993:71–72).

Indigenous use of fire for the cultivation of forested areas generally occurs on small parcels of land, maybe less than two hectares per household, and in this respect differs greatly from the fires that ran out of control in Roraima in 1998. In the practice of shifting cultivation, the dry leaves are burned to volatilize nitrogen. The tree trunks are not burned, but are left to decompose over perhaps ten years. The main causes of nitrogen loss are crops and denitrification, but rainfall and the planting of leguminous plants that fix nitrogen in the soil make up for some of the loss. After three years of cultivation at this scale, perhaps only 20 percent of nitrogen is actually lost (Moran 1993:73–75).

Traditional techniques also employ burning to eliminate sprouting plants and to slow down the return of successional forest. Hotter burns are better at retarding sprouting plants and weeds, they make the period of cultivation easier by requiring less work, and they produce larger yields. Mistakes in judging and carrying out the burning process mean that there will be more weeds, more pests, and fewer nutrients in the soil.

The main plants to benefit in the short term from burning are seed stocks and grasses. In a frequently disturbed environment, the ability to regenerate after burning is one of the most important attributes governing the survival of most woody plant species. Lianas, grasses, and herbs easily invade such sites, but only a few tree species can achieve regeneration (Preisinger et al. 1998). For that reason, burning in transitional and savanna areas is widely used to create an environment suitable for grazing animals (e.g., Moran 1993:73–75). Fire has been used in this way for centuries on the cerrados of Roraima and adjacent countries by both indigenous groups and incomers of mainly European origin.

Research from other Amazonian savanna areas has revealed significant changes in savanna vegetation, possibly created deliberately by indigenous techniques and environmental manipulation. In the savanna areas of eastern Amazonia, which are utilized by the Kayapó indigenous groups, islands of food production have been manipulated within the environment; up to 54 species, mainly of woody vegetation, are used. That kind of environment is also found in transitional areas between the cerrado and the forest. Although there is discussion about whether those woody plants are anthropogenic or a natural process of selection, it is clear that those islands do now provide important food sources both for plants and for animal habitats (Posey 1985; Parker 1992; Moran 1993). The Kayapó experience represents real expertise in successful tree planting, and as such could also provide important keys for the regeneration of areas severely affected by fire.

Use of fire for clearance depends on land being readily available. For shifting cultivators such as the Yanomami in Roraima, for example, it may provide for their traditional needs and practices, but when used by more intensive cultivators, or in areas where the cycle of rotation is too short, there is often neither time nor space for effective regeneration. Studies suggest that fire-return intervals of fewer than ninety years can eliminate rainforest species, and that returns of fewer than twenty years may eradicate trees entirely; fire-induced changes in the very short term, although taking several years to occur, are likely to make irreversible changes, transforming forest to savanna or grassland (e.g., Cochrane et al. 1999).

Short-term rotation and the use of fire for clearance can cause severe degradation in just a couple of decades. All of this indigenous experience should have been relevant in terms of responses to the 1998 crisis in Roraima, but was ignored: the indigenous peoples were not consulted. Indigenous knowledge should be built into any longer-term plans for land use and for preventing and controlling fire in Roraima and other parts of Brazilian Amazonia.

Fire Use in Roraima

In the grassland, or cerrado, area in Roraima, reports from the 1920s show that fire was sometimes used "two or three times a year, at random intervals" in order to improve pasture for the herds of cattle and horses; this was a practice borrowed from the Indians who used the technique for hunting (Hemming 1990:8). Fire is a controlling element in the environment and the predominance of fire-resistant species that have thick bark for protection is one indicator of long-established protection against burning.

Fire is also practiced by migrant settlers, cattle ranchers, and *colonos* (settlers), who rely on burning, planting, and harvesting for their own use and for the market. Through fire clearance, they also establish occupation of the land, confirm their claim, and increase the land's value.

Using fire to clear land creates large, open expanses of grasses for ranchers, who believe, sometimes mistakenly, that fire stimulates the growth of new vegetation. In 1967, for example, cattle ranchers in Roraima used the "practice of burning the vegetation cover in order to encourage the growth of new, tender shoots," though, in fact, this ended up reducing range quality because it produced "tougher, more fire resistant species" (Riviere 1972:10). That practice is still widely used by cattle ranchers and colonos who want to open up areas to settlement and production. Landsat imagery obtained in 1991 illustrated the pattern of "normal" fire use in Roraima. It showed that the incidence of fire was greater in cerrado than in forest areas and that, in general terms, fire use in Roraima increased from south to north (Schaefer 1994:14–15).

The use of fire by new settlers and ranchers in Roraima has been severely criticized. One expert in soil science defined it as in effect "mining the area" and called its purpose "purely speculative." Many of the migrants come from the arid northeastern part of the country (Maranhão, Ceará, and Piauí) as well as from the southern states of Paraná and Rio Grande do Sul; they have little or no experience with the type of forest frontier vegetation that occurs between savanna and humid forest. Accidental fires, those that burn out of control, result in damages for all of society; the fear of loss of investment (e.g., in fencing, buildings, and forest management programs) is a disincentive to sustainable forest management and agricultural intensification (IPAM 1998; Woods Hole Research Center 2000).

The process of migration and settlement has therefore substantially contributed to deforestation in Roraima. The average rate of deforestation declined marginally from 1990 to 1991, but even so, more than 210 square kilometers of forest have been destroyed each year since 1992 (IBAMA 1998b). The pressure of numbers and the greater proximity of fires used to clear cerrado and forest for permanent cultivation have set the scene whereby, in an unusually dry season, the normal burning of cleared vegetation can get out of hand. Fire then becomes the wild, devouring enemy, resulting in diminished cattle fodder, negligible biomass, and a very low regenerative potential.

The 1998 Fires

Within the framework of steadily intensifying settlement and changes in land use, the 1998 fires in Roraima became an uncontrolled, unregulated disaster that was broadcast to the world. It flared up during the "burning season"—an established part of the calendar for Amazonia. That is the time when small-scale cultivators and large-scale landowners alike take advantage of the dry season to burn off cleared vegetation. A serious drought had prevailed in the region since early September 1997 and was exacerbated by the effects of El Niño, which eliminated the "little rainy season" at the end of the year. In combination with man-made pressures on the land, these natural events led to what has been termed "the climatic event of the century" (McPhaden 1999). Usually, 600 millimeters of rain fall between September and December, but during that period in 1997 none fell. The failure of the little rainy season meant that by January the area was suffering severe, prolonged drought conditions that threatened to kill 100,000 head of cattle and 500,000 tomato plants (*Istoé* 1998a). Severe, recurring droughts are not unusual for this area, however. There is good oral evidence from indigenous groups that such *secas* occurred, for instance, during the 1940s and the late 1970s (Rival 1998).

What was different in 1998 was that the annual preparation for forest and savanna clearance had left larger areas, which when combined with drier conditions, were much more vulnerable to fire. As more fires were started and took firm hold, high temperatures and strong winds of up to 100 kilometers an hour increased the effect of burning, taking sparks and red-hot ashes high into the sky to ignite new fires elsewhere. Crisis swiftly turned into disaster. By the first of April, firefighters and military personnel were working on nine fronts: Serra de Pacaraima, Anajari, Mucajai, Apiau, Ajanari, Caracarai, Boqueiro, Confianca, and Ilha do Maracá. Overall, the fires extended from the north to the southeast along a 400-kilometer line (OCHA 1998b).

Although natural environmental forces were important factors during the buildup to the fires, it is now clear that the INCRA (National Institute for Colonization and Agrarian Reform) programs played a major role as well. As new settlements grew and spread, clearing and agricultural advances moved relentlessly into the more vulnerable fringe area lying between the cerrado and the humid forest, covering a larger area than ever before. Settlers operating in adjacent areas and seeking to increase production stimulated the burning that ran out of control (Nando.net and Reuters 1998); so too did settlers using fire to clear and manage their land (Alvarado 1997). Ranchers and loggers, who used fire irresponsibly by ignoring the potential consequences, also shared the blame (*El País Semanal* 1998). Thus the main cause of the fires was human impact, with groups of people operating in a tinderbox.

Even though the fires originated in different loci, they did not represent isolated or random incidents. Those raging fires, moving ever farther into the evergreen forest, were man made—sometimes, it is alleged, deliberately. They resulted in an environmental disaster unprecedented in Brazilian Amazonia, and represented a crisis that could have been averted.

As the fires were being started in Roraima, the impact of the prolonged drought was already being felt: the drought had caused the failure of the staple manioc crop for the Macuxi, Wapixana, and Taurepang in both Brazil and Guyana. It is not, then, surprising that the fires that began in January were, by March, already out of control, moving from the savanna area westward into the evergreen forest and threatening to destroy the forest cover of one group of Yanomami Indians. The fires entered the transition forest in the beginning of March and the dense forest by the middle of the month (IBAMA 1998b; *Istoé* 1998a:42; Nando.net and Reuters 1998; Rival 1998; SEJUP 1998b).

TABLE 4.1 Development of Fire in Roraima and Adjacent Areas, 1998

Date	Event
March 9	Fires in nonforest areas
March 11	Fires identified in forest in Guyana
March 12	Fires identified in forest in Brazil
March 15–18	Burning concentrated in this period
March 17	Peak of burning; dense plume of smoke covers 58,000 sq. km

FIGURE 4.3. Location of main fires burning in Roraima, Brazil, March 1998.

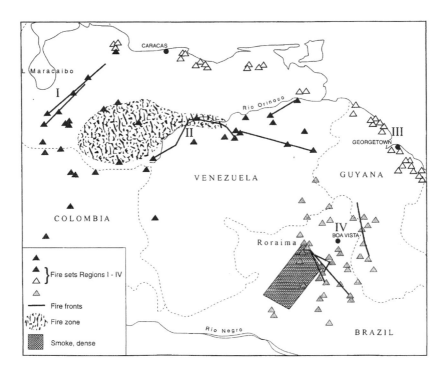

FIGURE 4.4. Fire fronts in northern Brazil, Colombia, Venezuela, and Guyana, March 1998.

In January, Neudo Campos, the state governor, became alarmed. He de clared a state of public emergency. Campos appealed to the people not to use slash-and-burn techniques during the drought and sought urgent help from the federal government. He received none. On March 15, federal officials were accused of delaying the release of money for firefighting. This threatened to allow further damage, a fear that was confirmed as the fires spread and the lack of resources to fight the flames became apparent to Brazilians and to the rest of the world (BBC News 1998a; Correio Brasiliense 1998; Independent on Sunday 1998; Istoé 1998b; Jornal do Brasil 1998a; LARR-Brazil Report 1998; Reuters 1998a, 1998b). The Roraima fires were now out of hand, to the increasing embarrassment of the federal government, which was charged with culpable neglect.

The start of the worst period in the spread and intensity of the fires seems to have been March 15. The Space Applications Institute of the Joint Research Centre European Commission (SAI-JRC) carried out monitoring and assessment of satellite imagery between March 9 and March 18 and reported the development of the fires as shown in table 4.1.

Satellite imagery clearly shows the foci of fires in the region on March 15 as dark pixels scattered along the savanna-forest edge; one day later, on March 16, an extensive plume of smoke could be seen spreading over most of the northern part of Roraima and into Guyana (figure 4.3). In addition, the sequence of satellite photos revealed that even in the nonforested areas the fires were not positioned at random. Many fires were detected at the edges of the forest or the fragmented forest area, suggesting that they were started for a purpose other than land clearing. Although the overall activity of the fires was concentrated in three regional sets around the Boa Vista region, the fires were not an independent event. They were linked to larger phenomena, which included burning in Guyana and on the border with Venezuela. At the same time, important fires were noted in the Venezuelan *llanos* (grasslands), on Andean slopes, on the coasts of Guyana and Suriname, and in Colombia and Peru (figure 4.4) (SAI-JRC/CLAIRE/LBA 1998a; see also *International Forest Fire News* 1998a).

Halting the Fires

It is difficult to know just when the drought in Roraima was overtaken in public and political consciousness by a perception of fire as being the more serious and immediate crisis. The dynamics of response and reaction are still too complex to disentangle, but it soon became clear that the infrastructure available to fight the fires was inadequate. The situation was then made worse by misperception of the scale of the problem and by delay and confusion in responding to it.

The drought, as noted, really began in September 1997, when some southern *municípios* alerted the state governor. It took the governor until January 1998, when the drought attacked 75 percent of the state and the fires 25 percent, to declare a state of public emergency and inform the president. There was then a long delay, and some 45 days later the federal funds requested still had not been released. Beginning in February, national newspapers, such as the *Folha de São Paulo*, began to publish dramatic reports of the situation in Roraima, provoking worldwide attention (Conselho Indigena de Roraima 1998; *Folha de São Paulo* 1998a).

International consciousness had been sharpened earlier by widespread reporting of appalling, disastrous fires in Indonesia and by that government's failure to respond. Editors seized on the news from Brazil and headlines appeared around the world. Television and radio newscasts and articles in the press now reported from Brazil on the dangers of environmental disaster, the threat to a Stone Age tribe, and a rain forest under destruction (e.g., Associated Press 1998; Cooper 1998; *Independent on Sunday* 1998; *Toronto Star* 1998). The Brazilian press, in turn, became loud in its demands for action and increasingly critical of the alleged lack of concern and the delayed response of President Fernando Henrique Cardoso and the federal government (*Jornal do Brasil* 1998a).

The apparent unconcern at the federal level even provoked the threat, by Deputado Socorro Gomes (representing the Communist Party of Brazil [PC do B] from the state of Pará), of opening an inquiry to prove Gustavo Krause, the Minister of the Environment, responsible for the crime of omission and negligence (*Jornal do Brasil* 1998a). Such criticism extended to President Cardoso; one of Brazil's most senior political columnists, Villas Boas Correa, noted dryly: "FH, que tanto viaja, não teve tempo de ir a Roraima" [Fernando Henrique, who travels so much, did not have time to go to Roraima][1] (*Jornal do Brasil* 1998b). In a lighter vein, though not without its own distinct wry twist, was the April Fool's Day report in a Brazilian newspaper of a broadcast from a Moroccan radio station; it said that "Brazil would not be participating in the World Cup because the money for the team would be used in the fight against the fires in Roraima" (*Istoé* 1998c). It was only on March 27, 1998, that the U.N. Office for the Coordination of Humanitarian Affairs (OCHA) could report that a team of five disaster-management specialists and two specialists from the U.N. Environment Program (UNEP) was being sent to Brasília. On Monday, March 30, a meeting was organized for representatives of the office of the president, the Brazilian army and air force, the Brazilian Institute for Environment and Renewable Natural Resources (IBAMA), the National Foundation for Assistance to Indigenous Peoples (FUNAI), and the Ministry of Foreign Affairs. On March 31, the U.N. team arrived in Boa Vista to consult the military command in charge of the operation and to look at priority needs for planes, helicopters, and firefighting equipment not available in Brazil (OCHA 1998b).

By that time an emergency operation, *Operação Roraima*, was being coordinated by the Brazilian Army Seventh Amazon Brigade with the support of the Brazilian civil defense and IBAMA. Firefighters from seven states (including a small, crack team of forest firefighters from Rio de Janeiro) were sent to the area, as were military personnel. Indigenous people helped speed

---

1. All translations from the Portuguese have been rendered by the author of this paper.

the firefighting teams, who had been dropped off by helicopters, through the forest to the fire foci. The group working in Roraima totaled 1,000 firefighters, including manpower and equipment from Argentina and Venezuela. Despite the use of helicopters and water-bombing equipment, the fires still raged out of control (OCHA 1998b).

Help from abroad and offers of further support, as and when needed, immediately reawakened and stirred always sensitive concerns over the questions of sovereignty and control of the Amazonian region; there were sharp reactions to any idea that Amazonia, including its rain forest, might in some way be an international as well as a Brazilian responsibility. Such sensitivities were understandable, legitimate, and deeply entrenched. They partly underlay the much publicized statement attributed to President José Sarney, the first civilian president in the restored democratic system (1985–1989), that "Amazonia is ours, even to destroy." That claim reflected an old legal concept of the most absolute possession or ownership, *ius alieni*. One of Sarney's first major initiatives after taking office in 1985 was the Calha Norte Project. This was essentially conceived by the Brazilian Armed Forces, especially the army, for the defense, security, and development of the Amazon region (Allen 1992b). "Development" and "security" were concepts central to military thinking as developed in the geopolitics of the Escola Superior da Guerra (ESG), the National War College founded in 1949. National security and national development were, for theorists of the armed forces and their civilian allies, inextricably interlinked. *Seguranca e Desenvolvimento* was the title of the journal of the ESG, in which geopolitical thinking was elaborated and debated.

These finely tuned nationalist sensibilities, which in the military perception of Amazonia stretched back as far as Marshal Candido Rondon and his work in the region, are not confined to the armed forces. Throughout the Amazon they find resonance, with strong regional overtones, among powerful political and economic forces, lobbies who often harbor an understandable suspicion of interference from "outside," however well meant. Such nationalist and regionalist perceptions partly explain the apparent reluctance to accept more extensive international assistance in fighting the fires in Roraima. This response was severely criticized in certain sections of the Brazilian press, but it forms an important part of the cultural and political tradition of "human impact" in Brazilian Amazonia and must be appreciated as such.

In the midst of growing controversy and increasingly desperate efforts to contain and put out the fires, other responses took a more traditional form. While plans were afoot to seed the clouds from aircraft to induce rainfall,

Yanomami and Kayapó shamans practiced their own professional skills, with some calling for rain in special rain-making ceremonies, while others prayed for release from the danger, the choking smoke, and the air pollution. Suddenly and unexpectedly the rains came, producing the solution that government and firefighters had failed to provide. The headlines said it all: "*Roraima faz festa cantando na chuva*" [Roraima has a party singing in the rain] (*Jornal do Brasil* 1998a); "*Chuva reduz fogo no Roraima*" [Rain reduces fire in Roraima] (*Jornal do Brasil* 1998a); "Rain douses Brazil's Amazon fires" (*BBC News* 1998b); "Medicine men save rain forest" (*The Times* 1998); "*Hechiceros extinguen incendio en Amazonia*" [Medicine men extinguish Amazon fires] (*Cronicalatina* 1998); "*Amazzonia, la fame dopo gli incendi*" [Amazonia: The hunger after the fires] (*Corriere della Sera* 1998). The inhabitants of Boa Vista came out to dance in the streets after nearly seven months of drought and fire, and everyone involved breathed a sigh of relief that this disaster had been stopped in its tracks.

The Impact of the Fires

People now could literally draw breath and start to count the costs of this disaster. Fire had attacked some 37,000 square kilometers, or about 16 percent of the state territory; two-thirds were cerrado or agricultural land and the rest (about 925,000 hectares) were rain forest (LARR-Brazil Report 1998:6). The National Institute for Amazonian Research (INPA), in a subsequent study, estimated the total area burned as 9,251.7 square kilometers, or 54.1 percent of the total area (some 17,000 square kilometers) affected by fires (OCHA 1998c).

Twelve of Roraima's municipalities were affected: Alto Alegre, Amajari, Apiau, Bomfim, Canta, Caracarai, Iracema, Maracá, Mucajai, Normandie, Pacaraima, and Uiramuta. The fires burning in the ecological station on Ilha Maracá had been put out, but there were still some fires burning in Apiau, Caracarai, and Pacaraima. It was claimed that as many as 12,000 people were affected by drought and fires, and some 7,000 were in serious condition owing to food and water shortages (OCHA 1998c). Many people who were not even in the direct line of the fires had suffered health problems from the fumes and the smoke, as the pall hung in huge swaths over the state and some of the most densely populated areas.

It was also reported that indigenous groups were in desperate need of help during the four-month period leading up to the harvest. The Yanomami Indians of the Mucajaí and Ajarani areas were surrounded by the fires and about 300 people had to flee to more remote forest areas. The fire moved 19 miles into the Yanomami area, driving back the Indians as well as their sources of food and game. Similarly, the Macuxi and Taurepang Indian groups in the northeastern part of the state had to find refuge. Their orange, tangerine, and manioc plantations had been totally destroyed by the drought and now they were exposed to the threat of fire, fumes, and smoke. Most of these indigenous peoples live in savanna and on cleared land, and the destruction of their subsistence crops of yucca and maize forced them to buy from the markets (Environment Brazil 1998; Kirchhoff and Escada 1998).

Survival International and OXFAM were urgently asked to supply seeds and food, and because 1998 was an election year, to help make sure that the situation was not used for political ends (Conselho Indigena de Roraima 1998). There were also claims of cattle killed and 700 barns destroyed; in addition 300 electric power posts were damaged, as were wells, bridges, schools, and health centers (OCHA 1998b).

The most urgent short-term needs were medical supplies, seeds, tools, the opening of 72 wells, and the provision of 104 housing units. Approximately 200 firefighters in the northwestern part of the state continued to put out the remaining fires and to ensure that reignitions were quickly dealt with. They were to be supported by six military helicopters (OCHA 1998b).

Assessment of the environmental situation and development of a policy for the future were also a priority. In the medium term, this included an estimate of the damage to different biomes, the acquisition of NOAA AVITAR antennae for fire monitoring, and the enforcement of tighter regulations concerning the use of fire (OCHA 1998b).

A report issued by IBAMA in August 1998 concluded that the fires in Roraima affected large areas of different types of forest, forming mosaics consisting of tracts that had been burned as well as those untouched by fire. The effects on forest undergrowth also varied. The smaller trees of the understory suffered most, especially saplings, palms, and vines. That damage would alter the future plant communities of the forest. Though the burning of the forest did not destroy all the vegetation, it left the remaining plants more susceptible to outbreaks of fire in dry periods (IBAMA 1998b].

Although the part of the forest most affected was the understory, the damage was not as disastrous as had first been feared. In many parts of the understory the possibility of recuperation appeared possible, as long as the vegetation did not suffer further burning. A preliminary evaluation showed that the fire affected 81 percent of the vegetation, and only in 1.5 percent of the area did the fire have a drastic effect. This suggested that 79.5 percent of the forest would have a substantial capacity for regeneration (IBAMA 1998b:3–4), an outlook certainly not envisaged at the height of the crisis.

As the ash settled and the smoke cleared, claims for compensation predictably appeared. This is a question of secondary importance to this paper but is real enough for those involved, and is certainly part of the pattern of human impact and ongoing levels of responsibility. The claims stated that Roraima should receive R$17 million, with an additional R$4 million to be spent on the costs of firefighting, and that the Macuxi and Wapixana indigenous groups should receive R$3 million for food and recuperation of dwellings. In addition the federal banks were to lend R$78 million to 4,000 cultivators whose land was affected by the fires (an average of about R$19,500 each) and the debts, particularly those of small-scale farmers and ranchers working with credit, would be renegotiated in the year 2000 (*Correio Brasiliense* 1998:21).

There were inevitably questions about some of these payments and financial waivers. It was claimed, for example, that some fires were started intentionally by landowners or their agents in order to claim compensation. In these cases, handed on to the federal police, it was said that landowners would be punished for lighting fires and that they would not be allowed to renegotiate their debts (*Folha de São Paulo* 1998a; *El País Semanal* 1998:82; SEJUP 1998a). There were similar questions over alleged fraudulent claims for loss of buildings and crops due to the fires. It was noted, for instance, that there were no crops to be lost on land being prepared for planting and that it was hard to find evidence that buildings and barns had burned.

From Roraima Onward

One of the positive results of the Roraima fires of 1998 is the stimulus they provide to thinking about human impact on Amazonia at a variety of levels, from the practical consequences of fires to political organization and response. There has been, for example, renewed discussion of the causes and prevention of wildfires and of how fire could be made most useful in clearing and cultivation. Use of fire will continue: as one poor farmer stated bluntly, "Ali, quem não faz queimada, não planta e quem não planta, não come" [Here, who doesn't burn, doesn't plant, and who doesn't plant, doesn't eat] (*Istoé* 1998b:26). Fire is such an important traditional tool in the management of land that it cannot simply be banned (Goldammer 1999a:1782). The need to use fire to clear land is not resolved simply by recommending mechanized clearing to reduce fire risk, as many cannot afford to do this and others may doubt the efficiency of such a clearance system (Nepstad et al. 1991; Nortcliff, chapter 8 this volume).

There are also the questions of why, in many cases, fires are started deliberately, and by whom. One of the main fears reported in Roraima was that the fires were started by landowners and loggers as an excuse to invade land reserved for the use of indigenous groups. In Roraima, indigenous areas account for nearly 55 percent of the state. At every stage of their demarcation they have been resisted by commercial and business sectors that want to see the land and its reserves of natural resources available for exploitation. This is an ongoing debate that no governor or candidate for political office can ignore and that may ultimately determine the pattern of human impact in this region.

By comparison with economic and political pressures, more technical questions, such as planting techniques, selection of fire-resistant vegetation, establishment of buffer zones in the transition forest, protection from grazing animals, and protection from erosion are of secondary importance, as is the application of technology based on traditional ecological knowledge. As research on the fires in Indonesia concluded: "The only useful way to address the problem of major fires ... is a sustained, long-term attack on the underlying causes" (CIFOR 1999:6).

Political priorities will continue to shape institutional and organizational responses to crises such as the 1998 fires in Roraima. That particular threat of environmental disaster led to unfortunate national and international reporting about the Brazilian government. In addition the subsequent fires that occurred later in the year have, arguably, had a direct impact. Under the Brazilian Constitution of 1988, firefighting became the responsibility of the *Corpos de Bombeiros* (fire brigades), but they were not trained or equipped to deal with the kind and size of fires that ran uncontrolled in Roraima. They could offer no rapid or effective response; nor were they expected to do so.

On April 10, 1989, under Decreto No. 97635, the federal government established a national system for the prevention and fighting of forest fires (PREVFOGO) as part of the IBAMA framework. That body, centered in Brasília, was charged with all aspects of organizing and implementing activities (e.g., forest clearance and preparation of sugarcane for cutting) that could affect forest fires and agricultural burning (PREVFOGO 1998).

In 1991, PREVFOGO entered into agreements with the Corpos de Bombeiros in the states of Rio de Janeiro, the Federal District, and Goiás; later, links were set up with the Corpos in Mato Grosso and Minas Gerais. But those efforts were essentially inadequate: "The techniques, the strategies, and the equipment for fighting fires in [the] forest in Brazil simply do not exist, with rare exceptions" (IBAMA 1998a). The fires in Roraima may help to speed up the training and equipping of this arm of IBAMA, as well as the educating and monitoring of communities—especially those in particularly vulnerable areas such as Amazonia and the cerrado. It has been proposed that local groups in indigenous areas be trained to fight fires more effectively, and that firefighting equipment be provided for them (Ministerio do Meio Ambiente 1999b).

As an immediate response to the Roraima crisis, steps to prevent uncontrolled fires were put in place, only, it was alleged, to be weakened by a president under political pressure. The Law of Environmental Crimes was passed in congress on February 13, 1998, and took effect at the end of March. It is said that, because of pressure from the important lobby group of the sugarcane industry—particularly that in the northeast of the country—President Cardoso vetoed a clause that would have imposed fines and prison sentences on those who set careless or intentional uncontrolled fires (LARR-Brazil Report 1998:7). This is a charge that needs closer scrutiny.

Further steps to prevent uncontrolled fires were laid out in the recommendations made by IBAMA after the Roraima fires. These included giving help to colonos in the mechanical preparation of ground for agriculture, promoting agroforestry and fruit-tree cultivation, and assisting in the regeneration of deforested areas. In addition, there were recommendations for wider legislation and fiscalization for environmental protection, promotion of community interest and involvement in revegetating river margins and ecological reserves, implementation of educational programs about fire prevention, and detection of fires by remote sensing. Finally, it was proposed that government agencies, groups of rural producers, community leaders, and others concerned with land use should have a greater role in the prevention of fires. All of this was to go hand in hand with better enforcement of protection for conservation areas (IBAMA 1998b).

Other positive responses that came in July 1998 were further evidence that the Cardoso government was aware of the seriousness of the question of human impact, and not just in Amazonia but throughout Brazil. Under *Proarco*, a project that was concerned with the whole area subject to deforestation in Brazilian Amazonia, an antideforestation firefighting team was set up as a rapid-response squad. It was to establish a permanent firewatch and have a team of 500 crack firefighters who could be mobilized within 48 hours.

The need for such response was all too clear. In August there were again reports of uncontrolled fires in Amazonia, this time destroying 172,000 acres (just under 70,000 hectares) of savanna in the states of Mato Grosso, Tocantins, and Pará, and again threatening forests and the land reserved for the use of indigenous peoples (*Guardian* 1998). In September 1998, severe fires also raged through the forests of the Biological Reserve of Sooretama in Espirito Santo, in the Chapada Diamantina in the interior of Bahia, in the Serra da Araras in Mato Grosso, in the state of Tocantins, and in the National Park of the Serra da Canastra; they even destroyed one-third of the Parque Nacional on the edge of Brasília in the heart of the Federal District (*Folha de São Paulo* 1998c).

This placed severe strain on firefighting resources: it was the third time in one year that the army had to be mobilized to fight fires in environmentally important areas. Some 100 men, for example, were sent to fight fires on the Ilha Bananal in the National Park of Araguaia and in the Parque Nacional (*Folha de Sao Paulo* 1998b; *Guardian* 1998; *Jornal do Brasília* 1998; *O Globo* 1998).

Another intriguing factor, and something of a paradox, is that reports from Roraima in July appeared under the headline, "*Roraima renasce: A chuva apaga as marcas do incendio e os agricultores esperam a maior safra da década*" [Rebirth of Roraima: The rains wash away the marks of the fire and the farmers expect the biggest harvest of the decade] (*Vejá* 1998).

From the beginning of April until the end of July 1998, it rained more in Roraima than it had in the whole of the preceding year, refilling the rivers, reducing the high temperatures, and in some places flooding cleared land. Yields of maize and rice, the main products of the state, exceeded the harvest of 1997 by 77 percent, an increase of 16,000 tons of grain (*Vejá* 1998).

And Now What?

By the time those later fires were being fought, a different kind of struggle was under way: elections were being held for president, congress, state assemblies, and state governors. The presidential election returned Fernando Henrique Cardoso to office for a second term, starting on January 1, 1999. By then the attention of the nation and the international community was focused on an altogether different form of "fire-fighting," as Brazil faced the most serious financial crisis of its modern history (Flynn 1999). That "fire" reached its height in January and February 1999, but went on smoldering and still threatens to break out.

This conflagration, too, had its costs and left a trail of destruction, including mounting criticism of the government for paying insufficient attention to social and environmental issues, and a consequent decline in support for President Cardoso and his government. Some of the fiercest critics have included state governors, thus threatening the close collaboration between federal, state, and municipal governments which is the necessary condition for effective environmental programs.

Some observers believe that nothing worthwhile has been learned from the experience of the Roraima fires and that little has been done to tackle the underlying causes of pressure and conflicting land claims in Roraima and throughout Amazonia. Such pressure is in fact increasing as internal migrants pour into the state of Roraima; many of them, disillusioned by the failure of exhausted lands, are from Rondônia, which is now losing population at a record rate as many of its earlier settlers arrive in Roraima daily by lorry (*Vejá* 1999). New settlers mean new pressures, which only intensifies the causes of fires such as those of 1998. Nevertheless the newcomers are welcomed by local authorities, especially by mayors eager to build up their municípios to take advantage of tax benefits conferred by the Brazilian Constitution of 1988.

Not surprisingly, in 1999 as in 1998 fires raged again throughout Brazil, including Amazonia, in some cases threatening long-distance transmission lines or water supplies to large cities. Once more, international assistance, including financial support, was offered and refused (Fleischer 1999). There have been some gestures of support from the federal government, but only time will show how real and effective they may be.

One such gesture includes the *Política Nacional Integrada para a Amazônia Legal* (Integrated Program for Legal Amazonia), which was laid out in Decree No. 1.541 of June 1995 by President Cardoso. Built on a charter of principles, it was signed on March 31, 1995, by all the governors of Amazonia, and approved (as Resolution No. 4) on July 14, 1995, by CONAMAZ, the *Conselho Nacional de Amazônia Legal* (National Council of Legal Amazonia) (Ministerio do Meio Ambiente 1995a, 1995b, 1995c; Nepstad et al. 1999). The formal commitment by the Brazilian state to coordinate programs and projects especially relevant to the environment in Amazonia has to be read alongside Cardoso's much more comprehensive program of *Avança Brazil* (Forward Brazil!) in order to get an idea of a possible blueprint for action. What those commitments will produce remains an open question, for the "political ecology of fire" does not always propel reforms, as was revealed in the cases of the Mexican and Indonesian wildfires (Pyne 1998; Barber and Schweithelm 2000).

Those programs are accompanied by another that will also need to be monitored: an agreement signed on July 20, 1999, by the president of IBAMA and the army's Head of Command for Terrestrial Operations (COTER), General Luciano Casales. The project was inaugurated by the new Minister of the Environment, José Sarney Filho, son of former President José Sarney who became a senator for Amapá after leaving office. COTER and IBAMA, it is said, will work together in monitoring both deforestation and fires in the Amazonian region.

The mission initially will be headed by 200 supervisors from IBAMA; they will be transported in army helicopters along with 70 military officers in charge of their security. The stated aim is to send up to 350 supervisors to the so-called Deforestation Arch, which runs across six states from Imperatriz in Maranhão to Humaitá in Amazonas (Brazilian Embassy 1999).

Those responses from the federal government—an integrated program for Amazonia and coordinated action by the armed forces, the Ministry of Environment, and agencies such as IBAMA—are not just a reaction to the 1998 fires in Roraima. It is, however, clear that the crisis that almost became a major environmental disaster did have substantial and lasting political resonance. Months after the event it was still being referred to by officials in Brasília as a defining moment or turning point in government thinking about Amazonia and the environment.

How effective that change of thinking has been remains to be seen. One cogent criticism of even the most recent government plans and initiatives is that they are still dealing with symptoms and effects rather than with causes, namely, the increasing human pressure and impact on fragile and vulnerable land.

The Cardoso government rejected such criticism, noting that in 1999 it earmarked US$550,000 for research into the social and economic causes of deforestation, and that in the two years from 1997 to 1999, 80 percent of all land distributed under the agrarian reform program has been in the Amazon region (LARR-Brazil Report 1998). It also pointed to renewed and expanded dialogue with the MST (*Movimento dos Trabalhadores Sem Terra*, or the Movement of Landless Rural Workers), and to the initiatives of IBAMA, INCRA, and other government agencies. But still, even the government itself would acknowledge it is not enough simply to douse flames and respond, often belatedly, to crises after they have started. What is needed, and urgently, is political fire prevention on a much greater and more ambitious scale than ever before.

Implications for Traditional Ecological Knowledge and Amazonian Fires

Looking to the future, one of the essential lessons to be drawn from the Roraima experience of 1998 is that the indigenous peoples of Roraima have rights—human, civil, and constitutional—that must be respected. In addition, one of the major priorities to be set is the establishment of mechanisms that show a greater sensitivity to the needs and rightful demands of the indigenous peoples in Roraima, and in Amazonia generally. After all, their whole environment and their very existence were threatened by fires started by others who had different interests and priorities.

It should be stressed that indigenous groups also have a contribution to make, one that is based on centuries of experience living in balance with their environment: this includes the use and control of fire. Little research has been done on this particular aspect of traditional knowledge, not just in Brazil but throughout the tropics. There has been much written about indigenous knowledge and forest management, but only now are foresters and forest officers taking real notice. Moreover little has been written on the specific issue of the traditional use of fire in tropical forest and savanna areas, except in terms of soil science and shifting cultivation. It is perhaps time to ask the right kinds of questions in the right kinds of ways to tap into the indigenous knowledge held on this subject.

From the point of view of indigenous peoples, such questions may seem strange, as the phenomenon of wildfires like those in Roraima is perhaps only a once-in-a-lifetime event. The great fear now is that the incidence of such fires may be much more frequent in the future. The utilization of indigenous knowledge may well hold important keys to protection of indigenous lands, to biodiversity protection, and to conservation of areas that have not yet been inundated with dense and permanent settlement.

Gradually, though time is running out, we are coming to realize that the incorporation and usage of indigenous knowledge about the conservation of biodiversity and sustainable forest management increase the active participation of indigenous communities and enhance the empowerment of indigenous peoples.

The fundamental barrier to such change lies in the reluctance and tardiness of governments to recognize indigenous knowledge in policy and implementation. Very rarely, it seems, are government interventions, at either state or national levels, based on the incorporation and full development of indigenous knowledge (see, for example, the report on merging indigenous forest-related knowledge with formal forest management in Pyne 1998 and Van Leeuwen 1998).

Now it seems that the Brazilian government, at various levels, may be starting to listen. Roraima has concentrated the federal mind to a notable degree. Structures are being put into place to counter both deforestation and destructive burning. IBAMA is strengthening its firefighting force, including helicopter monitoring, in nine of the Amazonian states that make up the *Arco de Desflorestamento* (Arc of Deforestation) (Ministerio do Meio Ambiente 2000b). There is some evidence that in these nine states, between September 1999 and September 2000, the number of outbreaks of fires fell by about 30 percent, from 22,101 to 15,730 (Ministerio do Meio Ambiente 2000a).

More significantly, in terms of the overall theme of this study, government officials at last say they are willing to listen to indigenous peoples and give them some share in determining policy for the lands on which they have lived for thousands of years. In June 1999, in the aftermath of the Roraima crisis, indigenous communities asked the Secretary for the Coordination for Amazonia, Mary Allegretti, for a fuller role in designing and carrying through the *Projeto Demonstrativo para Populações Indígenas* (Demonstration Project for Indigenous Populations–PDI). This project is part of the wider *Programa Piloto do Proteção das Florestas Tropicais do Brasil* (Pilot Program for the Protection of Brazilian Tropical Forests–PPG7), which aims to contribute to the sustainability of indigenous lands and the protection of their natural resources (Ministerio do Meio Ambiente 1999a).

This request to the government was extended to cover representation for Indians living in the Mata Atlântica, and it resulted in the first meeting ever between the Secretariat for Amazonian Coordination and indigenous peoples, which was held in Manaus in September 1999. Ms. Allegretti said the meeting was "of fundamental importance," and stressed the need for "direct participation" of local communities in any project for the region (Ministerio do Meio Ambiente 1999a). Not least, there is a strong, inherent argument in all this for fundamental changes to take place in the legal recognition and demarcation of indigenous territories in order to protect tropical forest and savanna ecosystems (e.g., Davis and Wali 1994).

However, what such a meeting and its attendant rhetoric really mean is still not clear. The indigenous peoples of Roraima and elsewhere in Brazil have a long-accumulated, but unrecorded, knowledge of forest and environmental management, including the use and control of fire for regeneration, reconstruction, and more efficient harvesting of their land, whether in the savanna or in tall forest. The knowledge of environmental management, like so much of long-garnered indigenous knowledge, is complex and may well be lost in a very short time. In any case, sadly but realistically, it does not afford effective protection against incomers eager to gain new ground at any cost, unconstrained by government control, and too often in league with vested and powerful interests. Traditional indigenous knowledge offers few defenses against rapacious interests when they are backed, literally, by superior firepower.

This, however, does not mean that indigenous knowledge and indigenous communities do not count in Roraima, or in Brazil, more than five hundred years after indigenous peoples first encountered European sailors, merchants, and priests. They have a voice and they have political power, as yet scarcely realized. They also have allies in Brazil and beyond, some specifically alerted by the fires in Roraima.

In Mexico, for example, where 60 firefighters died in that country's wildfires of 1998, it has been suggested that one path to fire prevention is the development of a "cultural consciousness of fire" that links pre-Columbian rituals and use of fire with new, modern technology (Pyne 1998). It is clear that strategies for combating fires must begin with local communities, indigenous as well as nonindigenous. Nepstad, Moreira, and Alencar (1999) and Woodwell (2000) argue that rules for a healthy, sustainable environment must be established locally when central government fails, as it evidently did in Roraima.

Conclusions

The most telling comparative experience for Roraima comes from Indonesia, where forest fires blazed at the same time as in Roraima, while the Indonesian government also failed to respond. The political context of these two conflagrations was importantly different, but the impact on indigenous communities was regrettably similar. One analysis of the Indonesian fires has pointers for Roraima. A detailed report of the World Resources Institute: Forest Frontiers Initiative starts by asking whether real forest reform is possible. Armed with this healthy, judicious skepticism, it then stresses the need to harness indigenous knowledge in managing forests. It sees the need to "recognize and legally protect forest ownership and utilization by indigenous and forest dependent communities" (Barber and Schweithelm 2000:43).

The report quotes the World Bank's estimate that "at least 30 million people are highly dependent on forests for important aspects of their daily livelihood" and also notes the World Bank's recommendation that any workable forest sector reform agenda "must give primacy to radically increased participation of forest dwelling and adjacent communities in the management, utilization and actual ownership of forests and forest lands" (Barber and Schweithelm 2000:43).

The fundamental question for Roraima, as in Indonesia and elsewhere, is this: How do indigenous peoples achieve sufficient political power to use their long experience and knowledge to survive, to ward off threats by fire or in other forms, and to contribute to the society that naturally they should enrich?

References

Allen, E. 1989. Brazil: Indians and the new constitution. *Third World Quarterly* 11(4):148–165.

Allen, E. 1992a. Brazilian Amazônia: Policy formation and democracy. Special edition, "Nachhaltinge Entwicklung in Amazonien: Konzept und Wirklichkeit," *Lateinamerika: Analysen-Daten Dokumentation* (Hamburg: Institut für Iberoaamerik-Kunde) 19 (November):15–27.

Allen, E. 1992b. Calha Norte: Military development in Brazilian Amazônia. *Development and Change* 23 (January):71–100.

Allen, E. 1992c. Jungle geopolitics. *Geographical* 64 (April):28–31.

Allen, E. 1994. Political responses to flood disaster: The example of Rio de Janeiro. In A. Varley, ed., *Disasters, Development and Environment*, pp. 99–108. Chichester: Wiley.

Alvarado, E. 1997. Fire on the Amazon-*Cerrado* transition. Special features forest list archive. msg00017. http://sol.cfr.washington.edu (accessed September 27, 2000).

Asbjømsen, H, L. E. Næss, and E. Torres. 1999. Fire in moist tropical forests: A progenitor or threat to biodiversity? *European Tropical Forest Research Network News* 29 (Autumn–Winter):36–39.

Associated Press. 1998. Approaching fire has Brazil natives nervous. *Chicago Tribune*, March 18:23.

Balée, W. 1989. The culture of Amazonian forests. In D. A. Posey and W. Balée, eds., *Resource Management in Amazonia: Indigenous and Folk Strategies*, pp. 1–21. Advances in Economic Botany 7. Bronx: New York Botanical Garden.

Barber, C. V. and J. Schweithelm. 2000. *Trial by Fire: Forest Fires and Forestry Policy in Indonesia's Era of Crisis and Reform*. Washington, DC: World Resources Institute, Forest Frontiers Initiative.

BBC News. 1998a. World: Americas; Amazon fires rage on. March 15. http://news.bbc.co.uk/hi/english/world/americas/newsid_65000/65696.stm.

BBC News. 1998b. World: Americas; Rain douses Brazil's Amazon fires. April 2. http://news.bbc.co.uk/hi/english/world/americas/newsid_72000/72811.stm.

Boyd, R., ed. 1999. *Indians, Fire and the Land in the Pacific Northwest*. Corvalis: Oregon State University Press.

Brazilian Embassy. 1999. Partnership IBAMA/army to fight deforestation in the Amazon. Press release. *Brazilian Briefing* 10 (August):28. London.

CCPY. 1995. *Update* 83 (December). Sao Paulo: Comissão pela Creação do Parque Yanomami.

CIFOR. 1999. Long-term solutions vital in combating forest fires. *CIFOR News* 22 (April/May):6–7. Jakarta: Centre for International Forestry Research.

Clark, K. and C. Uhl. 1987. Farming, fishing and fire in the history of the Upper Rio Negro region of Venezuela. *Human Ecology* 15:1–26.

Cochrane, M., A. Alencar, M. Schulse, C. M. Souza Jr, D. Nepstad, P. Lefebvre, and E. Davidson. 1999. Positive feedbacks in the fire dynamic of closed canopy tropical forests. *Science* 284:1832–1835.

Conselho Indigena de Roraima. 1998. *Noticias de Roraima*. Unpublished communication to Survival International, April 6.

Cooper, M. 1998. Through British eyes: UK print media coverage of Brazil 1993–1997. Paper presented at the Society for Latin American Studies annual conference, University of Liverpool, April 17–19.

Correio Brasiliense. 1998. Fogo queima 15% de Roraima. April 9:22.

Corriere della Sera. 1998. Amazzonia, la fame dopo gli incendi. April 5:27.

Cronicalatina. 1998. Hechiceros extinguen incendio en Amazonia. No. 132 (April 2):1. Londres.

Davis, S. H. and A. Wali. 1994. Indigenous land tenure and tropical forest management in Latin America. *Ambio* 33(8):485–490.

Eden, M. J. 1990. *Ecology and Land Management in Amazonia*. London: Belhaven Press.

El País Semanal. 1998. Amazonia: La selva herida. April 19:78–89.

Environment Brazil. 1998. Fire devours savannah and Amazon forests. *World News*, March 16. IPS homepage at http://www.oneworld.org/ips2/mar98/18_48_061. html (accessed September 27, 2000).

Fearnside, P. 1992. Greenhouse gas emissions from deforestation in the Brazilian Amazon. In W. Makundi and J. Sathaya, eds., *Tropical Forestry and Global Climate Change*. Berkeley, California: Lawrence Berkeley Laboratory, workshop proceedings.

Fleischer, D. 1999. *Brazil Focus*, Weekly Report, September 4–10 Brasília.

Flynn, P. 1999. Brazil: The politics of crisis. *Third World Quarterly* 20:287–317.

Folha de São Paulo. 1998a. Landowners may be punished for lighting fires. April 14. Quoted in SEJUP News from Brazil, no. 310. April 16.

Folha de São Paulo. 1998b. Combate a incêndio mobiliza Exército pela 3a. vez neste ano. September 26:3.

Folha de São Paulo. 1998c. O Vaivem do fogo pelo Brasil. September 26:3.

Furley, P. A., ed. 1994. *The Forest Frontier: Settlement and Change in Brazilian Roraima*. London and New York: Routledge.

Goldammer, J. G. 1999a. Forests on fire. *Science* 284 (June 11):1782–1783.

Goldammer, J. G. 1999b. Forest impoverishment by fire—the tropical and global context. *European Tropical Forest Research Network News* 29 (Autumn–Winter): 39–41.

Goldammer, J. G., ed. 1990. *Fire in the Tropical Biota: Ecosystem Processes and Global Challenges*. Berlin: Springer-Verlag.

Grimes, B., ed. 1992. *Ethnologue: Languages of the World*, 12th ed. Dallas: Summer Institute of Linguistics.

Guardian. 1998. South Amazon's farmers ignite season of fire. August 15:16.

Hecht, S. B. and D. Posey. 1990. Indigenous soil management in the Latin American Tropics: Some implications for the Amazon Basin. In D. A. Posey and W. L. Overal, eds., *Ethnobiology: Implications and Applications*, pp. 73–86. Proceedings of the First International Congress of Ethnobiology, vol. 2. Belém: Museu Paraense Emílio Goeldi.

Hemming, J. 1990. Roraima: Brazil's northernmost frontier. University of London Institute of Latin American Studies Research Papers, no. 20.

IBAMA [Brazilian Institute for Environment and Renewable Natural Resources]. 1998a. 4.3 Programa de Combate. http://www.ibama.gov.br/~prevfogo/pgcomb. html (accessed July 16, 1998).

IBAMA [Brazilian Institute for Environment and Renewable Natural Resources]. 1998b. Roraima: O Brasil aprendeu em 1998, a difícil lição de que não e exceção para o resto do mundo. August 3. http://www.ibama.gov.br/roraima/roraima.htm (accessed August 3, 1998).

IBGE. 1992. Roraima: Censo demografico 1991. Resultados preliminaries. Boa Vista: Instituto Brasileira de Geografia e Estatistica (IBGE).

Independent on Sunday. 1998. Rainforest fighters race to the rescue at the ends of the earth. March 29:17.

International Forest Fire News. 1998a. A brief history of forest fires in Mexico. IFFN, no. 19 (September). http://www.rut.uni-freiburg.de/fireglobe/iffn/country/mx/ mx_2.htm (accessed November 1, 2000).

International Forest Fire News. 1998b. Fire activity in the Guyana shield, the Orinoco and Amazon Basins during March 1998. IFFN, no. 19 (September). http:// www.ruf.uni-freiburg.de/fireglobe/iffn/country/br/br_5.htm (accessed November 1, 2000).

IPAM. 1998. *The Belém Letter on Accidental Fire in the Amazon*. April 30. Belém: Instituto de Pesquisa Ambiental da Amazônia (IPAM). http://www.ipam.org.br/ fogo/cartaen.num (accessed September 27, 2000).

Istoé. 1998a. Desastre: Seca na Amazonia; Roraima pode perder 1000mil cabeças de gado e 500mil pés de tomate se a estiagem continuar até abril. March 18:42.

Istoé. 1998b. Ensaio de apocalipse: Incéndio que se alastra Roraima desafia bombeiros e reflete o descaso governamental. April 1:24–26.

Istoé. 1998c. Almanaque: Pega na mentira; As principais lorotadas que correm pelo mundo no dia de 1° de abril. A Semana, April 8. http://www.zaz.com.br/istoe/ semana/148810a.htm (accessed June 27, 1998).

Jordan, C. F. 1989. *An Amazonian Rainforest: The Structure and Function of a Nutrient Stressed Ecosystem and the Impact of Slash and Burn Agriculture*. Man and the Biosphere Series, vol. 2. Carnforth and Park Ridge: Parthenon Publishing Group, UNESCO.

Jornal do Brasil. 1998a. Inquerito para Krause. April 1.

Jornal do Brasil. 1998b. Villas Boas Correa. April 1.

Jornal do Brasília. 1998. Parque nacional em chamas. September 29.

Kirchhoff, V. W. J. H. and P. A. S. Escada. 1998. *O Megaincêndio do Século—1998: The Wildfire of the Century*. São José dos Campos, SP: Transtec Editorial.

LARR-Brazil Report. 1998. Roraima rescued by heavy rains: But there could be worse to come, environmentalists warn. *Latin American Regional Report—Brazil*, LARR RB-98–04:6–7, April 28.

LARR-Brazil Report. 1999. Amazon/forest loss. *Latin American Regional Report—Brazil*, LARR RB-99–03:8, March 16.

MacMillan, G. 1995. *At the end of the rainbow? Gold, land, and people in the Brazilian Amazon*. London: Earthscan.

McPhaden, M. J. 1999. The child prodigy of 1997–98. *Nature* 398 (April 19):559–562.

Ministerio do Meio Ambiente. 1995a. Politica Integrada para Amazonia Legal, Apresentacao, Secretaria de Coordinação da Amazônia. http://www.mma.gov.br/port/SCA/polam/apresent.html (accessed August 19, 1999).

Ministerio do Meio Ambiente. 1995b. Politica Integrada para Amazonia Legal, Introdução, Secretaria de Coordinação da Amazônia. http://www.mma.gov.br/port/SCA/polam/intro.html (accessed August 19, 1999).

Ministerio do Meio Ambiente. 1995c. Politica Integrada para Amazonia Legal, Anexos, Secretaria de Coordinação da Amazônia. http://www.mma.gov.br/port/SCA/polam/anexos.html (accessed August 19, 1999).

Ministerio do Meio Ambiente. 1999a. SCA e índios reúnem-se em Manaus. InforMMA de hoje, September 21. http://www.mma.gov.br/port/ascom/imprensa/setembro99/informma134hum (accessed October 11, 2000).

Ministerio do Meio Ambiente. 1999b. Roraima terá atendimento itinerante de educação ambiental. InforMMA de hoje, November 10. http://www.mma.gov.br/port/ascom/imprensa/novembro99/informma165.html (accessed October 11, 2000).

Ministerio do Meio Ambiente. 2000a. Cai o número de focos de calor na Amazônia Legal, InforMMA de hoje, September 11. http://www.mma.gov.br/port/ascom/imprensa/linkinfor.cfm/idl = 235&idm = 85 (accessed October 11, 2000).

Ministerio do Meio Ambiente. 2000b. Sarney Filho reforça ações para evitar queimadas. InforMMA de hoje, September 25. http://www.mma,gov.br/port/ascom/ultimas/ultimas.clm?id = 50 (accessed October 11, 2000).

Moran, E. 1993. *Through Amazonian Eyes: The Ecology of Amazonian Populations*. Iowa City: University of Iowa Press.

Nando.net and Reuters News Service. 1998. Indians at risk as fires rage in Brazil. March 5. http://www.cgi2nandonet/newsroom/ntn/world/030598/world22_2196_noframes.html (accessed April 16, 1998).

Nepstad, D. C., A. Moreira, and A. A. Alencar. 1999. *Flames in the Rainforest: Origins, Impacts and Alternatives to Amazonian Fires*. Brasília: The Pilot Program for the Conservation of the Brazilian Rainforest, World Bank, Brasília.

Nepstad, D. C., C. Uhl, and E. A. S. Serrao. 1991. Recuperation of a degraded Amazonian landscape: Forest recovery and agricultural restoration. *Ambio* 20(6):248–255. Stockholm.

Nepstad, D. C., A. Verissiomo, A. Alencar, C. Nobre, E. Lima, P. Lefebvre, P. Sch-
lessinger, C. Potter, P. Moutinho, E. Mendoza, M. Cochrane, and V. Brooks.
1999. Large scale impoverishment of Amazonian forests by logging and fire. *Nature* 398 (April 8):505–508.

OCHA [U.N. Office for the Coordination of Humanitarian Affairs]. 1998a. Brazil-El
Niño forest fires. OCHA Situation Report No. 1, OCHA/GVA-98/0176, March
27. http://www.reliefweb.int (accessed April 10, 1998).

OCHA [U.N. Office for the Coordination of Humanitarian Affairs]. 1998b. Brazil-El
Niño forest fires. OCHA Situation Report No. 2, OCHA/GVA-98/0178, April 1.
http://www.reliefweb.int (accessed April 10, 1998).

OCHA [U.N. Office for the Coordination of Humanitarian Affairs]. 1998c. Brazil-El
Niño forest fires. OCHA Situation Report No. 3, OCHA/GVA-98/0184, April 9.
http://www.reliefweb.int (accessed April 10, 1998).

O *Globo*. 1998. O País: Incendio destrói parque ecológico e atrapalha o tráfego aérea
em Brasília. September 29. http://www.oglobo.com.br/pais/NAC60.htm (accessed September 29, 1998).

Parker, F. 1992. Forest islands and Kayapó resource management in Amazonia: A
reappraisal of the apêtê. *American Anthropologist* 94:406–428.

Posey, D. A. 1985. Indigenous management of tropical forest ecosystems: The case of
the Kayapó Indians of the Brazilian Amazon. *Agroforestry Systems* 3:139–158.

Posey, D. A. and W. Balée, eds. 1989. *Resource Management in Amazonia: Indigenous and Folk Strategies.* Advances in Economic Botany 7. Bronx: New York
Botanical Garden.

Preisinger, H., M. Skatulla, K. Richer, R. Lieberei, G. Gottsberger, R. da C. Araujo,
R. R. de Morais, L. Gasparotto, L. F. Coelho. 1998. Indicator value of anthropogenic vegetation in the Amazon. In R. Lieberei, K. Vob, and H. Bianchi, eds.,
Proceedings of the Third SHIFT-Workshop (Manaus, March 15–19, 1998), pp.
313–320. Hamburg: German Federal Ministry of Education and Research.

PREVFOGO. 1998. PREVFOGO: Sistema Nacional de Prevenção e Combate aos
Incendios Florestais. Ministerio de Ciencia e Tecnologia. http://www.mct.gov.br/
gabin/cpmg/climate/programa/port/prevfogo.htm (accessed July 16, 1998).

Pyne, S. J. 1998. The political ecology of fire: Thoughts prompted by the Mexican
fires of 1998. Guest editorial, *International Forest Fire News,* no. 19 (September).

Reuters. 1998a. Brazil seeks to limit settler damage to rainforest. March 19. Reuters,
Brasília.

Reuters. 1998b. Brazil sends more personnel to fight fires. March 23. Reuters, Brasília.

Rival, L. 1998. El Niño crisis in the northern savannahs of Guyana. APFT-Avenir
des Peuples des ForêtsTropicales Briefing Note, no. 13. Brussels: European Community IDG VIII and APFT (June).

Riviere, P. 1972. *The Forgotten Frontier: Ranchers of Northern Brazil.* Stanford University Case Studies in Cultural Anthropology. New York: Holt, Rinehart & Winston.

SAI-JRC/CLAIRE/LBA. 1998a. Satellite monitoring of vegetation fires. Performed by the Space Applications Institute of the Joint Research Centre European Commission (SAI-JRC) for CLAIRE (Cooperative LBA Airborne Regional Experiment)/LBA (Large Scale Biosphere Atmosphere Experiment). http://www.mtv.sai.jrc.it/parbo/Roraima_Web.html.

SAI-JRC/CLAIRE/LBA. 1998b. Map of fire locations in Roraima. Prepared by the Space Applications Institute of the Joint Research Centre European Commission (SAI-JRC) for CLAIRE (Cooperative LBA Airborne Regional Experiment)/LBA (Large Scale Biosphere Atmosphere Experiment). http://www.mtv.sai.jrc.it/parbo/Roraima2.jpg.

San Jose, J. J. and M. R. Farinas. 1971. Estudios sobre los cambios de la vegetacion protegida de la quema y el pastorero en las estacion biologico de los llanos. *Boletim de la Sociedad Venezoelana de Ciencias Naturales* 29:136–146.

Sanchez, P., D. Bandy, H. Villachica, and J. Nicholaides. 1982. Amazon Basin soils: Management for continuous crop production. *Science* 216:821–827.

Sanford, R. L., J. Saldarriaga, K. Clark, C. Uhl, and R. Herrera. 1985. Amazon rainforest fires. *Science* 227(4):53–55.

Schaefer, C. 1994. Landscape ecology and land use patterns in Northeast Roraima, Brazil. CEDAR Research Papers, no. 11 (November), Centre for Developing Areas Research-CEDAR, Department of Geography, Royal Holloway, University of London.

SEJUP [Servico Brasileiro de Justice e Paz]. 1998a. Landowners may be punished for lighting fires. *Folha de São Paulo*, April 14. Orig. in *News from Brazil*, no. 310:4, April 16.

SEJUP [Servico Brasileiro de Justice e Paz]. 1998b. Environment issues: Fire in Roraima destroys 15 km of Yanomami territory. *News from Brazil*, no. 306:3–4, March 11. Orig. in *Vejá*, March 11.

The Times. 1998. Medicine men save rainforest. April 2.

Toronto Star. 1998. Raging fires threaten stone-age tribe: Remote regions of Amazonia devoured by blaze that's burning out of control. March 17.

Uhl, C. and K. Clark. 1983. Selected seed ecology of selected Amazonian successional species. *Botanical Gazette* 144:419–425.

Van Leeuwen, L. 1998. Approaches for successful merging of indigenous forest-related knowledge with formal forest management: How can modern science and traditions join hands for sustainable forest management? Werkdocument IKC Natuurbeheer nr W-165. Wageningen: National Reference Centre for Nature Management, Ministry of Agriculture, Nature Management and Fisheries.

Vejá. 1998. Roraima renasce: A chuva apaga as marcas do incendio e os agricultores esperam a maior safra da década. July 22. Clipping Nacional, Ministerio das Relaçoes Exteriores. http://www.mre.gov.br/acs/clippings/vj8g2004.htm (accessed August 3, 1998).

Vejá. 1999. Vida Brasileira: O novo Eldorado — Roraima e o estado que mais recebe migrantes; E Rondonia, o que mais exporta gente. August 4:71-72.

Woods Hole Research Center. 2000. *Avança Brasil: The environmental costs for Amazonia. Report of the Scenarios Project—Feedback Cycle #1: Accidental fire and land improvements.* http://www.whrc.org/science/tropfor/roads/ciclo1en.htm (accessed September 27, 2000).

Woodwell, G. M. 2000. Regulation, not private enterprise, is the key to a healthy environment. *Nature* 405(8):613.

5 The Cerrado of Brazilian Amazonia

A Much-Endangered Vegetation

James A. Ratter, J. Felipe Ribeiro, and
Samuel Bridgewater

Attention in the Amazon Basin is almost always focused on the forested areas, the *terra firme* forest and the flooded *várzeas* and *igapós*, and it is often forgotten that *cerrado* and related Amazonian savannas are also important elements of the vegetation. However, the cerrado biome covers more than 700,000 square kilometers of Brazilian *Amazônia Legal* (Legal Amazonia), representing at least 15 percent of the total area of the region (figure 5.1). Most of it is made up of the parts of the core area of the biome lying in the states of Mato Grosso and Tocantins. In addition, areas of the Amazon Basin not included in Amazônia Legal are dominated by the cerrado biome; the principal of these is in the north and west of Goiás and covers more than 200,000 square kilometers. This means that cerrado is the second most important vegetation type of Amazonian Brazil, exceeded only by the terra firme Hylaean (Amazonian) forest; indeed in Tocantins cerrado is dominant, covering about 80 percent of the state, while in Mato Grosso it covers more than 40 percent. Thus nearly one-half of the two million square kilometers of the cerrado biome lie in Brazilian Amazonia sensu lato (i.e., in Amazônia Legal and the parts of the Amazon Basin lying outside it): cerrado is an extremely important vegetation of Brazilian Amazonia, and the southern part of Brazilian Amazonia is a vital part of the core area of the cerrado.

During much of the Pleistocene the cerrado biome probably covered far more of the Amazon Basin than it does at present. Evidence indicates great fluctuations between the relative sizes of the cerrado and Amazonian forest biomes during the alternating glacial and interglacial periods. Dynamic changes in vegetation distribution during the Pleistocene, bringing about

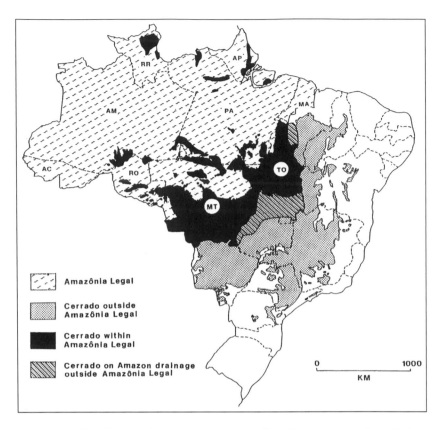

FIGURE 5.1 Distribution of cerrado vegetation in Brazilian Amazonia. State desig-
nations are as follows: AC, Acre; AM, Amazonas; AP, Amapá; MA, Maranhão; MT,
Mato Grosso; PA, Pará; RO, Rondônia; RR, Roraima; TO, Tocantins.

fragmentation of species ranges and isolation of populations, have probably
been a potent force in evolution and speciation in both the forest and cerrado biomes, producing much of their rich biodiversity (e.g., Prance 1973,
1982).

The Vegetation of the Cerrado Biome

The typical pattern of vegetation making up the landscape of the cerrado biome consists of cerrado (savanna woodland or sparser forms of scrub)
growing on the well-drained higher ground, and gallery forests following the
watercourses in the valleys. Between the cerrado and gallery forest there is

often a band of seasonally wet grassland where tall *Mauritia flexuosa* palms frequently occur. In addition, flat, ill-drained areas are common; they often consist of a pattern of regularly spaced tree-covered mounds of earth, borne on a seasonally wet *campo*—the so-called *campo de murundum*.

Cerrado is essentially a vegetation of nutrient-poor (dystrophic) soils, and more fertile areas within the cerrado biome are clothed with deciduous and semideciduous mesophytic forests. Such forests are very distinct from the Amazonian evergreen forest; in fact, in one area we studied in eastern Mato Grosso where the two forest types abutted each other, they had not a single species in common. The areas of mesophytic forest range from tiny patches where valleys have cut into base-rich rocks to such enclaves as the Mato Grosso de Goiás (on the drainage of the Rio Araguaia), estimated to have covered 40,000 square kilometers before its destruction for agriculture.

It is important to include the much-neglected cerrado–Amazonian forest transition vegetation in any consideration of the cerrado biome in Amazonia. There is a remarkable lack of research on this vegetation—literally only a handful of studies have been made (these are summarized and discussed in Ratter 1992).

Biodiversity of the Cerrado

The cerrado is a vegetation formation of great antiquity and there are even suggestions that it existed in prototypic form in the Cretaceous, before the final separation of the South American and African continents. The combination of the great age of the biome and the relatively recent (Pleistocene) dynamic phase in distribution patterns has probably led to its rich overall biodiversity, estimated by Dias (1992) as totaling about 160,000 species of plants, animals, and fungi.

Various estimates have been made of the number of taxa of vascular plants occurring in the cerrado biome. A recent compilation by Mendonça et al. (1998) lists 6,429 native species. This figure includes species in all habitats: cerrado sensu lato (i.e., all well-drained savanna-type vegetation), gallery forests, mesophytic forests, damp campos, montane campos (*campo rupestre*), and other less prominent habitats. The tree and large shrubby species of the cerrado sensu lato are relatively well known: a base list was provided by Rizzini (1963) and added to by Heringer et al. (1977). Those authors recorded 774 species belonging to 261 genera, and our work (Ratter et al. 1996, Biodiversity of the Cerrados project, unpubl.) has added some 60 species, so that the total species count for woody plants is now about 830.

The low vegetation (herbs and subshrubs) of the cerrado sensu lato contains many more species, probably at least four or five times more, than does the larger woody vegetation; there is also high species diversity of low-growing plants in the wet campos. The gallery forests also account for very high numbers of species: in areas where intensive studies have been carried out there are often higher numbers of tree species in the gallery forests than in the cerrado sensu lato.

The number of species actually endemic to the cerrado biome is an interesting question. A total of 336 of the 774 woody species of cerrado sensu lato recorded by Rizzini (1963) and Heringer et al. (1977) were regarded as endemic to the vegetation. The proportion of endemics among the herbs and subshrubs is certainly much higher; in fact it must represent the vast majority of the species. On the other hand, the number of endemic species in the gallery forests is much lower: the galleries of the western and northern cerrado region are largely composed of species of the Amazonian forest, while those in the central and southern cerrado have their floristic links with the semideciduous forests of southeastern Brazil (Oliveira-Filho and Ratter 1995).

The figures given above cover the entirety of the cerrado biome and at present it is not possible to estimate the number of species occurring only in the Amazonian part of it. However, surveys show that parts of Mato Grosso, Tocantins, and the Araguaia drainage of north and west Goiás are among the richest areas of the biome recorded. Particular biodiversity "hot-spots" have been noted on the Araguaia and Xingu drainage in eastern Mato Grosso and on the Chapada dos Guimarães (the latter in Amazônia Legal but on the Paraguay drainage). On the other hand, the isolated Amazonian savannas are rather species poor, in some cases having fewer than a dozen species of trees, in contrast to more than 100 at a single site in the richer communities of the core area (Ratter et al. 1996; Miranda 1997; Sanaiotti et al. 1997).

The Destruction of the Cerrado Biome

In the past, the cerrado domain was sparsely populated by Brazilian country people (backwoodsmen and Indians). Much of it was so remote that it only became incorporated into modern Brazilian life relatively recently, with the construction of railways and roads. The population practiced little more than subsistence agriculture based largely on low-density cattle-grazing in the cerrado vegetation, the raising of small crops in clearings in the gallery or mesophytic forests, charcoal burning (if there was an accessible market), and some hunting and fishing.[1] The native vegetation provided materials for housing (timber, palm thatch, etc.), seasonal fruits, fiber, firewood, and many other products for the rural economy. Recent research, however, has shown that Indian groups such as the Kayapó practiced a system of land use much more elaborate than was first obvious to the non-Indian observer (Posey 1984; Anderson and Posey 1989). Their agriculture centered on the creation of *apêtês* (forest islands), which were established by transporting forest soils enriched by humus and the remains of termite and ant nests into the cerrado, and by planting them with useful plants. This system was probably widespread in the Amazonian cerrados, and we were shown evidence of it in 1993 close to the Araguaia drainage near Jataí, Goiás, by an elderly *fazendeiro* (farmer) called Binômio da Silva. Binômio, a native of Jataí and a keen local naturalist, observed a remarkable constancy in the species composition and spatial distribution of species on the forest islands in the area. He confirmed his observations by measuring intertree distances, and discovered a repeated pattern that could only be explained by human activity. He therefore deduced that the forest islands had been planted by Indians (this was a completely independent conclusion: we discovered in discussions with him that he had never heard of other work on the subject). The Indian planters around Jataí were probably Kayapó, who once occupied a very extensive area in the state of Goiás.

All that has changed, however, and during the last twenty-five or so years the cerrados have been extensively developed for agriculture with the active encouragement of the Brazilian government. Such development is an important part of the policy to exploit the empty center of Brazil and incorporate it into the national economy. The best known elements of the same policy were the building of the new capital city, Brasília, right in the core of the cerrado area, and the construction of a vast system of national highways.

Encouragement for the agricultural development of the cerrados consisted of various forms of subsidy, extremely generous tax incentives, low

---

1. An exception to this was in the larger areas of mesophytic forest, where because of the fertile soils, large-scale agricultural development began much earlier. Such areas lie particularly in the Triângulo Mineiro (outside Amazonia) and in the Mato Grosso de Goiás (partially on the Araguaia drainage). The latter was opened for colonization by the construction of the railroad from Uberlândia to Anapolis in the 1920s and 1930s and became a center for cultivation of maize, upland rice, sugarcane, and coffee (Waibel 1948).

interest loans with no indexing (practically a donation in an economy suffering from hyperinflation during much of the period), and guaranteed prices. As intended, such inducements led to the establishment of a massive, highly mechanized, capital-intensive system of agriculture. Cerrado vegetation, with its small trees, is much more easily cleared than tall forest, and the soil is of a good structure for cultivation. However, before cultivation can take place a heavy application of lime and fertilizer is necessary to counteract soil acidity and to neutralize the aluminum that is present at toxic levels for virtually all cultivated plants. The arable crops planted are, in order of importance, soybeans, maize, rice, and manioc. Soybeans are grown largely for the export market and occupied 3.9 million hectares of the cerrado area in 1994, producing 8.8 million tons, while 4.9 million tons of maize were produced in the same year (Alho and Martins 1995). However, far more of the cerrado is exploited as improved pasture (planted with such exotic grasses as *Brachiaria* sp., *Hyparrhenia rufa*, and *Panicum maximum*) than as arable land (Alho and Martins 1995). It is difficult to find recent figures for the number of cattle raised on the cerrado, but in 1980 it was nearly 48 million and the present number is certainly much higher (Wagner 1985).

The system of cultivation of the cerrado is far from environmentally friendly. The employment of intensive mechanization requires huge tracts of monoculture with great areas of bare soil and the concomitant problems of erosion by rain and wind, while the legally required reserve areas tend to be kept in concentrated blocks so that the trees do not impede spraying aircraft. Clearly it would be better if such reserves were dispersed as a web to act as corridors for animals and as a more widespread seed source for recolonization. Of the many other environmental problems caused by cultivation, the heavy and often careless use of pesticides and the depletion of water reserves by giant rotating irrigators are among the most important.

In addition, the demand for charcoal by the Brazilian steel industry exerts pressure on the cerrado second only to agriculture. In fact, the two are often closely interlinked, as the trunks and roots of trees from the areas cleared for cultivation are usually used for charcoal production, partially offsetting the costs of preparation of the land. Today Brazilian steelmills are the largest charcoal users in the world—a fact one can readily appreciate when looking across the holocaust-like landscape of vast cleared areas with rows of domed charcoal kilns emitting dense smoke.

It is estimated that by 1994 some 695,000 square kilometers, representing 35 percent of the total cerrado area, had been changed to arable land or planted pasture, or was in altered fallow. Projections for the year 2000 indicate that the figure will have risen to 817,000–879,000 square kilometers, representing 41 to 44 percent of the total area (figures modified from Alho and Martins 1995). It is interesting to compare these data with the well-publicized figures for destruction of the Brazilian Amazonian rain forest. To date approximately 500,000 square kilometers of rain forest, representing some 13 percent of the original area, have been destroyed—an area far smaller in both absolute and relative terms than that of the cerrado.

Thus the level of destruction of the cerrado biome is alarming and it is important to know how much it affects the nearly half of its area which lies in Amazonia sensu lato. To some extent the enlightened Brazilian laws for land development should alleviate the situation: they require 50 percent of land to be kept under natural vegetation in Amazônia Legal as against 20 percent in the rest of Brazil (where, however, more than 200,000 square kilometers of the cerrado biome are situated in the Amazonian drainage of Goiás). This means that the extremes of environmental destruction seen in areas such as Mato Grosso do Sul, Minas Gerais, and in parts of Goiás should not be possible in Amazônia Legal. To some extent that appears to be the case, probably in part because of the 50 percent rule and in part because of the remoteness of many areas. However, even now, despite their distance from centers of population, agriculture in such areas as Humaitá, Amazonas, is being highly developed and this is likely to increase with improvements in the transport system. In particular, construction of the *hydrovias* (waterways produced by deepening the rivers to make them suitable for large vessels) will make the exploitation of distant areas very profitable. The Araguaia-Rio das Mortes and the Tocantins hydrovias will provide access to the cerrado areas of central and southeastern Amazonia, while the Paraguai-Paraná hydrovia will serve those of southwestern Amazonia.

Mato Grosso is the state of Amazônia Legal with the greatest area of cerrado and also the highest level of exploitation. Unfortunately, the figures available to us are approximate, but the original area of the cerrado biome in the state was about 400,000 square kilometers of which at least 60,000 square kilometers had been converted into arable and planted pastures by 1989, with over 17,000 square kilometers occupied by soybean plantations (figures modified from Sanchez 1992). If one considers that the legal limit for such altered agricultural areas in the state would be 200,000 square kilometers, allowing for an equal area in the 50 percent of reserves, this means that 30 percent of the Mato Grosso cerrados were already fully exploited by 1989, and the area has certainly greatly increased since that date.

It is impossible at present to produce figures for the level of destruction of cerrado–Amazonian forest transition vegetation; however, it is certain that they are extremely high. For instance, Ackerley et al. (1989), whose observations included the ecotonal cerrado around Sinop, Mato Grosso, emphasized that the elimination of that vegetation in southern Amazonia was proceeding at such a pace that observations might soon have only a historic interest, and Sanchez (1992) mentioned 6,000 kilometers of transitional forest converted to monocultures of perennial forage plants in the municipality of São Felix do Araguaia, Mato Grosso.

It is significant that most of the Amazonian areas identified as priorities for conservation by the internationally financed workshop *Áreas Prioritárias para a Conservação da Biodiversidade do Cerrado e Pantanal* (Cavalcanti 1999) are highly endangered and often already partly destroyed (e.g., the region of the Serra da Petrovina-Rondonopolis and the area around the municipalities of Riberão Cascalheira and Querência—all in Mato Grosso).

In conclusion, there is no doubt that cerrado is the most endangered major biome of Amazonia. As we have written before: "So much emphasis has been put on the emotive issue of the destruction of the rainforests that the world has largely forgotten the fate of their floristic cousins, the savanna woodlands!" (Ratter et al. 1997).

Conservation—The Way Ahead

The species-rich cerrado biome has always been undervalued as an important center of biodiversity. It stands alone among the major biomes of Brazil in not being recognized in the Brazilian constitution as a National Heritage, a status accorded the Amazon and Atlantic rain forests, the *Pantanal* (the enormous area of wetlands in Mato Grosso and Mato Grosso do Sul), and the coastal areas (Alho and Martins 1995). The cerrado has always been regarded as the poor relative of the Amazonian rain forests, and this is shown very strongly in the disproportionate allocation of international aid (e.g., the G7-financed Pilot Program for the Protection of Tropical Forests in Brazil [PPG7]). However, that attitude is now changing, and many important Brazilian politicians as well as the specialists of the Brazilian government–financed Center for Agricultural Research in the Cerrado (EMBRAPA-CPAC) and other scientists are well aware that urgent action must be taken to safeguard the existence and viability of the biome. In addition, Brazilian nongovernmental organizations (NGOs) are active in this field, and World Wildlife Federation (WWF) Brazil has recently produced a valuable assessment of the situation (Alho and Martins 1995) and is doing pioneering work in establishing extractive reserves.

One matter of urgency is to increase the number and area of conservation units. At present they represent less than 2 percent of the total area of the cerrado biome and Alho and Martins (1995) suggest that this should be at least tripled. The area of conserved cerrado in Amazônia Legal is tiny: for instance, the map in Cavalcanti (1999) shows only seven areas in Mato Grosso, only one of which is over 170,001 hectares, and five of these are in the Paraguay rather than the Amazon drainage. In contrast, 12 percent of the other Amazon ecosystems are in protected areas. The discrepancy extends to the size of conservation units, as well: while most of those in Amazonian forests are larger than 100,000 hectares, only 10 percent in cerrado surpass 50,000 hectares (Alho and Martins 1995).

However much research is required to choose the location of conservation units, as the cerrado biome is floristically very heterogeneous and constitutes a biological mosaic (Ratter and Dargie 1992; Castro 1994a, 1994b; Ratter et al. 1996). Teams from the University of Brasília, EMBRAPA-CPAC, and the Royal Botanic Garden Edinburgh have been collaborating on such research for a number of years, supported by Brazilian, European Community, and British funds. Their work has recently been expanded into a major Anglo-Brazilian initiative, Conservation and Management of the Biodiversity of the Cerrado Biome, with funding from the U.K. Department for International Development. The aim of the initiative is to survey the floristic patterns of cerrado vegetation and to discover representative areas and biodiversity hot spots. The method used covers the whole of the cerrado area with a grid subdivided into rectangles of $1°$ latitude $\times$ $1°30'$ longitude (the same method as established by Projeto Radambrasil in the early 1970s for the survey of the whole of Brazil) and plots the information for each rectangle in the grid. Priority is given to areas where there are either (a) few or no data, or (b) indications of exceptional diversity. Experienced field teams then carry out rapid surveys with techniques designed to provide the maximum amount of data in the minimum amount of time. A high percentage of the priority areas are in Amazonia, and recent surveys have concentrated on Tocantins, Mato Grosso, and Rondônia.

The internationally funded workshop *Áreas Prioritárias para a Conservação da Biodiversidade do Cerrado e Pantanal* (Cavalcanti 1999) brought together about 200 specialists from relevant disciplines to make recommendations on priority conservation areas. A remarkable consensus was achieved across the disciplines, and among the top priorities decided were about 10 Amazonian areas, including Mato Grosso sites such as the Serra da Petrovina region, the Riberão Cascalheira-Querência area (originally studied by the Xavantina-Cachimbo Expedition, 1967–1969 [Ratter et al. 1973, 1978]), and the Rio das

Mortes and Araguaia pantanals (much endangered by the potential effects of the Araguaia–Rio das Mortes hydrovia). The Riberão Cascalheira-Querência area would provide a reserve in the important cerrado–Amazonian forest eco-tone. If these recommendations are implemented they will provide the initial step toward the provision of a system of conservation units protecting the bio-diversity of the cerrado biome in Amazonia.

The Role of Traditional Ecological Knowledge in Conservation and Development

Traditional ecological knowledge has a vital part to play in conservation of the cerrados. Central Brazilian backwoodsmen understood the limitations of the soil and the dynamics of the vegetation to an extent that astonished me when I (J. A. R.) first met them more than thirty-five years ago. Since those days, research has always seemed to confirm the accuracy of their observa-tions, which were derived from long experience and so often unheeded by developers. The same can be said, probably to an even greater degree, about the knowledge of the cerrado Indians, at least some of whom had developed a land-use system completely in harmony with the environment (Posey 1984).

One extremely important factor well understood by native peoples is the use of controlled fire regimes to maintain soil fertility and the characteristics of the vegetation. Fires set more frequently than about once every three to four years result in impoverishment of the soil and great damage, particularly to the woody component of the vegetation. On the other hand, too infre-quent burning can lead to the establishment of a smothering thicket vegeta-tion, sometimes dominated by bracken fern (*Pteridium aquilinum*), and the occurrence of catastrophic damage if the accumulated mass of combustible material catches fire. Maintaining the right balance in reserves is difficult (e.g., Ratter et al. 1988) and the near-instinctive guidance of an experienced backwoodsman would often be of value. Similarly, such expertise would be invaluable in the recuperation of devastated cerrado areas, or in attempts to enhance cerrado by introducing or increasing economically useful species.

Conclusions

Although establishment and maintenance of conservation areas are ex-tremely important, it is unlikely that a sufficient amount of the biodiversity of the vast cerrado region could be protected by these activities alone. It is important therefore that a nondestructive system of agriculture be established over a large part of the region. Modern, highly mechanized systems of agriculture contrast so much with traditional land use in the cerrado that it seems impossible to combine their techniques: enormous arable monocultures and pastures of exotic grasses have little in common with low-density grazing of native vegetation and with subsistence cultivation. Similarly, the establishment of *Eucalyptus* or pine plantations to provide products like charcoal, wood chips, and paper pulp is very different from the exploitation of native woody species. However, at Fazenda Trijunção, at the meeting point of the states of Bahia, Minas Gerais, and Goiás, pioneering work is being carried out on the use of native pasture to provide profitable and sustainable land use of the cerrado. The scheme essentially involves refinement of existing techniques combined with some novel methods (e.g., in some cases cutting low vegetation rather than burning it). It even extends to breeding improved cattle, based on the Caracu race, to give high productivity on native pastures. The environmentally dedicated partners in the 20,000 hectare *fazenda* (farm), Srs José Roberto Marinho, Theodoro de Hungria Machado, and Neuber Joaquim dos Santos, aim to pioneer techniques that will safeguard the biodiversity and landscape of vast areas of the cerrado region while providing a profitable system of agriculture.

Acknowledgments

We wish to acknowledge the organizations that over the years have supported our research on the cerrado vegetation of Brazilian Amazonia. At present we are working as part of a major Anglo-Brazilian research initiative, Conservation and Management of the Biodiversity of the Cerrado Biome (U.K. Department of International Development/EMBRAPA Centro de Pesquisa Agropecuária dos Cerrados/Universidade de Brasília). In the past, the work has been supported by the European Community (financial contribution B92/4–3040/9304), the Baring Foundation, the Brazilian Academy of Sciences, the Conselho Nacional de Desenvolvimento Científico e Tecnológico (CNPq), the British Council, the Royal Society, and the Royal Geographical Society.

References

Ackerley, D. A., W. W. Thomas, C. A. Cid Ferreira, and J. R. Pirani. 1989. The forest-cerrado transition zone in southern Amazônia: Results of the 1985 Projeto Flora Amazônica expedition to Mato Grosso. *Brittonia* 41:113–128.

Alho, C. J. R. and E. de S. Martins, eds. 1995. *Bit by Bit the Cerrado Loses Space.* Brasília, DF, Brazil: World Wildlife Federation.

Anderson, A. B. and D. A. Posey. 1989. Management of a tropical scrub savanna by the Gorotire Kayapó of Brazil. In D. A. Posey and W. Balée, eds., *Resource Management in Amazonia: Indigenous and Folk Strategies*, pp. 159–173. Advances in Economic Botany 7. Bronx: New York Botanical Garden.

Castro, A. A. J. F. 1994a. Comparação florístico-geográfica (Brazil) e fitossociológica (Piauí-São Paulo) de amostras de cerrado. Ph.D. dissertation, Universidade Estadual de Campinas, SP, Brazil.

Castro, A. A. J. F. 1994b. Comparação florística de espécies do cerrado. *Silvicultura* 14(58):16–18.

Cavalcanti, R. B., coord. 1999. *Áreas Prioritárias para a Conservação da Biodiversidade do Cerrado e Pantanal.* Map of Unidades de Conservação (insert). Brasília, Brazil: Conservation International Brazil.

Dias, B. F. de S. 1992. Cerrados: Uma caracterização. In B. F. de S. Dias, ed., *Alternativas de Desenvolvimento dos Cerrados: Manejo e Conservação dos Recursos Naturais Renováveis*, ch. 2. Brasília, DF, Brazil: Funatura.

Heringer, E. P., G. M. Barroso, J. A. Rizzo, and C. T. Rizzini. 1977. A flora do cerrado. In M. G. Ferri, ed., *IV Simpósio Sobre o Cerrado*, pp. 211–232. São Paulo, Brazil: Editora Universidade de São Paulo.

Mendonça, R. de C., J. M. Felfili, B. M. T. Walter, M. C. da Silva Júnior, A. V. Rezende, T. S. Filgueiras, and P. E. Nogueira. 1998. Flora vascular do cerrado. In S. M. Sano and S. P. de Almeida, eds., *Cerrado: Ambiente e Flora*, pp. 289–556. Planaltina, Brazil: EMBRAPA-CPAC.

Miranda, I. de S. 1997. Flora, fisionomia e estrutura das savanas de Roraima, Brasil. Ph.D. dissertation. INPA, Manaus, AM, Brazil.

Oliveira-Filho, A. T. and J. A. Ratter. 1995. A study of the origin of central Brazilian forests by the analysis of plant species distribution patterns. *Edinburgh Journal of Botany* 52:141–194.

Posey, D. A. 1984. Keepers of the campo. *Garden* (USA) 8(6):8–12.

Prance, G. T. 1973. Phytogeographic support for the theory of Pleistocene forest refuges in the Amazon basin, based on evidence from distribution patterns in Caryocaraceae, Dichapetalaceae and Lecythidaceae. *Acta Amazônica* 3:5–29.

Prance, G. T. 1982. A review of the phytogeographic evidences for Pleistocene climate changes in the neotropics. *Annals of the Missouri Botanic Garden* 69:594–624.

Ratter, J. A. 1992. Transitions between cerrado and forest vegetation in Brazil. In P. A. Furley, J. Proctor, and J. A. Ratter, eds., *Nature and Dynamics of Forest-Savanna Boundaries*, pp. 417–429. London: Chapman & Hall.

Ratter, J. A., G. P. Askew, R. F. Montgomery, and D. R. Gifford. 1978. Observations on the vegetation of northeastern Mato Grosso. II. Forests and soils of the Rio Suiá-Missu area. *Proceedings of the Royal Society of London* B203:191–208.

Ratter, J. A., S. Bridgewater, R. Atkinson, and J. F. Ribeiro. 1996. Analysis of the floristic composition of the Brazilian cerrado vegetation. II. Comparison of the woody vegetation of 98 areas. *Edinburgh Journal of Botany* 53:153–180.

Ratter, J. A. and T. C. D. Dargie. 1992. An analysis of the floristic composition of 26 cerrado areas in Brazil. *Edinburgh Journal of Botany* 49:235–250.

Ratter, J. A., H. F. Leitão Filho, G. Argent, P. E. Gibbs, J. Semir, G. J. Shepherd, and J. Tamashiro. 1988. Floristic composition and community structure of a southern cerrado area in Brazil. *Notes from the Royal Botanic Garden Edinburgh* 45:137–151.

Ratter, J. A., J. F. Ribeiro, and S. Bridgewater. 1997. The Brazilian cerrado vegetation and threats to its biodiversity. *Annals of Botany* 80:223–230.

Ratter, J. A., P. W. Richards, G. Argent, and D. R. Gifford. 1973. Observations on the vegetation of northeastern Mato Grosso. I. The woody vegetation types of the Xavantina-Cachimbo expedition area. *Philosophical Transactions of the Royal Society of London* B226:449–492.

Rizzini, C. T. 1963. A flora do cerrado: Análise florística dos savanas centrais. In M. G. Ferri, coord., *Simpósio sobre o Cerrado*, pp. 127–177. São Paulo, Brazil: Editora Universidade de São Paulo.

Sanaiotti, T. M., S. Bridgewater, and J. A. Ratter. 1997. A floristic study of the savanna vegetation of the state of Amapá, Brazil, and suggestions for its conservation. *Boletim Museu Paraense Emílio Goeldi*, sér. Bot., 13:3–29.

Sanchez, R. O. 1992. *Zoneamento Agroecológico do Estado de Mato Grosso: Ordenamento Ecológico-Paisagístico do Meio Natural e Rural.* Cuiabá, Mato Grosso, Brazil: Fundação de Pesquisas Rondon.

Wagner, E. 1985. Desenvolvimento da região dos cerrados. In W. J. Goedert, ed., *Solos dos Cerrados: Tecnologias e Estratégias de Manejo*, pp. 19–31. São Paulo and Brasília, Brazil: EMBRAPA (CPAC)/Nobel.

Waibel, L. 1948. Vegetation and land use in the planalto central of Brazil. *Geographical Review* (New York) 38:529–554.

6 A Review of Amazonian Wetlands and Rivers

Valuable Environments Under Threat

Christopher Barrow

Traditional Exploitation of Floodlands and Possible Improvements

During the twentieth century, Amazonian wetlands (tables 6.1 and 6.2) and rivers have been exploited at relatively low intensity (Hall 1989:262). Rubber production, even at its height in the early 1920s, occurred mainly through the exploitation of wild trees; fiber crops like jute (*Corchorus capsularis*) and the similar malva (*Sida rhombifolia*) never became established enough to do much damage to the environment. However, in recent decades there has been a marked trend toward intensive *várzea* (floodland) agriculture, livestock production, timber extraction, palm-heart gathering, commercial fishing, gold mining, industrial pollution, *terra firme* (nonfloodland) agriculture that has contaminated runoff with agrochemicals, and hydroelectric projects (Eden 1990; Goodman and Hall 1990). Amazonian wetlands and rivers might well sustain intensified development, but caution is needed, for careless exploitation will destroy valuable potential crops, timber, and aquatic species.

There is historical and archaeological evidence that Amazonian wetlands and rivers supported large groups of people in the past. Dark-colored soil layers indicative of human activity ("black earths") have been found throughout Amazonia—often on bluffs between terra firme and wetlands—and are interpreted as indications of widespread pre-Columbian settlement (Denevan 1966, 1996; Meggers 1971; Roosevelt 1989; Alcorn 1990:212; Hiraoka 1992; Grenand and Grenand 1993; Macedo and Anderson 1993). After the conquests of the fifteenth and sixteenth centuries, *caboclos* settled Amazonia's wetlands (*caboclo* and *ribeirinho* are names applied to mixed-blood settlers who have adapted to the environment; the term *ribereño* is used more in Peru); they made their livings by fishing; extracting forest products, especially rubber (*Hevea* spp.); and shifting cultivation (Moran 1974; Eden and Andrade 1988; Nugent 1993; Brondizio 1996). Between the 1930s and 1970s many caboclos included jute or malva in their rotation crops, producing an income with little environmental damage; but in the 1970s the market failed, mainly because plastic replaced natural fibers. Rubber extraction from Amazonia's wild floodland trees still supports approximately 400,000 caboclo tappers.

Caboclo populations have remained low, so Amazonian várzeas have sustained relatively little degradation. The *aviamento* (sharecropper-patron) system that controls much of Amazonian wetland extraction and agriculture (Gray1990) and insecure land ownership have also deterred wetland development. This contrasts with the increasing settlement of terra firme areas.

Várzea Agriculture

Since the 1950s there have been calls for agricultural development of várzeas (Lima 1956; Hiraoka 1989, 1993; Prance 1990:899; Cleary 1991:122; Serrao 1994). Although várzeas are potentially capable of productive, sustainable agriculture it would be wise to research their ecology and traditional resource exploitation thoroughly before undertaking any major development (Goulding 1989; Demerona 1993; Posey 2000). Large scale irrigated rice production has been established along the Rio Jarí (Hall 1989:8), lower Rio Araguaia, and a few other rivers. However, rice production has been discouraged by land ownership problems as well as by volatile grain prices and competitive savanna rice production (Madeley 1993; author's field visits to Roraima in 1987).

Rice yields from várzeas can be good (Barrow and Patterson 1994; Barrow 1996) and some of the arable floodlands get regular depositions of silt, making fertilizers unnecessary. With careful drainage and soil management, compaction, erosion, and salinization can be avoided, herbicide use could be minimized, and one or more harvests a year might be sustained indefinitely (although várzeas in the lower Amazon are more likely to have salinity problems, and some of the clay soils become compacted or prone to acid-sulphate toxicity if cultivated). We can hope that opportunities will not be snatched from local peoples by a few large producers supported by government incentives.

TABLE 6.1 Types of Amazonian Wetlands

Wetlands on terra firme	

Wet hollows/swamps

Areas where road construction has caused water logging and accumulation of runoff

Várzantes — Areas periodically (often seasonally) flooded by runoff, not by river floods.

Floodlands (inundated by floodwater or tides)	

Várzeas

Areas formed by deposition of river sediments (see table 6.2). Four types of várzeas are recognized in Brazil, depending on cause of flooding: rivers, tides, sea, and rainfall. It is more practical to recognize *várzeas baixas* ("lower") and *várzeas altas* ("upper") — the latter are exposed first and are easier to irrigate, but may be difficult to drain. Often várzeas baixas merge into floating meadows (*campos de várzeas*) that are formed by mats of grasses like *Paspalum repens*, an excellent cattle forage.

In western Amazonia flooding is a consequence of rainfall/snowmelt in the Andes (occurs January–May).

In eastern Amazonia floods peak May–June; there is additional inundation as tides back up river water — near Belém this happens ca. every 15 days for two days in a row (significant tides are felt as far away as Santarém). There is some risk of unexpectedly high/untimely floods in most areas.

Cultivated várzeas of the lower Tocantins can flood for up to five months; periods of flooding seem to have lengthened in the last twenty-five years.

Igapós	Semipermanently flooded forests; often "backswamps" behind river-fringing várzeas. Some use term *igapó* for areas flooded by blackwater rivers and *várzea* for those flooded by whitewater rivers.
Restingas	Forests on levees/várzeas altas; flooded periodically at high water.
Paranás	Side channels of main streams, often fringed with forests.
Várzea lakes	Lakes that lie in the floodlands; can be a few ha to several hundred km^2 in size; seldom more than 4 m deep at low water; rich in plankton and weeds, they support important fisheries and act as breeding/nursery areas.
Saltmarshes and mangrove swamps	These may sometimes be *várzea* formations; regularly inundated when tides back up fresh water.

Note: It is difficult to assess the extent of várzeas, because of cloud and vegetation cover (side-scan radar has helped), and because the gently sloping land is subject to considerable variation in level, length, and frequency of flooding

Sources: Smith 1981:122; Eden 1990:17; Anderson and Ioris 1992a, 1992b:337.

TABLE 6.2 Informal Classification of Amazon Rivers

	"White" (turbid) Rivers	"Black" Rivers	"Clear" Rivers
pH	6.2–7.2	3.8–4.9	4.5–7.8
River	Madeira	Negro	Tapajós
	Purus	Preto	Xingú
	Solimoes	Urabú	Juruena
	Juruá	Uatuma	Cururu
	Napo		
	Caquetá-Japura		
	Branco		
	Ucayali		
	Guamá		
	Caeté,		
	Gurúpi		
	Jarí		
	Tocantins[1]		
	Araguaia[1]		

Note: Várzeas are best developed along whitewater rivers. There are no good long-term re-
cords of sediment loads, and monitoring is limited. Depositional conditions are highly variable
so generalization is difficult. Rivers draining Andean regions carry more sediment and tend to
deposit more fertile silt to form várzeas than do rivers from the Brazilian or Guyanan shields,
which tend to form infertile várzeas.
[1] Probably better classed as clear, although they carry sediment during floods and deposit
várzea-forming silt. Whitewater river-floodwater contains significant amounts of salts, so care
may be needed to avoid salinization of farmland.

Várzea Agroforestry and Extraction

There are more than 25,000 square kilometers of wetland forest in Amazonia. Traditionally, exploitation has been through agroforestry rather than extensive felling. To counteract the damage from increased logging, improved
agroforestry has been promoted as an alternative. Smith et al. (1995:258) argue that agroforestry could be widely adopted in Amazonia, especially in the
middle Amazon where there has been extensive deforestation, as a means
of reforestation. Certain fruit-bearing trees, if encouraged in várzea forests,

could improve food availability for fish during floods and yield sustainable timber (for a review of agroforestry potential, see Smith 2000).

Amazonia's palms are extensively used by local peoples and show promise for wider use; for example, açaí (*Euterpe oleracea* and *E. precatoria*), *pupunha* (*Bactris gasipaes*), *babaçu* or *babassu* (*Orbignya phalerata* [*Attalea* spp.]); *pataúa* (*Oenocarpus bataua*), *jauari* (*Astrocaryum jauari*), *burití* (*Mauritia flexuosa*), and *tucumá* (*Astrocaryum vulgare*) (Barrow 1990:376; Kahn 1991; Kahn and de Granville 1992). Collection and sale of fruit from açaí and palm hearts from burití and açaí in some regions provide the bulk of people's earnings (Anderson 1988; Anderson and Jardim 1989; Lopez Parodi and Freitas 1990). Better açaí management could improve and sustain livelihoods for many more people and help reduce forest clearance for other crops (Hiraoka 1995; Pollak, Mattos, and Uhl 1995). Nugent (1991:148) noted a rapid expansion of açaí production, with markets mainly in Belém and Manaus. Further expansion of the market is limited by the fruit's quick rate of fermentation and the possibility that people outside Amazonia may not acquire a taste for the fruit. Thus the key to opening wider markets for açaí and other extractive produce may lie in better processing and marketing.

Some Amazonian palms are productive where soils are unlikely to support other crops. Babaçu, which is of less value for wetlands because it prefers better-drained land, is widely used as part of linked floodland–terra firme livelihood strategies (Hall 1989:265; Alcorn 1990:207). As a source of thatch, food, oil, charcoal, and fibers, babaçu provides a major part of the income for about two million people in Brazil, especially in Maranháo and Goaís. It might be possible to expand production on degraded lands in Amazonia — and in a huge area south and northeast of the eastern Amazonian forest—to help divert development away from wetlands and rivers (Anderson, May, and Balick 1991).

Some Brazilian organizations and academics have expressed considerable interest in tolerant forest management (TFM), which is essentially an "improved" traditional agroforestry/extraction system that reduces clearance, provides income, and offers some biodiversity protection. TFM involves establishing "extractive reserves" by encouraging useful species without clearing other plants: the understory may be cleared a little to improve access, but the tree canopy is left largely intact so that many nonutilized species can survive (Anderson et al. 1985; Anderson 1990a; Clüsener-Godt and Sachs 2000:125–148). Examples of profitable and apparently sustainable wetland TFM have been described by Anderson and Jardim (1989), Anderson and Ioris (1992a, 1992b), and Padoch and De Jong (1992). One example is Combú Island, located approximately 1.5 kilometers from Belém. Inundated by seasonal and tidal flooding, this area of about 15 square kilometers supports TFM, which secures a year-round income (vital for smallholders) without the need for much cash or labor investment. There is still about 95 percent forest cover, and many little-disturbed areas survive. Family groups cultivate gardens, extract from a zone of managed forest surrounded by unmanaged forest, and fish or catch prawns. Forest management consists of opening paths to get at and promote the growth of useful trees (Anderson 1992:71–73). Average incomes from sales of nontimber products were good in the early- to mid-1980s (Anderson and Ioris 1992a, 1992b; figures quoted by Serrao 1994:27 for similar activities in Amazonas were less impressive). Ecotourism has been integrated with this TFM near Belém and Manaus to provide a diversified and improved means of earning a living for some local peoples.

Brondizio et al. (1994) compared caboclo groups that retained traditional TFM-type practices with those that turned to clearance and mechanized agriculture (all in the same eastern Amazonian locality). The groups having the greatest impact on the forest made no better a living than those who stayed with TFM-type extraction.

Várzea Livestock Production

Traditionally cattle have been grazed on várzeas during low water and moved to high ground or to floating barns during floods. McGrath et al. (1993:181) reported that Brazilian Amazonia's cattle herds had more than doubled between 1974 and 1984; at least half of floodland families now have some cattle, and grazing is widespread on the várzeas, especially in eastern Amazonia. Cattle are seen to be a wise investment, and water buffalo are increasing as a percentage of the total stock because they are better suited to wetlands and produce more milk. Increased grazing means that those practicing floodland cultivation now often have to prevent damage to their crops from these grazers. Larger herds are kept near Manaus and on Marajó Island (Faminow 1998).

River turtles (*Podocnemis unifilis* and *P. sextuberculata*) are collected and may have aquaculture potential (Mittermeir 1978; Smith 1981). There have also been suggestions that capybara (*Hydrochaeris hydrochaeris*), which are traditionally hunted, could be farmed in wetlands. Responding to a strong demand for skins, coboclos have hunted caiman to near extinction in many areas. Alligator and crocodile farming are well developed in Florida and Australasia and might be promoted in Amazonia, if native alligator or caiman (*Caiman crocodilus crocodilus*, *Melanosuchus niger*, or *Palaeosuchus trigonatus*) are used. However, aggressive exotics like the Nile crocodile (*Crocodylus niloticus*) could be a hazard to humans and wildlife if they escaped. Also, mercury pollution may hinder these developments as well as projects involving fishing and aquaculture improvements.

There is little potential for the captive breeding of river dolphins (*Inia geoffrensis* and *Sotalia fluviatalis*) because of their need for space and because of the many protective local myths relating to them. The much hunted manatee (*Trichechus inunguis*) might be bred, although it reproduces slowly (one calf every three years) (Best 1984), as might the iguana (*Iguana iguana*) (Werner 1991).

Impacts and Potential Impacts of Floodland Exploitation

Important tree crops such as cocoa and rubber originate from Amazonian wetlands, and many others may yet be developed (Smith 2000). Amazonian biodiversity is thus valuable for renewing existing tropical crops and as a source of new ones. Wetlands are a crucial part of the wider Amazonian ecosystem: many plants are dispersed by floods or require floodwater to grow; these areas provide feeding, breeding, and refuge areas for fish and other animals, and help regulate river flows. The primary productivity of many Amazonian rivers (notably "blackwaters" and "clearwaters") is poor; consequently animals depend a lot on flooded forests, so it is vital that areas under tree cover be left intact. Alternatively, one might establish useful tree species as "flooded plantations" or simply plant such species among natural floodland forest trees. Caboclos have a long tradition of making a living from such seasonally flooded environments, with little outside aid (Brondizio et al. 1994).

In 1981 várzeas yielded about 60 percent of Amazonian timber (FAO 1993:17). The main species exploited were *ucuúba* (*Virola surinamensis*), *andiroba* (*Carapa guianensis*), *sumaúma* (*Ceiba pentandra*), and *assaçu* (*Hura crepitans*) (Macedo and Anderson 1993). Ucuúba is used for plywood, particleboard, paper pulp, and veneer, and by the late 1980s was Brazil's second timber export after mahogany. Controls on its logging have been poorly enforced and sustainable stands are being lost, with nearby várzea forests damaged as a consequence; yet it is fast growing and could be easily managed sustainably (Prance 1990:894).

Impact of Agricultural Development

Wetlands may be drained and cultivated or may provide water for pump irrigation of terra firme land. Várzea flood recession and terra firme irrigation may increase as the use of motorized pump sets spreads; the falling water levels and return flows contaminated with sediment can have serious impacts on aquatic ecosystems. The ideal várzeas for rice cultivation are those receiving fertile silt with each flooding and needing little or no pump irrigation. In areas where less fertile silt is deposited there may still be opportunities for agricultural development if farmers can afford the necessary chemical fertilizers—but that is likely to be environmentally damaging if fertilizer runoff causes pollution (Junk 1980; Hiraoka 1993; Barrow and Patterson 1994; Barrow 1995). Probably one of the greatest threats to conservation of biodiversity and sustainable development in Amazonia is posed by the careless use of agrochemicals, which can cause widespread wildlife damage, disrupt fisheries and aquaculture, and threaten human health. At present, mercury pollution associated with gold mining is a major problem, but agrochemicals are likely to become the greater problem as settlement progresses and gold mining passes its peak activity. During visits to várzea smallholdings downstream of Tucuruí (author's field notes, 1985), I found that small-scale farmers were applying large amounts of pesticides like Aldrin, Dimecron-50, and Mirex. Herbicides are still seldom used, because flooding discourages most weeds that cannot can be cut and burned. Caboclos and indigenous groups are especially affected by agrochemicals and mercury river pollution because they have to depend on local—contaminated—food sources.

Impacts and Potential Impacts of Riverine Resources Exploitation

Fisheries

Fish are crucial to nutrition in Amazonia, and there is tremendous potential for providing new species for aquaculture if they are not lost through overfishing, pollution, or wetland degradation before they can be recognized. Smith (1981:18) noted that the Congo River had roughly 560 fish species; the Amazon has at least 1,300, of which only about 300 were exploited by 1967. Too little is known about the productivity and vulnerability of Amazonian fisheries and the potential of species that are not traditionally exploited.

There has been marked expansion of commercial fishing in Amazonia (Flores et al. 1990; McGrath 2000). This seems to be in response to the growth of várzea ranching, the decline of jute cultivation, and the improved possibilities for transport and sale of fish. Better boats and engines, nets, plastic food coolers, and refrigeration allow fishermen to range farther from the market and packing stations. Commercial fishing has caused problems in eastern Amazonia. For example, Manaus now has a large commercial fleet that operates up to 1,700 kilometers from the city and often comes into conflict with traditional fishermen (Serrao 1994; Clüsener-Godt and Sachs 2000:175–204). Commonly caught species include *tucunaré* (*Cichla temensis* and *C. ocellaris*), *pirarucu* (*Arapaima gigas*), *aruana* (*Osteoglossum bicirrhosum*), *filhote* (*Brachyplatystoma filamentosum*), *dourada* (*B. flavicans*), *tambaqui* (*Colossoma macropomum*), *piranha* (*Serrasalamus* spp.), and *pescada* (*Plagioscion squamosissimus*) (for Peruvian names see Hiraoka 1989:94). The best fish, such as the pirarucu, have been badly overfished to feed local populations and to sell outside Amazonia (Goulding 1981).

In a number of parts of Amazonia, declining fish and prawn stocks have been reported (Chapman 1989 noted a 23 percent fall in catches between 1970 and 1975), and the problem should be monitored. Research has shown that prawns (*Macrobrachium amazonicum*) and some fish may depend on extensive rafts or patches of floating macrophytes as "nursery areas," which are easily disturbed by some development activities (Collart and Moreira 1993). The same researchers suggested mid-Amazonia's prawn potential is underutilized.

Fisheries are also damaged by terra firme exploitation (leading to silting, pollution, and altered river flows), hydroelectric development, várzea disturbances, and urban and industrial pollution. The breakdown of traditions that discouraged overfishing now calls for better agreements or legal controls to conserve stocks—for example, through the use of permits, quotas, and closed seasons (Smith 1981:122). Increased commercial fishing has caused conflicts and has led some communities to try to control their local várzea and river/lake fisheries. Those community management developments may provide a foundation for better management (McGrath et al. 1993; Elchart 1997; McDaniel 1997). Commercialization might help break down aviamento, which pressures producers to increase catches yet offers few benefits to the fishermen, who become more marginalized.

There is potential for improving Amazonian fisheries and for aquaculture, but efforts must be carefully managed to avoid environmental damage and to ensure that local people benefit (Demerona 1990; Flores et al. 1990; McGrath et al. 1993). Also it is unwise to consider agriculture, extraction, and fisheries separately, because they are often closely interrelated (Serrao 1994).

Prawn and Fish Aquaculture

Prawns are an important source of income, especially in eastern Amazonia. So far they have mainly been netted or caught in wooden traps, but in many countries "tiger prawns" (such as *Penaeus monodon*) are farmed in ponds. Intensive aquaculture has serious impacts: floodland deforestation to build ponds; discharge of effluent; lowering of water tables; saltwater intrusion into aquifers (where groundwater is pumped into ponds); and the marginalization of traditional producers (Primavera 1991). There is also a possibility that farmed species could escape and damage native prawn fisheries and other wildlife. If intensive prawn (pond) aquaculture is carelessly adopted in Amazonia, the negative impacts on wildlife and local economies could be considerable.

Dam Construction

A number of Amazonian rivers have been impounded and more will be in the future. There has been success in developing Tucuruí Reservoir fisheries—catches in 1993 exceeded 3,000 tons (Boonstra 1993), and Balbina Reservoir near Manaus has also established a tucunaré fishery. However, Curu-Una Reservoir near Santarém has proved unproductive.

The first major Amazonian dam, Tucuruí (8,000 megawatts) has provided power since 1984, but flooded roughly 2,100 square kilometers of floodlands. The Balbina Reservoir flooded 2,430 kilometers to generate only about 250 megawatts (Fearnside 1989). After decades of experience, efforts to reduce dam-related impacts seem to have had limited success (Fearnside and Barbosa 1996). Water released downstream of a dam may be anoxic, acidic, and contaminated with toxic compounds; unless discharged with caution it can injure fish and other aquatic life (Fearnside 1989:410). By trapping silt, a dam damages the food supplies of aquatic life downstream (Magee 1989) and reduces deposition of várzea silt. Prawn fisheries downstream of Tucuruí have suffered, but it is difficult to establish to what extent that is due to dam construction (author's field notes, 1985; Demerona, De Cavalho, and Bittencourt 1987; Odinetz-Collart 1987).

Floodland forests and their wildlife suffer if dams control flows too much. Fish and other animals are often unable to migrate past a dam and can be deprived of the flow variation needed to trigger breeding (Reeves and Leatherwood 1994). Commercially important Amazonian fish that migrate include *jaraqui* (*Semaprochilodus* spp.) and dourada. Smith et al. (1991:319) reported the downstream loss of important fish species following closure of the Tucuruí and the Samuel dams. Studies have been made of the problems dams cause to the land use of várzeas along the São Francisco (Bolivian Amazonia) and the Guaporé (Rondônia) rivers (Diegues 1992).

Mineral Exploitation

Much of Amazonia's mining—whether the large-scale mining of bauxite, copper, tungsten, kaolinite, and iron, or the widespread small-scale gold mining by bands of *garimpeiros*—is alluvial and interacts with rivers and wetlands. In Rondônia, tin mining has silted streams (Smith 1981:131). Aluminum processing in eastern Amazonia contaminates rivers and causes widespread acid deposition (Anderson 1990b) that damages vegetation and mobilizes toxic compounds in soils. Although abandoned gold-mining pools may one day offer opportunities for wildlife and aquaculture, they are now contaminated with mercury (and may stay contaminated for a long time) and often support mosquito breeding. Many of the garimpeiros carry strains of malaria to which local people have little resistance, and so with increased pond breeding of mosquitoes there is a growing threat to caboclo and indigenous peoples.

There has been an Amazonian "gold rush" since 1979; by the early 1990s there were around 650,000 garimpeiros in Brazilian Amazonia alone (Smith et al. 1991; Greer 1993). Mercury used by garimpeiros to extract gold particles poisons the riverine and wetlands food webs. Perhaps 80 percent of Brazilian Amazonia's gold is sold to unofficial dealers, so it is difficult to make accurate estimates of production, and thus of pollution. Malm et al. (1990) suggested that more than 100 tons of mercury got into the Madeira River between 1979 and 1985; more would have entered other tributaries (De Lacerda et al. 1989, 1990; Smith et al. 1991; Greer 1993). Garimpeiros are most active along the Madeira, Branco, Moju, Tapajós, Garupi, and Tocantins rivers; their activity in the Mato Grosso contaminates the wildlife-rich Pantanal wetlands southwest of Amazonia (Cleary 1990:xxiv; Porvari 1995). Levels of mercury found in samples of river water, in hair from people living alongside rivers, and in fish are often well above World Health Organization safety limits and thus threaten the health and survival of fish, wildlife, and humans (Pfeiffer et al. 1991; Nriagu et al. 1992; Cleary et al. 1994). In parts of Amazonia people increasingly avoid fish they traditionally consumed. Toxic mono-methyl mercury will be released from riverbed and várzea sediments (some of which have 1,500 times the normal background levels) by microbial action for decades—even if mercury use could be immediately controlled (Greer 1993).

The drug trade "launders" profits by buying gold—and miners are driven by poverty to work in the mines—so control of mining practices may prove difficult. Smith et al. (1991:314) noted that in Colombia in colonial times miners used a foaming plant sap instead of mercury; it is worth researching this (and modern detergents or high-density fluids) to find something cheaper, more effective, and less of a health risk than mercury—if found, its use might spread fast. Cleary (1990:225) noted that, although garimpeiros could work without mercury, that would require skill and would reduce gold recovery. The European Community has funded research to develop ways of reducing mercury pollution caused by the garimpeiros' mining (The Times 1994). It might also be worthwhile to offer the "gold barons" who control production inexpensive devices for recovering gold and mercury from sluices and from vapor emitted when the gold-mercury amalgam is heated (Coghlan 1994).

Other Impacts on Floodland and Riverine Environments

Urban-Industrial Developments

There has been considerable industrial development in Amazonia since the 1970s, especially in the vicinities of Belém, Manaus, Santarém, Tucuruí, and Altimira (Lucarelli et al. 1994). Those centers also cause sewage pollution, acid deposition, and traffic-related pollution, which are felt well downwind. The cutting of forest to provide charcoal for iron smelting in the Grand Carajás Program region has caused serious environmental degradation of eastern Amazonia's rivers and wetlands (Anderson 1990b). Urban pollution has damaged fisheries near Manaus and Belém, below Marabá and Boa Vista. Acid deposition especially threatens soils that have high aluminum content and rivers that have acidic or neutral pH levels (i.e., clearwaters and blackwaters).

Small rum distilleries cause only localized river pollution, but widespread damage will occur if production of alcohol-based motor fuel spreads into Amazonia. Such fuel distilleries discharge effluent that pollutes rivers; their demand for cassava, yam, or sugar feedstock depletes these resources while marginalizing smallholders and causing forest clearance.

So far, wood pulp production is mainly limited to the Jarí Project, which was supposed to be supplied by approximately 200,000 hectares of fast-growing exotic-tree plantations. Jarí pulp mills cause river and air pollution and use trees from natural forests as well as from the plantations (Halperin 1980; Hoppe 1992:232).

Impacts of Petroleum Exploitation

The greatest impact has been in western Amazonia—particularly Ecuador, Colombia, and Peru—where road building has made land accessible, and oil spills or fires at wellheads have caused pollution (Thompson and Dudley 1989). At least one Brazilian company has found gas in the Jurúa River basin and oil in the Ucurú and Tapajós basins, and is prospecting on Marajó Island and in the Carajás region; so far there has been little noticeable impact. The petrochemical industry will probably develop existing refineries close to Manaus or Belém, where they will create acid deposition and river pollution.

The Drug Trade

The production of coca (*Erythroxylum* spp.) in the upper Amazon—especially in Peru, Bolivia, and Colombia—has had considerable environmental impact. The huge profits from "crack" and cocaine make it difficult to deter the shift from traditional production for local consumption to more intensive and environmentally damaging cultivation. Processing drugs in forest "factories" poisons rivers with an estimated 38,000 tons per year of toxic waste (kerosene, sulphuric acid, acetone, toluene, and filter paper). These enter Amazonian headwaters where many fish and other aquatic species breed and feed. Antidrug squads have used herbicides like 2,4-D and tebuthurion on coca fields; tebuthurion can remain in aquatic environments for up to five years, killing birds, fish, and other organisms (Redclift and Sage 1994:178).

The Impact of El Niño–Related Events

It seems likely that recent fires in Roraima and Rondônia are related to El Niño. There are indications that past El Niño events caused serious environmental change in Amazonia (Meggers 1994). Perhaps as many as 40,000 square kilometers of Roraima have been affected by fire. That is likely to have an impact on river flows, and on riverine and floodland nutrient regimes, and may have meteorological effects at the regional scale. Whether the impact on wetlands and rivers will be negative and how long it will be felt is not yet apparent.

The Threat of Global Warming

Amazonia's wetlands (especially in eastern Amazonia) lie at a low elevation, so even a slight rise in sea level would have a marked effect. The periods of inundation, extent of flooding, character of sediments deposited, and risks of salt accumulation in floodland soils and floodplain groundwater could change (Fearnside 1995). Smallholders can probably adapt, as they already cope with considerable year-to-year variations; larger-scale development of floodplain areas may be less flexible.

Conclusions

Development in and around wetland and riverine environments will cause serious loss of biodiversity unless careful monitoring and effective measures are applied. Wherever possible, development should take place only after a thorough assessment of likely impacts, and existing livelihood strategies should be studied to see whether they offer a foundation for improved production. Local strategies have evolved to fit difficult conditions and can suggest the most likely ways to achieve secure and sustainable livelihoods. The involvement of local people in development is important but the outcome will only be successful if the community can see a significant benefit to the proposed changes. Much of the research and advocacy focused on improving Amazonian livelihoods and environmental management has missed the point of "social capital"—that is, the institutions, traditions, and social obligations (e.g., exchanges of labor and other mutual assistance) that allow groups of people to be more resilient, adaptive, and innovative. During fieldwork along the Tocantins, I observed that villages with seemingly comparable resource endowment varied a great deal in apparent affluence; it would be interesting to study whether those variations reflected different endowments of social capital. If social capital proves important in determining the success of livelihood development and environmental management, it would be worth supporting measures to prevent its breakdown, and wherever possible to work to strengthen it.

According to Barber (1993), potential wetlands offer five categories of value (in addition to agricultural development):

1. *Direct uses* that consume resources: grazing, cultivating, cutting wood for fuel
2. *Nonconsumptive uses*: tourism, TFM
3. *Indirect uses*: storm protection, flood mitigation, fish feeding/breeding
4. *Nonuse*: intrinsic value, moral obligation to conserve
5. *Reservoirs of biodiversity*

Carefully selected wetlands, like Jau National Park, should be set aside and protected. Larger areas could be used for TFM, which would still preserve some biodiversity. There is a pressing need to control mercury pollution. The development of floodland for agriculture must be controlled and monitored, especially for the use of agrochemicals. Monitoring and management of fisheries, the drug trade, floodland logging, and urban and industrial pollution must be improved. Wetland conservation could take the form of extensive reserves linked by riverside belts of forest that would be kept secure from disruption by dam building, terra firme development, or sea-level change. Riverside reserves could provide biodiversity conservation, breeding and feeding sites for fish and other animals, riverbank protection, opportunities for TFM, and moderation of river flows. Selected lower várzeas could be used for crops and grazing. Some of these land uses are compatible with tourism: already "jungle walks" have been organized along TFM trails.

What is needed is a plan for the integration of Amazonian wetland development, river use, conservation, and tourism. There is also a need for monitoring to ensure that terra firme development does not damage wetlands or rivers. Without good coordination, development efforts are likely to disrupt adjoining interdependent ecosystems. Fearnside (1997) suggested that zoning would be useful for achieving sustainable development in Amazonia. But for zoning to be effective, there needs to be much more information about the effects of development on the environment, land-use systems, and local societies.

Traditional exploitation strategies—many of which can be learned from archaeology—have much to offer in support of sustainable agriculture and biodiversity conservation (Posey 1989; Posey and Balee 1989; Goodman and Redclift 1991; Roosevelt 1992). Coboclo strategies have evolved to suit Amazonian wetlands and so are potential sources of workable models (for example, TFM draws on traditional knowledge). It makes sense to involve local people to support production and conservation, as they stand to gain livelihoods and can help exclude more exploitative settlement. Small-scale traditional land uses are probably going to be more adaptable to future environmental and socioeconomic changes.

It should be easier to sustain production in wetland than in terra firme environments. Furthermore, rivers provide protein and more appropriate transport than Amazonian roads. Because wetlands, rivers, and terra firme are interdependent, development must maintain a locally adapted mix of conservation, cultivation, TFM, aquaculture, tourism, and sustainable forestry. Ideally, each locality should develop a local livelihood strategy, which will often draw on traditional knowledge and be appropriate for existing conditions. Strategies will need to be coordinated to ensure compatibility with other development efforts, to minimize conflicts with environmental and social requirements, to maximize mutual support between communities, and to encourage improvement and innovation. These local strategies will form an overall Amazonian "mosaic." Once a local component strategy is determined a success, it should be duplicated somewhere else to multiply benefits and to ensure security in case there is a local failure elsewhere, in which case a counterpart component could be used to supply help and genetic resources. A mix of locally adapted components would give people a diverse and probably more secure livelihood, and should better guarantee biodiversity conservation. With the marketing and transport challenges faced by small-scale producers in Amazonia, it makes sense for them to seek adequate subsistence based on diversity, using a variety of cash crops to avoid glutting the markets. However, a diversity of crops that may be little known beyond Amazonia also presents challenges for processing, transporting, and marketing.

There is growing interest in the value of traditional knowledge as a foundation for development when combined with in situ biodiversity conservation (IUCN Inter-Commission Task Force on Indigenous Peoples 1997). Traditional knowledge has much to offer wetland agriculture. For example, in preconquest Mexico the Aztecs had over 100,000 hectares of wetland around Lake Texcoco under *chinampas* agriculture (raised-bed and channel horticulture); this sustained more than 100,000 people (Roggeri 1995:208). If that could be accomplished at high altitude, then Amazonian wetlands, with their warmer climate and periodic receipts of fertile flood silt, should be able to reward efforts to develop chinampas-type strategies that would feed and employ local people (and could be integrated with pisciculture). Traditional Amazonian livelihoods should be researched and ways of enhancing them explored.

References

Alcorn, J. B. 1990. Indigenous agroforestry systems in the Latin American tropics. In M. A. Altieri and S. B. Hecht, eds., *Agroecology and Small Farm Development*, pp. 203–213. Boca Raton: CRC Press.

Anderson, A. B. 1988. Use and management of native forests dominated by açaí palm (*Euterpe oleracea* Mart.) in the Amazon estuary. In M. J. Balick, ed., *The Palm — Tree of Life: Biology, Utilization and Conservation*, pp. 144–154. Advances in Economic Botany 6. Bronx: New York Botanical Garden.

Anderson, A. B. 1990a. Extraction and forest management by rural inhabitants in the Amazon estuary. In A. B. Anderson, ed., *Alternatives to Deforestation: Steps Toward Sustainable Use of the Amazon Rain Forest*, pp. 65–85. New York: Columbia University Press.

Anderson, A. B. 1990b. Smokestacks in the rainforest: Industrial development and deforestation in the Amazon Basin. *World Development* 18(9):1191–1205.

Anderson, A. B. 1992. Land-use strategies for successful extractive economies in Amazonia. In D. C. Nepstad and S. Schwartzman, eds., *Non-Timber Products from Tropical Forests: Evaluation of a Conservation and Development Strategy*, pp. 67–77. Advances in Economic Botany 9. Bronx: New York Botanical Garden.

Anderson, A. B., A. Gely, J. Strudwick, G. L. Sobel, and M. Das Cracas C. Pinto. 1985. Uma sistema agroflorestal na várzea do estúario amazônica (Ilha das Onças, Município de Barcarena, Estado do Pará). *Acta Amazônica* (suppl.) 15(1–2):195–224.

Anderson, A. B. and F. M. Ioris. 1992a. The logic of extraction: Resource management and income generation by extractive producers in the Amazon estuary. In K. H. Redford and C. Padoch, eds., *Conservation of Neotropical Forests: Working from Traditional Resource Use*, pp. 175–202. New York: Columbia University Press.

Anderson, A. B. and E. M. Ioris. 1992b. Valuing the rain-forest: Economic strategies used by small-scale forest extractivists in the Amazon Estuary. *Human Ecology* 20(3):337–369.

Anderson, A. B. and M. A. Jardim. 1989. Costs and benefits of floodplain management by rural inhabitants in the Amazon estuary: A case study of açaí palm production. In J. O. Browder, ed., *Fragile Lands of Latin America: Strategies for Sustainable Development*, pp. 114–129. Boulder, Colorado: Westview.

Anderson, A. B., P. H. May, and M. J. Balick. 1991. *The Subsidy from Nature: Palm Forests, Peasantry, and Development on the Amazon Frontier*. New York: Columbia University Press.

Barber, E. B. 1993. Sustainable use of wetlands: Valuing tropical wetland benefits; Economic methodologies and applications. *The Geographical Journal* 159(1): 22–32.

Barrow, C. J. 1990. Environmentally appropriate, sustainable small-farm strategies for Amazonia. In D. Goodman and A. Hall, eds., *The Future of Amazonia: Destruction or Sustainable Development?*, pp. 360–382. London: Macmillan.

Barrow, C. J. 1995. Agricultural production in Brazilian Amazonia: Problems and prospects. *Journal of Sustainable Agriculture* 7(1):19–37.

Barrow, C. J. 1996. Environmental impact of resource use on wetland and riverine habitats in Amazonia. In M. J. Eden and J. T. Parrey, eds., *Land Degradation in the Tropics: Environmental and Policy Issues*, pp. 177–189. London: Pinter.

Barrow, C. J. and A. Patterson. 1994. Agricultural diversification: The contribution of rice and horticultural producers. In P. A. Furley, ed., *The Forest Frontier: Settlement and Change in Brazilian Roraima*, pp. 153–184. London: Routledge.

Best, R. C. 1984. The aquatic mammals and reptiles. In H. Sioli, ed., *The Amazon: Limnology and Landscape Ecology of a Mighty Tropical River and Its Basin*, pp. 371–412. Dordrecht: W. Junk.

Boonstra, T. E. 1993. Commercialisation of the Tucuruí reservoir fishery in the Brazilian Amazon. *TCD Newsletter* 28 (December):1–4. Gainesville: Center for Latin American Studies, University of Florida.

Brondizio, E. S. 1996. Forest farmers: Human and landscape ecology of caboclo populations in the Amazon Estuary. Ph.D. dissertation, Indiana State University, Bloomington.

Brondizio, E. S., E. G. Moran, P. Mausel, and Y. Wu. 1994. Land-use change in the Amazon Estuary: Patterns of caboclo settlement and landscape management. *Human Ecology* 22(3):249–278.

Chapman, M. D. 1989. The political ecology of fisheries depletion in Amazonia. *Environmental Conservation* 16(4):331–337.

Cleary, D. 1990. *Anatomy of the Amazon Gold Rush*. London: Macmillan.

Cleary, D. 1991. The greening of the Amazon. In D. Goodman and M. Redclift, eds., *Environment and Development in Latin America: The Policies of Sustainability*, pp.116–140. Manchester: Manchester University Press.

Cleary, D., I. Thornton, N. Brown, G. Kazantzis, T. Delves, and S. Washington. 1994. Mercury in Brazil. *Nature* 369(6482):613–614.

Clüsener-Godt, M. and I. Sachs, eds. 2000. *Brazilian Perspectives on Sustainable Development of the Amazon Region*. New York: Parthenon Publishing Inc. (for UNESCO).

Coghlan, A. 1994. Midas touch could end Amazon's pollution. *New Scientist* 141(1916):10.

Collart, O. O. and L. C. Moreira. 1993. Fishing potential of *Macrobrachium amazonicum* in central Amazonia (Careiro Island): Abundance and size variation. (In French.) *Amazoniana Limnologia et Oecologia Regionalis Systemae Fluminis Amazonas* 13(3–4):399–413.

De Lacerda, L. D., F. C. F. De Paula, A. R. C. Ovalle, W. C. Pfeiffer, and O. Malm. 1990. Trace metals in fluvial sediments of the Madeira River watershed, Amazon, Brazil. *Science of the Total Environment* (1990):525–530.

De Lacerda, L. D., W. C. Pfeiffer, A. T. Ott, and E. G. da Silveira. 1989. Mercury contamination in the Madeira River, Amazon—Hg inputs to the environment. *Biotropica* 21(1):91–93.

Demerona, B. 1990. Amazonian fisheries—general characteristics based on two case studies. *Interciencia* 15(6):461–468.

Demerona, B. 1993. Ecological conditions of the production in a floodplain island of central Amazonia: A multidisciplinary project. (In French.) *Amazoniana Limnologia et Oecologia Regionalis Systemae Fluminis Amazonas* 12(3–4):353–363.

Demerona, B., J. L. De Cavalho, and M. M. Bittencourt. 1987. Les effets immediats de la fermature du barrage de Tucuruí (Brésil) sur l'ichtyofaune en aval. *Revu de Hydrobiologie Tropicale* 20(1):73–84.

Denevan, W. M. 1966. A cultural-ecological view of the former aboriginal settlement in the Amazon region. *Professional Geographer* 18(3):346–351.

Denevan, W. M. 1996. A bluff model of riverine settlement in prehistoric Amazonia. *Annals of the Association of American Geographers* 86(4):654–681.

Diegues, A. C. S. 1992. Sustainable development and people's participation in wetland ecosystem conservation in Brazil: Two comparative studies. In D. Ghai and J. M. Vivian, eds., *Grassroots Environmental Action: People's Participation in Sustainable Development*, pp. 141–158. London: Routledge.

Eden, M. J.1990. *Ecology and Land Management in Amazonia*. London: Belhaven.

Eden, M. J. and A. Andrade. 1988. Colonos, agriculture and adaptation in the Colombian Amazon. *Journal of Biogeography* 15(1):79–85.

Elchart, G. 1997. Sustainable resource management in the Brazilian Amazon: The case of the community of Tiningu. *Coastal Management* 25(2):205–226.

Faminow, M. D. 1998. *Cattle, Deforestation and Development in the Amazon: An Economic, Agronomic and Environmental Perspective*. Wallingford: CAB International.

FAO 1993. *Management and Conservation of Closed Forests in Tropical America*. FAO Forestry Paper, no. 101. Rome: Food and Agriculture Organization of the U.N. (FAO).

Fearnside, P. M. 1989. Brazil's Balbina Dam: Environment versus the legacy of the pharaohs in Amazonia. *Environmental Management* 13(4):401–423.

Fearnside, P. M. 1995. Potential impacts of climatic change on natural forest and forestry in Brazilian Amazonia. *Forest Ecology and Management* 78(1–3):51–70.

Fearnside, P. M. 1997. Human carrying capacity estimation in Brazilian Amazonia as a basis for sustainable development. *Environmental Conservation* 24(3):271–282.

Fearnside, P. M. and R. I. Barbosa. 1996. The Cotingo Dam as a test of Brazil's system for evaluating proposed developments in Amazonia. *Environmental Management* 20(5):631–648.

Flores, H. G., F. A. Bocanegra, J. M. Garcia, and H. S. Riveiro. 1990. Fisheries in the Peruvian Amazon. *Interciencia* 15(6):469–475.

Goodman, D. and A. Hall, eds. 1990. *The Future of Amazonia: Destruction or Sustainable Development*. London: Macmillan.

Goodman, D. and M. Redclift, eds. 1991. *Environment and Development in Latin America: The Policies of Sustainability.* Manchester: Manchester University Press.

Goulding, M. 1981. *Man and Fisheries on an Amazon Frontier.* Dordrecht: W. Junk.

Goulding, M. 1989. *Amazon: The Flooded Forest.* London: BBC Publications.

Gray, A. 1990. Indigenous people and the marketing of the rainforest. *The Ecologist* 20(6):223–227.

Greer, J. 1993. The price of gold: Environmental costs of the new gold rush. *The Ecologist* 23(3):91–96.

Grenand, F. and P. Grenand. 1993. Historical stages of the várzea settlement in the Amazon. (In French.) *Amazoniana Limnologia et Oecologia Regionalis Systemae Fluminis Amazonas* 12(3–4):509–526.

Hall, A. 1989. *Developing Amazonia: Deforestation and Conflict in Brazil's Carájas Programme.* Manchester: Manchester University Press.

Halperin, D. T. 1980. The Jarí Project: Large-scale land and labor utilisation in the Amazon. *Geographical Survey* 9(1):13–21.

Hiraoka, M. 1989. Agricultural systems on the floodplains of the Peruvian Amazon. In J. O. Browder, ed., *Fragile Lands of Latin America: Strategies for Sustainable Development,* pp. 75–101. Boulder, Colorado: Westview.

Hiraoka, M. 1992. Caboclo and ribereño resource management: A review. In K. H. Redford and C. Padoch, eds., *Conservation of Neotropical Forests: Working From Traditional Resource Use,* pp. 134–157. New York: Columbia University Press.

Hiraoka, M. 1993. Sustainable resource management in the Amazon floodplain: Report on the first stage of the "Várzea Project." Newsletter of the UNU-PLEC Project, H. Brookfield, ed. Canberra: Australian National University. *PLEC News and Views* 1 (July):11–13.

Hiraoka, M. 1995. Land-use changes in the Amazon Estuary. *Global Environmental Change: Human and Policy Dimensions* 5(4):323–336.

Hoppe, A. 1992. The Amazon between economy and ecology. *Natural Resources Forum* 16(3):232–234.

IUCN Inter-Commission Task Force on Indigenous Peoples. 1997. *Indigenous Peoples and Sustainability: Cases and Actions.* Gland, Switzerland: International Union for the Conservation of Nature and Natural Resources (IUCN).

Junk, W. J. 1980. Areas inundáveis—um desafio para limnologica. *Acta Amazônica* 10(4):775–795.

Kahn, F. 1991. Palms as key swamp forest resources in Amazonia. *Forest Ecology and Management* 38(3–4):133–142.

Kahn, F. and J.-J. de Granville 1992. *Palms in Forest Ecosystems of Amazonia.* Berlin: Springer-Verlag.

Lima, R. R. 1956. A agricultura nas várzeas do estuario do Amazonas. *Boletim Téchnico do Instituto Agronômico do Norte,* no. 33. Belém: Instituto Agronômico do Norte.

Lopez Parodi, J. and D. Freitas. 1990. Geographical aspects of forested wetlands in the lower Ucayali, Peruvian Amazonia. *Forest Ecology & Management* 33–34(1–4): 157–168.

Lucarelli, F., P. Destefano, L. G. Napolitano, P. Murino, and R. Vigliotti. 1994. Brazilian Amazonia: Industrial development and environmental monitoring. *Environmental Management* 18(4):597–604.

Macedo, D. S. and A. B. Anderson. 1993. Early ecological changes associated with logging in the lower Amazon floodplain. *Biotropica* 25(2):151–163.

Madeley, J. 1993. Raising rice in the savannas. *New Scientist* 138(1878): 36–39.

Magee, P. 1989. Peasant political identity and the Tucuruí Dam: A case study of the island dwellers of Pará, Brazil. *Latin Americanist* 24(1):6–10.

Malm, O., W. Pfeiffer, C. M. M. Souza, and R. Reuther. 1990. Mercury pollution due to gold mining in the Madeira River basin, Brazil. *Ambio* 19(1):11–15.

McDaniel, J. 1997. Communal fisheries management in the Peruvian Amazon. *Human Organisation* 56(2):147–152.

McGrath, D. G. 2000. Avoiding a tragedy of the commons: Recent developments in the management of Amazonian fisheries. In A. Hall, ed., *Amazonia at the Crossroads: The Challenge of Sustainable Development*, pp. 171–187. London: Institute of Latin American Studies.

McGrath, D. G., F. De Castro, C. Futemma, B. D. de Amaral, and J. Calabria. 1993. Fisheries and the evolution of resource management on the lower Amazon floodplain. *Human Ecology* 21(2):167–195.

Meggers, B. J. 1971. *Amazonia: Man and Culture in a Counterfeit Paradise*. Chicago: Aldine.

Meggers, B. J. 1994. Archaeological evidence for the impact of mega-El Niño events on Amazonia during the past 2 millennia. *Climate Change* 28(4):321–338.

Mittermeir, R. A. 1978. South American river turtles: Saving their future. *Oryx* 14(3):220–230.

Moran, E. F. 1974. The adaptive system of the Amazonian caboclo. In C. Wagley, ed., *Man in the Amazon*, pp. 136–159. Gainesville: University of Florida Press.

Nriagu, J. O., W. C. Pfeiffer, O. Malm, and C. M. M. de Souza. 1992. Mercury pollution in Brazil. *Nature*, vol. 368, pt. 66368, p. 389.

Nugent, S. 1991. The limitations of environmental "management": Forest utilisation in the lower Amazon. In D. Goodman and M. Redclift, eds., *Environment and Development in Latin America: The Policies of Sustainability*, pp. 141–154. Manchester: Manchester University Press.

Nugent, S. 1993. *Amazonian Caboclo Society: An Essay on Invisibility and Peasant Economy*. Berlin: Berg.

Odinetz-Collart, O. 1987. La pêche crevettère de *Macrobrachium amazonicum* (Palaemonidae) dans le Bas-Tocantins, après la fermature du barrage de Tucuruí (Brésil). *Revue de Hydrobiologie Tropicale* 20(2):131–144.

Padoch, C. and W. De Jong. 1992. Diversity, variation, and change in ribeireño agriculture. In K. H. Redford and C. Padoch, eds., *Conservation of Neotropical Forests: Working From Traditional Resource Use*, pp. 158–274. New York: Columbia University Press.

Pfeiffer, W. C., O. Malm, C. M. M. de Souza, L. Drude de Lacerda, E. G. Silveira, and W. R. Bastos.1991. Mercury in the Madeira River ecosystem, Rondônia, Brazil. *Forest Ecology and Management* 38(3):239–245.

Pollak, H., M. Mattos, and C. Uhl. 1995. A profile of palm heart extraction in the Amazon Estuary. *Human Ecology* 23(3):357–385.

Porvari, P. 1995. Mercury levels of fish in Tucuruí hydroelectric reservoir and in River Moju in Amazonia, in the State of Pará, Brazil. *Science of the Total Environment* 175(2):109–117.

Posey, D. A. 1989. Alternatives to forest destruction: Lessons from the Mêbêngôkre Indians. *The Ecologist* 19(6):241–244.

Posey, D. A. 2000. Biodiversity, genetic resources and indigenous people in Amazonia: (Re)discovering the wealth of traditional resources of Native Amazonians. In A. Hall, ed., *Amazonia at the Crossroads: The Challenge of Sustainable Development*, pp. 188–204. London: Institute of Latin American Studies.

Posey, D. A. and W. Balee, eds. 1989. *Resource Management in Amazonia: Indigenous and Folk Strategies*. Advances in Economic Botany 7. Bronx: New York Botanical Garden.

Prance, G. 1990. Future of the Amazonian rainforest. *Futures* 22(9):891–903.

Primavera, J. H. 1991. Intensive prawn farming in the Philippines—ecological, social and economic implications. *Ambio* 20(1):28–33.

Redclift, M. and C. Sage, eds. 1994. *Strategies for Sustainable Development: Local Agendas for the Southern Hemisphere*. Chichester: Wiley.

Reeves, R. R. and S. Leatherwood. 1994. Dams and river dolphins: Can they coexist? *Ambio* 23(3):172–175.

Roggeri, H. 1995. *Tropical Freshwater Wetlands: A Guide to Current Knowledge and Sustainable Management*. Dordrecht: Kluwer.

Roosevelt, A. C. 1989. Lost civilisation of the lower Amazon. *Natural History* 98(2):74–83.

Roosevelt, A. C. 1992. Secrets of the forest—an archaeologist reappraises the past and future of Amazonia. *Sciences* (New York) 32(6):22–28.

Serrao, E. A. 1994. The Amazon floodplain: The next major frontier for food production. UNU-PLEC Project publication, H. Brookfield, ed. Canberra: Australian National University. *PLEC News and Views* 2 (February):25–28.

Smith, N. J. H. 1981. *Man, Fishes and the Amazon*. New York: Columbia University Press.

Smith, N. J. H. 2000. Agroforestry development and prospects in the Brazilian Amazon. In A. Hall, ed., *Amazonia at the Crossroads: The Challenge of Sustainable Development*, pp. 151–170. London: Institute of Latin American Studies.

Smith, N. J. H., P. de-T. Alvim, A. Homma, I. C. Falesi, and A. E. S. Serrao. 1991.

Environmental impacts of resource exploitation in Amazonia. *Human and Policy Dimensions* 1(4):313–320.

Smith, N. J. H., T. J. Fik, P. de-T. Alvim, I. C. Falesi, and E. A. S. Serrao. 1995. Agroforestry developments and potential in the Brazilian Amazon. *Land Degradation & Rehabilitation* 6(4):251–263.

The Times. 1994. Higher Education Supplement, S4/3/94. March 4:9

Thompson, K. and N. Dudley. 1989. Transnationals and oil in Amazonia. *The Ecologist* 19(6):219–224.

Werner, D. 1991. The rational use of green iguanas. In J. G. Robinson and K. H. Redford, eds., *Neotropical Wildlife Use and Conservation*, pp. 181–201. Chicago: University of Chicago Press.

7 Fragility and Resilience of Amazonian Soils

Models from Indigenous Management

Peter A. Furley

Despite the increased availability and dissemination of information, perceptions of Amazonian soils are still polarized. At one extreme there persists a utopian view of rich fertility (colored by the luxuriance of the tropical forest cover). At the other extreme there exists an excessively restrictive view of an extremely precarious resource, vulnerable to rapid degeneration at the slightest disturbance (a perception based on the extreme effects of deforestation). The reality lies somewhere in between and reflects the variety of soils associated with diverse plant communities and environments. Many Amazonian soils are indeed fragile (that is, easily broken or destroyed), while others exhibit surprising resilience (that is, the power to recover their original character after disturbance). The realization of such diversity has only emerged with intensification of the scale of research. What was once assumed to be a largely homogeneous and stable soil cover over extensive tracts of land is now recognized as being remarkably varied and dynamic, even at the level of a clearing or small farm. Consequently when a bureaucratic and mechanistic allocation of land has occurred without an understanding of the variety of its soil components, there has frequently been a highly inequitable distribution of soil resources.

The object of this paper is to explore those concepts further by considering the nature and pattern of soils at regional and local scales, by concentrating on key soil properties and soil-plant dynamics, and consistent with the theme of this volume, by examining whether indigenous peoples have managed their soil resources better than subsequent colonizers and whether we have something to learn from their stewardship of the land.

Soil Resources, Scale, and Dynamic Change

The sheer size of the Brazilian Amazon (approximately six million square kilometers) has inevitably resulted in broad generalizations on the nature and distribution of its soils. As a starting point, however, the regional level of the catchment basin offers a useful overview. At that level, geomorphological and landscape groups stand out with their constituent floodplains (*várzeas*), drier raised interfluvial plains and residual sedimentary landscapes (*terra firme*), higher plateaus (*altoplanos*), and submontane/montane environments—the last group being less evident in Brazil than in neighboring Amazonian countries. The alluvial várzeas and base-rich igneous intrusive soils form the only significant naturally fertile tracts in a region that, from a land management point of view, possesses generally infertile and problematic soils.

At that broad scale, as understood from the Projeto Radambrasil surveys in the early 1970s and onward (Cochrane and Sanchez 1982; Almeida 1984; Cochrane et al. 1985), the predominant soil orders, occupying around 75 percent of the area, are (using the Brazilian classifications) Latosols and red-yellow Podsols (Carmargo, Klamt, and Kaufman 1987), or (using taxonomic classifications) Oxisols and Ultisols (Soil Survey Staff 1996) (figure 7.1). Equivalent terms from the U.S. Department of Agriculture (USDA) and the Food and Agriculture Organization of the U.N. (FAO) are given in a number of texts, including Sanchez (1976), Eden (1990), Furley (1990), and Nortcliff (chapter 8 this volume). This widespread group of soils comprises deep, highly weathered, acidic, well-drained soils of low nutrient fertility, which have significant limitations for agriculture no matter what the level of technology. They have characteristic deficiencies of essential plant nutrients such as phosphorus, potassium, and calcium, and of micronutrients such as zinc and boron, while often possessing inhibiting or toxic levels of aluminum (Serrão and Homma 1993). It has been estimated that 88 percent of Legal Amazonia's soils are dystrophic (that is, acidic with low nutrient fertility and high-to-toxic levels of aluminum) and are mostly located over the terra firme areas (Nascimento and Homma 1984). Of the remaining 12 percent of more fertile (mesotrophic and eutrophic) soils, about half are hydromorphic, or seasonally wet, Inceptisols and Entisols and are found in the várzeas (Alfaia and Falcão 1993), and half are more fertile upland soils such as the *terra roxas*, or Alfisols, that are usually associated with base-rich parent materials. Mesotrophic and eutrophic soils possess natural fertility that is capable of sustaining intensive agriculture, but they require considerable management skills to deal with the seasonal water excesses (in the várzeas) or the problems of erosion (on steep slopes in the uplands).

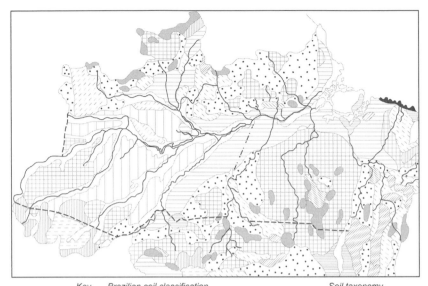

Key	Brazilian soil classification	Soil taxonomy
LA	Yellow Latosol	Oxisol
LV	Red-Yellow Latosol (Red-Yellow Podzol)	Oxisol
LE	Dark Red Latosol and Reddish Brunizem Soilic	Oxisol
PV1	Red-Yellow Podzol (Red-Yellow Latosol)	Ultisol
PV2	Red-Yellow Podzol and Plinthosol	Oxisol
PVe	Eutrophic Red-Yellow Podzol, and Eutrophic Cambisol (Terra Roxa Estruturada)	Alfisol
CL	Concretionary Laterite	Oxisol or Ultisol
TE	Terra Roxa Estruturada and Red Latosol (Red-Yellow Eutrophic Podzol)	Alfisol
Ce	Eutrophic Cambuisol and Eutrophic Red (Yellow Podzol)	Alfisol
P	Plinthosolic (Hydrromorphic Gleys)	Inceptisol or Entisol
PH	Hydromorphic Podzol and Quartz Sand (Red-Yellow Podzol)	Inceptisol or Entisol
AQ	Quartz Sand (Hydromorphic Podzol)	Entisol
HG	Hydromorphic Gley	Inceptisol or Entisol
SM	Mangrove Soil	Entisol
R	Lithosol (rock outcrop)	Entisol

FIGURE 7.1 Soil types in Brazilian Amazonia.
Source: After Furley 1990.

The concept of soil fertility is a human construct and malleable. It has been viewed in the developed world as a term to represent—in quantitative, scientific terms—those natural resources capable of supporting a commercial return and sustaining continuous production. In indigenous terms, that interpretation is sometimes of less relevance given the variety of crops, trees, shrubs, and other plants that are utilized at different times and intensities and that cover a spectrum of soil conditions (Sillitoe 1998). Generally, but as will be shown not always, the demands of indigenous systems are less specific and less constant in the exploitation of the soil resource.

A regional scale perspective is helpful in understanding the far-reaching nature of soil limitations for large-scale agricultural development (table 7.1), but does not provide an adequate view of local soil diversity, for example at the level of the municipality or village. At the latter scale, land terrain or landscape units are often more valuable in revealing the pattern of slope and drainage or parent-material change. Finally, at the farm scale (say at 100 hectares or even within one hectare), subtle changes in gradient, minor differences in water availability or periodic excess, and differences in soil-rooting depth or texture may make a great deal of difference to the resources available to a shifting agriculturist or smallholder with limited capital. There are also small patches and a few larger tracts of mesotrophic soil with better chemical characteristics scattered throughout the Amazon, their distributions are not at all clear from generalized data (Furley 1990; Ratter 1992).

Soil properties can be conveniently divided into *physical* (depth of rooting, structure, texture, aeration, and moisture status), *chemical* (nutrient status, pH and conductivity, ion exchange capacity, toxicities, and deficiencies), and *biological* (organic matter—labile and stable fractions, living plant roots, macrofauna, and microorganisms). However there is a great deal of overlap and interconnectivity between them. It is also important to consider the dynamics of soil change, which closely relate to the concepts of fragility and resilience. Some properties, such as soil texture, may remain stable for great lengths of time and may prevail over large areas, whereas others, such as soil moisture, soluble ions, and easily decomposable organic fractions, can vary significantly over small areas and are less resistant to disturbance.

In addition to spatial variability, temporal change is significant and is especially marked in properties associated with the soil surface. Some properties are extremely dynamic and have a high variability—for example, electrical and saturated hydraulic conductivity, soil moisture status, labile organic matter, and soluble or rapidly complexed ions. Others, such as the clay fraction

TABLE 7.1 Estimates of Major Soil Constraints to Crop Production

Constraint	% of Amazon Basin
N deficiency	90
P deficiency	90
Al toxicity	79
K deficiency	78
Ca deficiency	62
S deficiency	58
Mg deficiency	58
Zn deficiency	48
Poor drainage/Flood hazard	24
P fixation	16
Low CEC	15
High potential erosion hazard	8
Steep slopes (> 30%)	6
Laterization hazard	4
Shallow soils (< 50cm)	< 1

Source: Adapted from Nicholaides et al. 1984.

or bulk density, develop over long periods and are characterized by intermediate or low variability. Some properties (e.g., nitrogen, phosphorus, potassium, and carbon) are inversely related to increasing disturbance, whereas
bulk density increases with greater pressure from trampling or mechanization, especially in the surface horizons. Other properties (e.g., calcium and
magnesium) may benefit from disturbance and can remain at residually
high levels (McNabb et al. 1997).

There is increasing evidence that human-induced enrichment of soil
organic matter (SOM) can be maintained over long periods, for example
through the sequestering of carbon in resistant forms (Sombroek, Nachtergaele, and Hebel 1993; Glaser et al. 2001; Glaser, Lehmann, and Zech
2002; Lehmann, Gebauer, and Zech 2002). Although such patches of *terra*

preta, or black soils, are relatively small (often only a few hectares in size), there are numerous pockets scattered throughout the Amazon, particularly on the higher bluffs of land overlooking floodplains. Essentially the soils are formed by a concentrating process whereby organic matter in the form of litter, branches, and woody material is gathered over wide areas and burned to provide mulch. Organic matter and waste matter accumulate in middens, particularly around settlements. The resultant black carbon and associated organic compounds are highly resistant to the normal processes of weathering and decay, resulting in deep, organic chemical complexes capable of retaining nutrients (such as phosphorus, nitrogen, potassium, and calcium) and moisture. Similar dark soils (*terra mulata*, or brown soils) have been reported to surround settlement areas (McCann, Woods, and Meyer 2001) and behave in a similar way, retaining nutrients and moisture though to a lesser degree. Analogous soil patches have also been described outside forest boundaries, for example at savanna edges (Anderson and Posey 1989; Hecht and Posey 1989; Furley 1996). These dark soils represent some of the few examples where intensive human use that exploits a considerable area around settlements has resulted in a localized increase in fertility within a region more notable for the poverty of its soils.

The most fragile soil properties are at their most vulnerable within 10 to 20 centimeters of the surface. There is therefore a vertical as well as spatial dimension to soil variability. Most of the key soil properties in humid tropical soils are associated with the nutrient circulation and moisture retention of the organic matter. Below that dynamic surface horizon the soils are frequently infertile and inappropriate for agricultural use, so the disturbance that normally accompanies clearance or even selected logging can radically affect soil resources. As far as the vertical profile is concerned, therefore, the nature and maintenance of the surface properties are essential. Spatial variation in humid tropical soils has also been studied, at different scales from molecular to landscape (Dijkerman 1974), and for purposes ranging from agriculture to engineering (Young 1976; van Wambeke 1992; Syers and Rimmer 1994). For the present discussion, soil properties are simplified into groups whose characteristics affect primary land use, specifically plant growth.

Somewhere between 10 and 20 soil properties are commonly analyzed in soil and land-use surveys, although the total number of properties relevant to the present topic might run to a hundred or more (Landon 1991). Because many properties are autocorrelated (such as the proportions of sand, silt, and clay) or are closely related (such as carbon and nitrogen), it is justifiable to target key groups of properties that reflect the total soil character in order to make sense of an extremely complex set of interrelationships. That is particularly true where little is known of a biological group or property; for instance Borneman and Triplett (1997) found that nearly 20 percent of the bacterial sequences they were studying in eastern Amazonia could not be classified into any known bacterial kingdom. Within those groups it may be further necessary to target "indicator" properties, for example, termites to reflect soil fauna or carbon to represent organic matter.

Management of Soil Properties

Our understanding of the management of soil properties tends to be highly site specific. Several physical properties illustrate the diversity. Micro-differences in drainage can radically affect moisture availability and aeration as well as the length of time that water spends in the rooting zone. Extremes of soil temperature are uncommon in the humid tropics, but where they exist they can be alleviated by protecting the soil surface by retaining a cover of trees and shrubs, or by organic mulching, a practice that also inhibits weed growth, checks water loss, and increases fertility. Some Oxisols, despite their chemical limitations, have the advantage of possessing deep profiles and stable aggregates that are coated with oxides and organic matter. Such aggregate properties can render heavy clay into a soil that acts more like sand (through the production of microaggregates). That helps drainage, reduces compaction problems, and minimizes the risk of erosion. On the other hand, leaching is intense and there can be deficits of available moisture during dry spells. Ultisols and Alfisols have different management problems: with cultivation the sandy topsoils facilitate compaction, runoff, and erosion. However, by way of compensation they usually have sufficient water in the argillic (clay-rich) subhorizons. There are also limited occurrences of Vertisols (soils that are 30 percent or more clay), which have problems associated with swelling and shrinking clays, making soil management difficult despite their naturally high organic levels and fertility. As a broad generalization, farmers tend to use simpler implements when soils are more fertile. With increasingly difficult soil physical properties, more intensive and complex technology is often required to maintain productivity. Over the long span of indigenous occupation and settlement in the Amazon and elsewhere in the Neotropics, the areas of natural soil fertility, especially in alluvial river valleys, are generally well known and have been exploited.

Management of soil chemical properties also varies with site, although a number of generalizations can be made according to major soil groups and the type of land use. As indicated earlier, humid tropical areas with heavy precipitation suffer from extreme eluviation (the combined effect of leaching and the mechanical washing out of solids, suspended materials, and solutes). In the Amazon, that leads to the widespread occurrence of acid soils, often accentuated by high or toxic levels of aluminum. Whereas native plants are adapted to such acidity, exotic plants are vulnerable and require careful control or amelioration (Smyth and Cravo 1992) to avoid the inhibition of root development that occurs in aluminum-rich subsoils. Deforestation inevitably intensifies the soil degradation processes and depletes the nutrient balance (de Koning et al. 1997), and also affects solute dynamics (Williams, Fisher, and Melack 1997). The importance of organic matter to the supply of the major plant nutrients (nitrogen, phosphorus, and sulphur) is a clear indication of the critical need to manage organic resources in soil.

In shifting-cultivation systems, which are characteristic of numerous indigenous and smallholder groups, temporary clearings are succeeded by fallow periods that allow regeneration and restoration of soil fertility, and the soils may therefore be regarded as resilient under that management system (e.g., Holscher et al. 1997). Shifting cultivation tends to disturb topsoils in savanna areas (occupying up to 20 percent of Legal Amazonia) more than in forested areas by removing roots and by ridging or in other ways physically affecting the surface. In addition, the cropping period in savanna areas tends to be longer, weed growth is a serious problem, and there is a long dry fallow season where the unvegetated surface is susceptible to erosion (Sanchez 1976). In the more widespread forested areas, maintenance of soil fertility depends on the effectiveness of nutrient cycling, and low-impact, small-scale clearances permit relatively rapid regeneration (Furley 1990). The few areas of more fertile soil in the Amazon (usually Alfisols or Inceptisols) contain significant quantities of nutrients in the top 30 centimeters of soil, whereas Oxisols and Ultisols have most of their nutrients lodged in the biomass and organic matter (Jordan 1985; Proctor 1989). The extreme oligotrophic conditions on deep sandy soils (Spodosols, for example) support plant cover through a remarkably tight nutrient-cycling process dependent upon the effectiveness of a dense, absorbent root mass and mycorrhizal activity (Klinge and Herrera 1978; Herrera 1985; Jordan 1985; St. John 1985). Those soils would be incapable of supporting even shifting agriculture and are usually colonized by specialized plant communities (Amazonian *campinas*, also known as *caatingas* or *campinaranas*, that consist of a range of low woody savanna types with tortuously branched trees). Burning accelerates the process of nutrient release through overland flow, erosion, and solute flow (Cerri, Volkoff, and Andreaux 1991; Williams, Fisher, and Melack 1997), but also increases nutrient loss from gaseous emissions and leaching. Yield decline is most marked in acidic soils of forested areas. Methods of land clearance also affect the nutrient availability, as well as the physical properties of soils, particularly near the surface (Lal, Sanchez, and Cummings 1986).

In some cropping systems, such as shifting agriculture and some forms of multiple cropping, the apparent randomness and disorder is counterbalanced by a closer mimicry of natural forest processes. That probably results in greater ecological efficiency and sustainability—though at the expense of short-term economic productivity. Smallholder farm systems are typified by the restricted area (a few hectares), limited mechanization, and the growing of several crops simultaneously or sequentially (Young 1998). Data on the effects of multiple cropping are increasingly available, though the value of maintaining soil organic constituents has been appreciated for many years (Mulongoy and Merckx 1993; Lehmann, Gebauer, and Zech 2002). Generally, intercropping does seem to result in higher yields than monocrop cultivation over a sustained period of time (Fearnside 1995; Correa and Correa 1996). It is difficult to get accurate information from such a mosaic of plant types and to quantify the effects of "advanced" technologies in such situations. Soil-plant relationships (for example, nutrient tapping from overlapping and successive species, competition for water, interactions between plants, and effectiveness of fertilization) are still not well understood in multiple-crop areas. More research is available for commercial arable cultivation where high and regular rates of fertilization seem essential (Cahn, Bouldin, and Cravo 1993; Cahn, Bouldin, Cravo, and Bowen 1993) and lime is a requirement to counteract subsurface acidity. Low tillage systems seem to offer a means of minimizing disturbance in areas where continuous cultivation is considered viable (Landers 1998).

Pasture represents one of the most important targets for forest conversion, yet it is one of the least appropriate forms of land use for most Amazon soils typified by severe fertility and soil moisture limitations. According to some ecologists (e.g., Goodland 1980), cattle ranching is, from an ecological point of view, one of the least acceptable forms of land use. It can equally be argued that developing any system of harvesting monospecific plants over large areas in humid tropical environments such as the Amazon Basin is inimitable to good soil management. Destruction of the natural vegetation cover and its associated surface organic litter eliminates many of the nutrient-cycling mechanisms and weakens the "buffering effect" against atmospheric damage caused by such factors as higher insolation, lower moisture, and humidity. Organic matter is reduced through lower organic inputs at the surface and higher decomposition rates accompanying increased insolation and higher temperatures (Neill et al. 1997). Trampling can also reduce porosity and permeability and can increase bulk density (Koutika et al. 1997). Despite those constraints, livestock rearing is clearly of great national importance in Brazil and reflects high demand. With the exception of low-density, small-scale enterprises, livestock systems necessitate a commercial level of management with expensive controls and fertilizer practices (Smith et al. 1995). However cattle can often be reared at low densities, on soils too poor for commercial crop production, and most contemporary research is concentrated on improving and maintaining cattle on existing pastures (Faminow 1998). The fact that the cattle operations are often marginal is highlighted by the high rates of abandonment (Bushbacher, Uhl, and Serrão 1988; Serrão and Toledo 1990).

From this brief outline of the key soil properties and their relationship to current land-use systems, it should be evident that SOM is a major integrating factor. Understanding and manipulating SOM has been a feature of indigenous, smallholder, and some commercial enterprises. It has been suggested that the fluctuations of organic matter through the soil organic pools during the course of a year have often been considerably underestimated (Trumbore 1993). The site character, as well as the amount, distribution, and dynamics of organic matter, relate to physical, chemical, and biological soil characteristics. For instance, on a *physical* level, the presence of organic matter lowers bulk density, ameliorates structure, and improves water retention. The *chemical* properties of SOM increase the surface area, augment ion exchange sites, improve ion exchange, add nutrients, and buffer losses. Finally, in terms of soil *biology*, SOM acts as a food source for organisms and is the basis of the food chain for some soil fauna as well as microorganisms (Coleman, Oades, and Uehara 1989; Mulongoy and Merckx 1993), especially in the highly active surface and litter layers (Hofer, Martius, and Beck 1996; Lips and Duivenvoorden 1996a, 1996b). Consequently land-use systems that conserve organic matter or control the speed of decomposition are more likely to sustain fertility.

As alluded to earlier, the dynamic nature of many soil properties is another major consideration. Dynamic change is an essential aspect of the inherent fragility or resilience of soil. Fragile soils are those whose properties react sharply to disturbance and are vulnerable to nutrient loss and/or erosion and structural degradation. All soils are in a state of constant change, but the more resilient ones are those in which the intrinsic properties are capable of resisting disturbance and have a buffering capacity against short- and even medium-term exploitation. Appropriate soil management is therefore the process of channeling land use in the direction of sustained plant growth along pathways guided by the soil's inherent properties. For example, it might be possible to accelerate nutrient restoration in managed fallows as indicated by Szott and Palm (1996). The further the deviation from natural pathways, the more the intervention of other technologies becomes necessary and the more costly the maintenance or recuperative costs are likely to become.

Relationships Between Soils and Plant Communities

It has long been believed that there is little correlation between soil character and plants in the humid parts of the tropics. That belief stems from the notion that the considerable age of (many) tropical soils tends to produce homogeneous properties as a result of long and continued weathering, leaching, and soil development. At the same time, however, plant communities seem to be adapting and changing more rapidly in response to climatic and other controls. The widespread occurrence of Oxisols and Ultisols underlying both forest and savanna ecosystems in the Amazon might appear to support the contention that soils are relatively similar. Yet it is now known that, at scales applicable to farming enterprises and village communities (say 1:25,000 or larger), soils can be extremely heterogeneous as a result of differences in

- *age*—many are younger and more dynamic than appreciated earlier (e.g., Costa 1997; Costa and Moraes 1998)
- *slope and drainage* (e.g., Tiessen, Chacon, and Cuevas 1994; Botschek et al. 1996)
- *parent material*—subtle differences that are often textural and structural rather than chemical, as shown by Lips and Duivenvoorden (1996a, 1996b) in the Colombian Amazon
- *human disturbances*, which are now appreciated as having been far more pervasive than formerly realized

Thus at a more detailed level of investigation and as a result of the influence of those pedogenic factors, plant communities and even species or individual plants can often reveal close pedological affinities. That has been demonstrated for large-scale ecosystems—for example, the *cerrado* (Brazilian savanna) and northern Amazon forest margin (Furley 1996, 1998)—and for individual plant species such as pteridophytes (Tuomista and Poulsen 1996), nurse plants (Kellman 1979), and cone-rooted palms (Furley 1975). It is almost certain that there are numerous analogous relationships that have not yet been thoroughly explored. The significance of plant and soil distributions at microscales (one hectare down to one square meter or less) has not been explored in detail, but research on it is underway elsewhere.

Where soils have been utilized for agriculture over long periods of time, either permanently or periodically, local people have an unrivaled knowledge of microvariation and it is to their experience that we may now turn.

Have Indigenous Peoples and Traditional Methods Managed Soil Resources Successfully?

Indigenous use of land in the Neotropics can be characterized by the small scale of activities, generally though not universally at low intensity and pressure, and the long-term perspective. The small scale of operation (a few hectares per family), even when amplified in a communal society, promotes rapid restoration of soil fertility by permitting forest regeneration; encourages rapid recolonization of vegetation through successional phases from adjacent seed banks; and avoids exposing the ground surface, thereby limiting desiccation, excess runoff, and damage to natural soil recovery mechanisms (such as nutrient cycling through mycorrhizal and soil faunal activities). The low intensity of use, which employs hand tools or simple tools associated with what has been termed intermediate technology, is less damaging than modern commercial methods to the plant cover, to the underlying soils, and particularly to the sensitive plant-soil processes located in the surface horizons. The short period of occupancy means less pressure and disturbance, and normally leaves useful trees and shrubs. Such forms of management necessitate considerable labor inputs and a lack of population pressure on the land.

However there have been times in the past when population pressures have been greater, especially in more favorable locations; that is illustrated by the widespread occurrence of anthropogenic soils in the Amazon, notably along water courses (Smith 1980; Eden et al. 1984), or by evidence of past agricultural landscapes throughout the Neotropics (Turner 1983; Farrington 1989; Roosevelt 2000; Denevan 2001; Whitmore and Turner 2001). As the demand for food increases, so the level of disturbance accelerates, and the requirement to put back harvested nutrients and regulate soil moisture inevitably increases proportionately (Lunaorea and Wagger 1996). That ratchets up the demands on land-use management, and in some indigenous societies pressure on land has eventually contributed to a breakdown of community social structures, indicating that the pressures put upon land management determine its success or failure. Nowhere is that more evident than in the

example of the Central American Maya, who will be examined as an analogy
to the studies of Amazonian groups such as the Kayapó.

Soil Management Under Conditions of Population Pressure

The Land Husbandry of the Maya

The historical record of the Maya imprinted on the landscape of Central
America seems to epitomize the best practice, and at the same time some
of the potential dangers, of indigenous land management in the tropics. On
the one hand, Mayan management techniques were among the most sophisticated of any native agricultural system and appear to have been sustainable at high levels of production over long periods. On the other hand, their
highly organized societies collapsed, and it has been argued that population
pressure on available land resources, probably in association with a period of
climatic aridity and political upheaval, contributed to the decline in Mayan
fortunes. What is evident today, from careful reconstruction of the landscapes
of the Mayan period (Harrison and Turner 1978; Turner 1983; Whitmore
and Turner 2001), is the elaborate control those people had of the different
soil environments of the region. Their management techniques ranged from
drained and raised fields in wetlands (*chinampas*) to irrigated fields around
wet depressions (*bajos*) to slash-and-burn rotations in the form of shifting
cultivation (*milpas*) to the construction of extensive terraces that allowed the
manipulation of soils and water in the drier environments of sloping karstic
hills. Allied to the intensive use of crops and other plant products were a
highly advanced forest management system and an extensive utilization of
fresh-water and marine resources. Mayan agrosilvicultural techniques can
be grouped into a number of characteristic features:

- *Hydraulic systems* for storing, regulating, and distributing water (e.g.,
  Turner and Harrison 1983; Farrington 1989; Gomez-Pompa 1991).
- *Raised field and drained field systems* (chinampas) (Gliessman 1990;
  Siemens and Puleston 1972; Darch 1983).
- *Terraces and rotational practices*, usually in conjunction with milpa
  farming (Lundell 1940; Turner 1983).
- *Forest gardens* (present today and assumed to be widespread in
  the past) (Wiseman 1978), with raised beds for seed germination
  (*caanches*) and/or tree planting for fruit, shade, fuelwood, and seeds
  (figure 7.2a).

- *Successional stages* of natural forest growth and regrowth (figure 7.2b).
- *Conservation of forest patches for specific purposes*, with planting and removal of selected species (*pet kot*) (Gomez-Pompa, Flores, and Sosa 1987; Gomez-Pompa 1991).

Although they were consummate managers of the environment, the Maya could not cope with rising population pressures in a context of increased climatic dryness and political instability. Population densities have been reported at between 40–70 people and more than 200 people per square kilometer (Turner 1976; Santley, Killon, and Lycett 1986; Rice and Culbert 1990) and may have reached 1,000 people per square kilometer in some locations. Such densities impose immense demands upon soil management, which, it could be argued, indigenous practices cannot support. So although the milpa (the predominant Mayan land husbandry) was ecologically benevolent, sustainable over long periods, and in many respects conservationist, it is unlikely that it was able to cope with sustained, unremitting demands on soil resources without a sufficient fallow period for natural restoration.

Best Practice as Evidenced in the Soil Management of the Kayapó in Eastern Amazonia

It seems reasonably certain that different indigenous groups in Amazonia succeeded in significantly manipulating the soils and plants of their territories and had a profound influence upon their environment. It has been estimated that at least 12 percent of the terra firme might have been managed (Balée 1989). Although population densities may have been locally high in favorable areas, giving rise to anthropogenic soils (notably the black soils), the pressure was light overall. Balée (1989) quotes figures of around 0.2 people per square kilometer over large areas of Amazonia, although those figures may not reflect soil constraints alone. A number of characteristic forms of soil management have been reported; they illustrate considerable sophistication in some tribal groups, one of which—the Kayapó—has been intensively studied (Anderson and Posey 1989; Hecht and Posey 1989; Posey and Balée 1989). Kayapó soil management activities range from very slight disturbance of the forest, involving harvesting at a light level and protection of favored species, to total intervention and clearing (Alcorn 1983), as shown below.

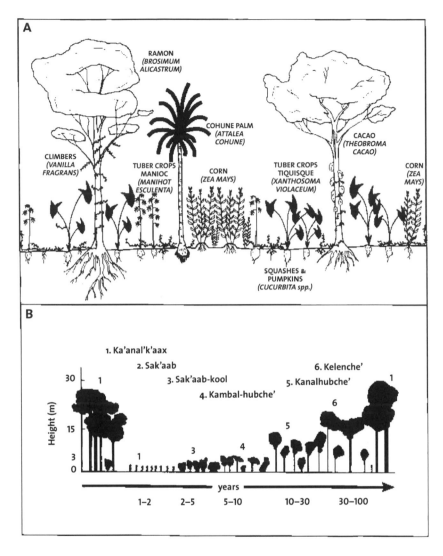

FIGURE 7.2 Mayan land husbandry: (a) the spatial distribution of crops, shrubs, and trees in an "artificial forest" or forest garden (based on work by Wiseman 1978); (b) vegetation classification based on successional growth stages from initial cutting and/or burning (based on Gomez-Pompa 1991).

- *Understanding and managing small-scale environmental and ecological diversity*, which includes the development of a complex mosaic of land uses: home gardens, forest and cerrado plantings, plantings along trails, and use of successional phases of forest regrowth (Anderson and Posey 1989).
- *Local mounding of soil; improving drainage and friability* (Carneiro 1983).
- *Raising and draining fields*; ridging to cope with seasonal or periodic water excess (Denevan and Zucchi 1978).
- *Accumulating and transferring soils to form mesotrophic conditions.* For example, "islands," or moundings of earth surrounded by water (*apêtês*), of up to one hectare in area (Posey 1985; Anderson and Posey 1989; Hecht and Posey 1989). Similar features have been recorded in the cerrados of the planalto in central Brazil, above gallery forest on sloping ground (Furley 1996).
- *Enriching local soil*, which includes manipulating organic matter. For example, use of termite and ant nests; mulching with organic litter that has been beaten and macerated with sticks (Hecht and Posey 1989).
- *Manipulating plant species* that deliberately or inadvertently have a beneficial effect on soil properties. For example, planting leguminous trees such as *Inga* spp. (Denevan et al. 1984); retaining palms such as *Orbignya cohune*, which is related to the *babaçu* in Central America (Furley 1975).
- *Using biological controls.* For example, *Azteca* ants, to protect against leafcutters and other horticultural pests (Overal and Posey 1990).
- *Using fire* to reduce weeds, clear land of vegetation and pests, and add ash for fertilization (increases availability of potassium and nitrogen, and lowers acidity by counteracting aluminum).

The soil management techniques of Amazonian indigenous groups illustrate a detailed knowledge of the diversity and local scale variety of environments. The recognition of the value of organic matter, the specific uses of different plant species, the understanding of the need for soil restoration through additions of ash or through forest fallows, the awareness of site-specific requirements (especially those concerned with wetness and dryness), and often an excellent appreciation of the synchronization between crop moisture requirements and available soil moisture (Sanchez 1976)—all indicate a high level of environmental awareness. Whether that can inform contemporary agriculture depends to some extent on land pressures; furthermore, intensive agricultural practices necessitate high labor inputs, at least periodically. Those constraints may inhibit the widespread adoption of indigenous methods of soil management, even though Indian principles of land management are shown to be concordant with findings from modern research into the sustainable use of tropical soils: zero tillage or disturbance, water control, and management of organic matter and nutrient balances.

Conclusions

The soils of the Amazon Basin are complex, exhibiting high spatial and temporal variability. Their sustained management is therefore unlikely to be achieved by any one technique or strategy. Indigenous peoples have developed long-term methods for dealing holistically with their land. They contrast with Western techniques that attempt to simplify ecosystems, concentrate upon single land uses, and demand mechanization and capital-intensive inputs, but can also achieve high production levels. Indigenous systems of land use offer alternatives for conservation, supported by their long-term perspective, lower intensity, smaller scale, and better understanding and use of regenerative stages that permit vegetation to recover. Such indigenous strategies encourage biodiversity at low levels of disturbance and are arguably more appropriate in areas of fragile soil. However they do not, by themselves, offer solutions where greater productivity is required—where for example pressure on land creates greater demand, necessitating more intense use of the more resilient soils. The mosaic of soil types found within the Amazon Basin requires a multiple approach to soil and land-use management.

One direction for resolving that dilemma might be to combine some aspects of the strategies adopted by indigenous peoples and more recent land colonizers. The advantages of indigenous land management include the marshaling of available labor at critical periods throughout the year, multiple cropping, and wide-ranging resource use. Some of the advantages offered by Western technology include a knowledge of nutrient replacement to cope with increased harvest pressure; liming to address problems of soil acidity; the ability to control pests and weeds; and better storage systems. At present, however, indigenous management of the soil and commercial farming sit uncomfortably together and serve different purposes.

References

Alcorn, J. 1983. *Huastec ethnobotany.* Austin: University of Texas Press.

Alfaia, S. S. and N. P. Falcao. 1993. Study of nutrient dynamics in floodplain soils of Careiro Island, Central Amazonia. *Amazonia Limnologia et Oecologia Regionalis Systemae Fluminis Amazonas* 12:485–493.

Almeida, A. L. S. de. 1984. *Realizações.* Projeto Radambrasil. Salvador, Bahia: Ministério das Minas e Energia.

Anderson, A. B. and D. A. Posey. 1989. Management of a tropical scrub savanna by the Gorotire Kayapó of Brazil. In D. A. Posey and W. Balée, eds., *Resource Management in Amazonia: Indigenous and Folk Strategies,* pp.159–173. Advances in Economic Botany 7. Bronx: New York Botanical Garden.

Balée, W. 1989. The culture of Amazonian forests. In D. A. Posey and W. Balée, eds., *Resource Management in Amazonia: Indigenous and Folk Strategies,* pp. 1–21. Advances in Economic Botany 7. Bronx: New York Botanical Garden

Borneman, J. and E. W. Triplett. 1997. Molecular microbial diversity in soils from eastern Amazonia: Evidence for unusual microorganisms and microbial population shifts associated with deforestation. *Applied and Environmental Microbiology* 63(7):2647–2653.

Botschek, J., J. Ferraz, M. Jahnel, and A. Skowronek. 1996. Soil chemical properties of a toposequence under primary rainforest in the Itacoatiara vicinity (Amazonas, Brazil). *Geoderma* 72:119–132.

Bushbacher, R., C. Uhl, and E. A. S. Serrão. 1988. Abandoned pastures in eastern Amazonia. II. Nutrient stocks in the soil and vegetation. *Journal of Ecology* 76:682–699.

Cahn, M. D., D. R. Bouldin, and M. S. Cravo. 1993. Amelioration of subsoil acidity in an Oxisol of the humid tropics. *Biology and Fertility of Soils* 15:153–159.

Cahn, M. D., D. R. Bouldin, M. S. Cravo, and W. T. Bowen. 1993. Cation and nitrate leaching in an Oxisol of the Brazilian Amazon. *Agronomy Journal* 85:334–340.

Carmargo, M. N., E. E. Klamt, and J. H. Kaufman. 1987. *Sistema brasileiro de clasificação de solos.* Campinas: Sociedade Brasileiro de Ciência de Solo.

Carneiro, R. 1983. The cultivation of manioc among the Kuikuru of the upper Xingu. In R. B. Hames and W. T. Vickers, eds., *Adaptive Responses of Native Amazonians,* pp.65–111. New York: Academic Press.

Cerri, C. C., B. Volkoff, and F. Andreaux. 1991. Nature and behaviour of organic matter in soils under natural forest and after deforestation, burning and cultivation, near Manaus. *Forest Ecology and Management* 38:247–257.

Cochrane, T. T. and P. A. Sanchez. 1982. Land resources, soils and land management in the Amazon region: A state of knowledge report. In S. B. Hecht, ed., *Amazonia: Agriculture and Land Use Research,* pp. 137–209. Cali, Colombia: Centro Internacional de Agricultura Tropical (CIAT).

Cochrane, T. T., L. G. de Azevedo, J. A. Porras, and C. L. Garver. 1985. *Land in*

Tropical America. 3 volumes. Cali, Colombia: Centro Internacional de Agricultura Tropical (CIAT).

Coleman, D. C., J. M. Oades, and G. Uehara, eds. 1989. *Dynamics of Soil Organic Matter in Tropical Ecosystems.* Honolulu: University of Hawaii Press.

Correa, J. C. and A. F. G. Correa. 1996. Nutrient cycling in an intercropped plantation of Jacaranda da Baia (*Dalbergia nigra* Fr. Allem.) with Desmodium (*Desmodium ovalifolium* Wall.). *Pesquisa Agropecuaria Brasileira* 31:467–472.

Costa, M. L. da. 1997. Laterization as a major process of ore deposit formation in the Amazon region. *Exploration and Mining Geology* 6:79–104.

Costa, M. L. da. and E. L. Moraes. 1998. Mineralogy, geochemistry and genesis of kaolins from the Amazon region. *Mineralium Deposita* 33:283–297.

Darch, J. P., ed. 1983. *Drained Field Agriculture in Central and South America.* International Series 189. London: British Archeological Reports.

De Koning, G. H. J., P. J. van de Kop, and L. O. Fresco. 1997. Estimates of subnational nutrient balances as sustainability indicators for agro-ecosystems in Ecuador. *Agricultural Ecosystems and Environments* 65:127–139.

Denevan, W. M. 2001. *Cultivated Landscapes of Native Amazonia and the Andes.* New York: Oxford University Press.

Denevan, W. M., J. M. Treacy, J. B. Alcorn, C. Padoch, J. Denslow, and S. F. Paitan. 1984. Indigenous agroforestry in the Peruvian Amazon: Bora Indian management of swidden fallows. *Interciencia* 9:346–357.

Denevan, W. M. and A. Zucchi. 1978. Ridged field excavations in the central Orinoco Llanos, Venezuela. In D. L. Bowman, ed., *Advances in Andean Archeology,* pp. 235–245. The Hague: Mouton.

Dijkerman, J. C. 1974. Pedology as a science: The role of data, models and theories in the study of natural soils systems. *Geoderma* 11:73–93.

Eden, M. J. 1990. *Ecology and Land Management in Amazonia.* London: Belhaven Press.

Eden, M. J., W. Bray, L. Herrera, and C. McEwan. 1984. Terra preta soils and their archeological context in the Caqueta Basin of southeast Colombia. *American Antiquity* 49:25–40.

Faminow, M. D. 1998. *Cattle, Deforestation and Development in the Amazon.* New York: Oxford University Press.

Farrington, I., ed. 1989. *Prehistoric Intensive Agriculture in the Tropics.* International Series 232. London: British Archeological Reports.

Fearnside, P. M. 1995. Agroforestry in Brazil's Amazon development policy. In M. Clüsener-Godt and I. Sachs, eds., *Brazilian Perspectives on Sustainable Development of the Amazon Region,* pp. 125–148. Carnforth, Lancashire: UNESCO-Parthenon Press.

Fernandes, E. C. M., P. P. Motavalli, and L. Mukurumbira. 1997. Management control of soil organic matter dynamics in tropical land use systems. *Geoderma* 79:49–67.

Furley, P. A. 1975.The significance of the cohune palm, *Orbignya cohune* (Mart.) Dahlgren on the nature and in the development of the soil profile. *Biotropica* 7:32–36.

Furley, P. A. 1990. The nature and sustainability of Brazilian Amazon soils. In D. Goodman and A. Hall, eds., *The Future of Amazonia*, pp. 309–359. London: Macmillan.

Furley, P. A. 1996. The influence of slope on the nature and distribution of soils and plant communities in the central Brazilian cerrado. In M. G. Anderson and S. M. Brooks, eds., *Advances in Hillslope Processes*, vol. 1, pp. 327–346. Chichester: Wiley.

Furley, P. A. 1998. Soil properties and plant communities over the eastern sector of the Ilha de Maracá. In W. Milliken and J. A. Ratter, eds., *Maracá: The Biodiversity and Environment of an Amazonian Rainforest*, pp. 415–430. Chichester: Wiley.

Glaser, B., G. Guggenberger, L. Haumaier, and W. Zech. 2001. Persistence of soil organic matter in archeological soils (terra preta) of the Brazilian Amazon region. In R. Rees, B. Ball, C. Campbell, and C. Watson, eds., *Sustainable Management of Soil Organic Matter*, pp. 190–194. Farnham: Commonwealth Agricultural Bureau.

Glaser, B., J. Lehmann., and W. Zech. 2002. Ameliorating physical and chemical properties of highly weathered soils in the tropics with charcoal—a review. *Biology and Fertility of Soils* 35:219–230.

Gliessman, S. R. 1990. *Agroecology: Researching the Basis for Sustainable Development.* New York: Springer-Verlag.

Gomez-Pompa, A. 1991. Learning from traditional ecological knowledge: Insights from Mayan silviculture. In A. Gomez-Pompa, T. C. Whitmore, and M. Hadley, eds., *Rain Forest Regeneration and Management*, pp. 335–342. Carnforth, Lancashire: UNESCO-Parthenon Press.

Gomez-Pompa, A., E. Flores, and V. Sosa. 1987. The "pet kot": A man-made tropical forest of the Maya. *Interciencia* 12:10–15.

Goodland, R. J. A. 1980. Environmental ranking of Amazonian development projects. In F. Barbira-Scazzocchio, ed., *Land, People and Planning in Contemporary Amazonia*, pp. 1–26. Cambridge: Centre of Latin American Studies.

Harrison, P. D. and B. L. Turner, eds. 1978. *Pre-Hispanic Maya Agriculture.* Albuquerque: University of New Mexico Press.

Hecht, S. B. and D. A. Posey. 1989. Preliminary results on soil management techniques of the Kayapó Indians. In D. A. Posey and W. Balee, eds., *Resource Management in Amazonia: Indigenous and Folk Strategies*, pp. 174–188. Advances in Economic Botany 7. Bronx: New York Botanical Garden.

Herrera, R. 1985. Nutrient cycling in Amazonian forests. In G. T. Prance and T. E. Lovejoy, eds., *Key Environments: Amazonia*, pp. 95–108. Oxford: Pergamon.

Hofer, H., C. Martius, and L. Beck. 1996. Decomposition in an Amazonian rain forest after experimental litter addition in small plots. *Pedobiologia* 40:570–576.

Holscher, D., B. Ludwig, R. F. Moller, and H. Folster. 1997. Dynamics of soil chemical parameters in shifting agriculture in the eastern Amazon. *Agriculture Ecosystems and the Environment* 66:153–163.

Jordan, C. 1985. *Nutrient Cycling in Tropical Forest Ecosystems: Principles and Applications in Management and Conservation.* Chichester: Wiley.

Kellman, M. 1979. Soil enrichment by neotropical savanna trees. *Journal of Ecology* 7:565–577.

Klinge, H. and R. Herrera. 1978. Biomass studies in Amazon caatinga forest in southern Venezuela. *Tropical Ecology* 19:93–110.

Koutika, L. S., F. Bartoli, F. Andreux, C. C. Cerri, G. Burtin, T. Chone, and R. Philippy. 1997. Organic matter dynamics and aggregation in soils under rain forest and pastures of increasing age in the eastern Amazon Basin. *Geoderma* 76:87–112.

Lal, R., P. A. Sanchez, and R. W. Cummings., eds. 1986. *Land Clearing and Development in the Tropics.* Boston: A. A. Balkema.

Landers, J. N. 1998. *Zero Tillage Development in Tropical Brazil.* External contribution to the Annals of Zimbabwe Conservation Tillage Workshop FAO/GTZ/ZFU. Harare, Zimbabwe.

Landon, J. R, ed. 1991. *Booker Tropical Soil Manual.* London: Longman.

Lehmann, J., G. Gebauer, and W. Zech. 2002. Nitrogen cycling assessment in a hedgerow intercropping system using N15 enrichment. *Nutrient Cycling in Agroecosystems* 62:1–9.

Lips, J. M. and J. F. Duivenvoorden. 1996a. The regional patterns of well drained upland soil differentiation in the middle Caqueta Basin of Colombian Amazonia. *Geoderma* 72:219–257.

Lips, J. M. and J. F. Duivenvoorden. 1996b. Fine litter input to terrestrial humus forms in Colombian Amazonia. *Oecologia* 108:138–150.

Lunaorea, P. and M. G. Wagger. 1996. Management of tropical legume cover crops in the Bolivian Amazon to sustain crop yields and soil productivity. *Agronomy Journal* 88:765–776.

Lundell, C. L. 1940. *The 1936 Michigan-Carnegie Botanical Expedition to British Honduras.* Publication 522:1–57. Washington, DC: Carnegie Institute.

McCann, J. M., W. I. Woods, and D. W. Meyer. 2001. Organic matter and Anthrosols in Amazonia: Interpreting the Amerindian legacy. In R. Rees, B. Ball, C. Campbell, and C. Watson, eds., *Sustainable Management of Soil Organic Matter,* pp. 180–189. Farnham: Commonwealth Agricultural Bureau.

McNabb, K. L., M. S. Miller, B. G. Lockaby, B. J. Stokes, R. G. Clawson, J. A. Stanturf, and J. N. M. Silva. 1997. Selection harvests in Amazonian rain forests: Long term impacts on soil properties. *Forest Ecology and Management* 93:153–160.

Mulongoy, K. and R. Merckx, eds. 1993. *Soil Organic Matter Dynamics and Sustainability of Tropical Agriculture.* Chichester: Wiley.

Nascimento, C. N. B. and A. K. O. Homma. 1984. *Amazônia: Meio ambiente e tec-*

nologia agricola. Documento 27. Belém, Brazil: Brazilian Enterprise for Agricultural Research—Center for Agroforestry Research of the Eastern Amazon.

Neill, C., J. M. Melillo, P. A. Steudler, C. C. Cerri, J. F. L. de Moraes, M. C. Piccolo, and M. Brito. 1997. Soil carbon and nitrogen stocks following forest clearing for pasture in the southwestern Brazilian Amazon. *Ecological Applications* 7:1216–1225.

Nicholaides, J. J., D. E. Bandy, P. A. Sanchez, and J. H. Villachica. 1984. From migratory to continuous agriculture in the Amazon Basin. In FAO Soils Bulletin, *Improved Production Systems as an Alternative to Shifting Agriculture*, pp. 141–168. Rome: FAO/UN.

Overal, W. L. and D. A. Posey. 1990. Uso de formigas *Azteca* spp. para controle biológico de pragas agrícoles entre os Indios Kayapó do Brasil. In D. A. Posey and W. L. Overal, eds., *Ethnobiology: Implications and Applications*. Proceedings of the First International Congress of Ethnobiology. 1988. Belém, Pará: Museu Paraense Emílio Goeldi.

Posey, D. A. 1985. Indigenous management of tropical forest ecosystems: The case of the Kayapó Indians of the Brazilian Amazon. *Agroforestry Systems* 3:139–158.

Posey, D. A. and W. Balée, eds. 1989. *Resource Management in Amazonia: Indigenous and Folk Strategies.* Advances in Economic Botany 7. Bronx: New York Botanical Garden.

Posey, D. A. and S. B. Hecht. 1990. Indigenous soil management in the Latin American tropics: Some implications for the Amazon Basin. In D. A. Posey and W. L. Overal, eds., *Ethnobiology: Implications and Applications*, pp. 73–86. Proceedings of the First International Congress of Ethnobiology. 1988. Belém, Pará: Museu Paraense Emílio Goeldi.

Proctor, J., ed., 1989. *Mineral Nutrients in Tropical Forests and Savanna Ecosystems.* Oxford: Blackwell Scientific.

Ratter, J. A. 1992. Transitions between cerrado and forest vegetation in Brazil. In P. A. Furley, J. Proctor, and J. A. Ratter, eds., *Nature and Dynamics of Forest-Savanna Boundaries*, pp. 417–429. London: Chapman & Hall.

Rice, D. C. and T. P. Culbert. 1990. Historical contexts for population reconstruction in the Maya lowlands. In T. P. Culbert and D. C. Rice, eds., *Pre-Columbian Population History in the Maya Lowlands*, pp. 1–36. Albuquerque: University of New Mexico Press.

Roosevelt, A. 2000. The lower Amazon: A dynamic human habitat. In D. L. Lentz, ed., *Imperfect Balance: Landscape Transformations in the Pre-Columbian Americas*, pp. 391–453. New York: Columbia University Press.

Sanchez, P. A. 1976. *Properties and Management of Soils in the Tropics.* New York: Wiley-Interscience.

Santley, R. S., T. W. Killon, and M. T. Lycett. 1986. On the Maya collapse. *Journal of Anthropological Research* 42:123–159.

Serrão, E. A. S. and A. K. O. Homma. 1993. Brazil. In *Sustainable Agriculture and the Environment in the Humid Tropics*, pp. 263–351. Washington, DC: National Academy Press.

Serrão, E. A. S. and J. M. Toledo. 1990. The search for sustainability in Amazonian pastures. In A. R. Anderson, ed., *Alternatives to Deforestation*, pp. 195–124. New York: Columbia University Press.

Siemens, A. H. and D. E. Puleston. 1972. Ridged fields and associated features in southern Campeche: New perspectives on the lowland Maya. *American Antiquity* 37:228–239.

Sillitoe, P. 1998. Knowing the land: Soil and land resource evaluation and indigenous knowledge. *Soil Use and Management* 14:188–193.

Smith, N. J. H. 1980. Anthrosols and human carrying capacity in Amazonia. *Annals of the American Association of Geographers* 70:553–566.

Smith, N. J. H., E. A. S. Serrão, P. T. Alvim, and I. C. Falesi. 1995. *Amazonia: Resiliency and Dynamism of the Land and Its People*. Tokyo: United Nations University Press.

Smyth, T. J. and M. S. Cravo. 1992. Aluminium and calcium constraints to continuous crop production in a Brazilian Amazon Oxisol. *Agronomy Journal* 84:843–850.

Soil Survey Staff. 1996. *Keys to Soil Taxonomy*. 7th ed. Washington, DC: USDA-NRCS.

Sombroek, W. G., F. O. Nachtergaele, and A. Hebel. 1993. Amounts, dynamics and sequestering of carbon in tropical and sub-tropical soils. *Ambio* 22:417–426.

St. John, T. V. 1985. Mycorrhizae. In G. T. Prance and T. R. Lovejoy, eds., *Key Environments: Amazonia*, pp. 277–283. Oxford: Pergamon Press.

Syers, J. K. and D. L. Rimmer, eds. 1994. *Soil Science and Sustainable Land Management in the Tropics*. Wallingford, Oxford: CAB International.

Szott, L. T. and C. A. Palm. 1996. Nutrient stocks in managed and natural humid tropical fallows. *Plant and Soil* 186:293–309.

Tiessen, H., P. Chacon, and E. Cuevas. 1994. Phosphorus and nitrogen status in soils and vegetation along a toposequence of dystrophic rain forests on the upper Rio Negro. *Oecologia* 99:145–150.

Trumbore, S. E. 1993. Comparison of carbon dynamics in tropical and temperate soils using radiocarbon measurements. *Global Biogeochemical Cycles* 7:275–290.

Tuomisto, H. and A. D. Poulsen. 1996. Influence of edaphic specialization on Pteridophyte distribution in neotropical rain forests. *Journal of Biogeography* 23:283–293.

Turner, B. L. 1976. The population density in the Classic Maya lowlands: New evidence for old approaches. *Geographical Review* 66:73–82.

Turner, B. L. 1983. *Once Beneath the Forest*. Boulder, Colorado: Westview Press.

Turner, B. L. and P. D. Harrison. 1983. *Pulltrouser Swamp: Ancient Maya Habitat, Agriculture and Settlement in Northern Belize*. Austin: University of Texas Press.

Van Wambeke, A. 1992. *Soils of the Tropics: Properties and Appraisal*. New York: McGraw-Hill.

Whitmore, T. M. and B. L. Turner. 2001. *Cultivated Landscapes of Middle America on the Eve of Conquest.* New York: Oxford University Press.

Williams, M. R., T. R. Fisher, and J. M. Melack. 1997. Solute dynamics in soil water and groundwater in a central Amazon catchment undergoing deforestation. *Biogeochemistry* 38:303–335.

Wiseman, F. M. 1978. Agricultural and historical ecology of the Maya lowlands. In P. D. Harrison and B. L. Turner, eds., *Pre-Hispanic Maya Agriculture*, pp. 63–116. Albuquerque: University of New Mexico Press.

Young, A. 1976. *Tropical Soils and Soil Survey.* Cambridge: Cambridge University Press

Young, A. 1998. *Agroforestry for Soil Conservation.* 2nd ed. Wallingford, Oxford: CAB International.

8 Is Successful Development of Brazilian Amazonia Possible Without Knowledge of the Soil and Soil Response to Development?

Stephen Nortcliff

The land resources of the Amazon are estimated as 484 million hectares, with approximately 35 percent comprising true moist tropical forest, 57 percent semievergreen forest, and 7 percent savanna (Cochrane and Sanchez 1982). The landscape under this vegetation is generally of low relief and flat to gently rolling, frequently with short, relatively steep slopes separating flat and poorly drained low-lying lands from flat to gently rolling crest areas. Cochrane and Sanchez (1982) estimated that approximately 23 percent of the land surface was flat and poorly drained, and that the remainder consisted of well-drained slopes and free to moderately well-drained flat to gently rolling upland areas, with 73 percent of the area having slopes less than 8 percent. The humid tropical forest ecosystem is a unique environment and serves as a reservoir of sequestered carbon; a repository of biological diversity of plants and animals; and a source of food, timber, medicine, and other products for people. In the context of humid tropical ecosystems the Amazon Basin is recognized as particularly important, and it is estimated that as much as 40 percent of the world's moist tropical forests are found within its boundaries (Lanly 1982).

The Natural Forest Ecosystems

Probably the most important characteristic of the moist tropical forest is its tremendous lushness, which has attracted developers looking for land to exploit. The rain forest has a substantial biomass, often greater than that of any other natural ecosystem, and also has a high net primary productivity. The ill-informed interpretation of this ample biomass and productivity has often been that this system is supported by rich, fertile soils that offer great development potential. As will be seen, this is not the case; rather the soils are inherently poor and the system is sustained by an almost closed organic matter–based nutrient cycle (Went and Stark 1968; Stark 1971; Jordan 1985). The litter at the soil's surface is rapidly decomposed and the nutrients released are almost immediately taken up by plants. There is thus only very limited loss of nutrients from the natural ecosystem. Since the vast majority of the nutrients are held in the biomass, removal of the biomass results in a major depletion of the nutrient pool in the system. Given the apparent efficiency and sustainability with which the natural system functions, it might be appropriate to understand the mechanisms that enable this system to function so well, and if possible to adapt and apply these mechanisms to the introduction of sustainable land-management strategies. There is some evidence that this ecosystem has supported a diverse indigenous population for many hundreds of years (Hemming 1978). Those indigenous populations acknowledged the close relationships between the components of the ecosystem, and in particular recognized the need to allow time for the forest to regenerate between periods of occupation and periods of exploitation. In more recent times developers have used the forest ecosystem with no recognition of the close interrelationships, and the consequences have often been disastrous for the system.

Soils of the Amazon Basin

Unlike much of the temperate lands of Europe and North America, where much of the soil development has taken place in the relatively recent geological past (i.e., during the tens of thousands of years since the retreat of the ice sheets), soil development in the tropics may have been in progress, albeit under a possibly wide range of environmental conditions, for many hundreds of thousands of years. It is not surprising therefore that there are many contrasting soils found within the Amazon Basin. Sanchez and Cochrane (1980) suggested that at least eight of the 10 orders of Soil Taxonomy occupy significant portions of this area (Soil Survey Staff 1975), with Histosols, Aridisols, and Andisols being identified as the exceptions. Table 8.1 presents summary information on the most widespread soil orders.

Oxisols are the most widespread soil, occupying almost 46 percent (ca. 220 million hectares) of the land area, principally on the freely draining upland areas. The second most common order of soils is the Ultisols, occupying slightly

TABLE 8.1 Soil Orders Found Within the Amazon Basin
(Most Significant Orders Shown)

Order	Area (ha $\times$ 10^6)	Percentage of Amazon Basin
Oxisols	219.9	45.6
Ultisols	141.7	29.3
Entisols	72.0	14.9
Alfisols	19.8	4.1
Inceptisols	16.0	3.3
Spodosols	10.5	2.3
Mollisols	3.7	0.8
Vertisols	0.5	0.1
Total	484.1	100.4

Source: Soil Survey Staff 1975.

more than 29 percent (ca. 142 million hectares) of the land area. Ultisols are found in a wider range of landscape conditions than Oxisols, being common in both freely drained upland and poorly drained landscape positions.

Although Soil Taxonomy is probably the most widely used soil classification in the Americas, there is a second widely used international classification, which in its latest form is known as the World Reference Base for Soil Resources (Deckers, Nachtergaele, and Spaargaren 1998), but which owes much of its structure and class definition to the earlier Revised Legend to the Soil Map of the World (FAO/UNESCO 1984). In addition, within Brazil (Carmargo, Klamt, and Kaufman 1987) there is a national soil classification that incorporates some of the characteristics and criteria included in the international systems as well as criteria considered to be locally important in differentiating soils. Table 8.2 presents some of the major soils found in the Amazon Basin and attempts to cross-reference the three classifications. The terminology of Soil Taxonomy is used in this paper.

Soil Constraints to Development

The soil constraints to agronomic development may be broadly grouped into chemical and physical restrictions. The major chemical limitations to

TABLE 8.2 Classification of Major Soils Found in the Amazon Basin

Soil Taxonomy	World Reference Base	Brazilian Classification
Oxisols	Ferralsols	Latossolos
Ultisols	Acrisols	Podzolicos vermelho-amarelo
Aquents	Gleyic arenosols	Areias hidromorficos
Aquepts	Dystric gleysols	Hidromorficos distroficos

agricultural development are soil acidity, phosphorus deficiency, and low cation exchange capacity. It is estimated (Cochrane and Sanchez 1982) that 81 percent of the basin has native pH values of less than 5.3 in the topsoil, and that 82 percent of the area has native pH values of less than 5.3 in the subsoil. Aluminum toxicity is associated with low pH conditions; therefore if 60 percent aluminum saturation in the top 50 centimeters of soil is assumed to be toxic to aluminum-sensitive plants (such as many crop and forage plants), it is estimated that 73 percent of the area will be affected. A major consequence of high levels of aluminum is stunted root growth and function; this often limits a plant's ability to exploit the available nutrient and water resources with the soil volume, exacerbating the problems of inherently poor soils. Given the poor levels of nutrients available for plant growth, if agriculture is to be pursued on a sustained basis there is a need to provide additional nutrients, ameliorate the soil pH, and increase the ability of the soils to retain cations (Cochrane and Sanchez 1982). The importance of organic-matter cycling to maintaining the productivity of the nearly closed system of the natural forests was stressed above. Recent work in Africa (Wong et al. 1995) has shown how the incorporation of organic residues from certain agroforestry trees (e.g., *Calliandra alothyrsus*, *Grevillea robusta*, and *Leucaena diversifolia*) appears to mitigate the effects of aluminum toxicity on an Oxisol in Burundi. The effects of the residues varied, but the best responses showed increases in maize yields from 0.9 to 2.2 metric tons per hectare and for bean yields from 0.2 to 1.2 metric tons per hectare.

The major physical constraints relate to the potential for erosion and, perhaps surprisingly, to moisture stress. Given the rainfall characteristics in the region, bare soils are likely to be erodible, but the soils are also frequently characterized by high infiltration rates due to the very stable microstructures common in these soils (e.g., Nortcliff and Thornes 1989). Cochrane and Sanchez (1982) estimate that only 6 percent of the soils in the rain forest

and 10 percent of the soils in semievergreen seasonal forest are highly erodible, but more recently, higher percentages have been suggested. Earlier,
reference was made to the relatively flat topography in the Amazon Basin
and the low occurrence of steep slopes. Under conditions of level or low
slope angles, erosion will be of limited extent even when the forest cover is
removed. Although erosion may not occur, the exposure of bare soil to high-
intensity rainfall may lead to a breakdown of the aggregates at the soil surface
and the rapid deterioration of the soil's relatively good physical properties.
There is increasing evidence that Oxisols are less susceptible than Ultisols
to both erosion and structural deterioration owing to the stronger structural
development in Oxisols.

Despite the relatively high annual rainfall in the region, moisture stress is
another physical limitation (Cassel and Lal 1992). Even though mean total
rainfall exceeds mean total evapotranspiration for eight months or more each
year in the moist tropical zone, there are often one or more short periods
during the year when plants undergo moisture stress severe enough to affect
crop production. That may in part be related to the limited rooting volume
exploited by the crops because of subsoil acidity and associated aluminum
toxicity. Thus the markedly pronounced dry season may have the effect of
producing drought conditions for plants having these rooting limitations.

Oxisols and Ultisols

Oxisols and Ultisols, the dominant soils of the Amazon Basin, are considered to be mildly to severely infertile (in terms of agricultural potential).
Oxisols are the result of strong weathering and desilication in humid, freely
draining environments, and are characterized by an oxic subsurface horizon
in which the clay fraction is dominated by kaolinite and/or gibbsite. The
soils are characteristically acidic (pH < 5.0) throughout the profile with low
cation exchange capacities (< 10 cmolc kg^{-1} clay). Much of the retention of
cations is associated with soil organic matter principally found in the surface
horizons. Although Al^{3+} is the dominant cation on the exchange complex,
the absolute amount is small compared to that found on Ultisols because the
cation exchange capacity is so low; hence the severity of aluminum toxicity
may be less and the potential for remedy greater.

Ultisols are characterized by a subsurface horizon of increased clay content (an argillic horizon) that forms because clay is removed from upper soil
horizons, translocated down through the soil profile, and deposited in a lower
horizon where it accumulates. The clay fraction in these soils is dominated

TABLE 8.3 Major Soil-Based Constraints to Agricultural Development
in Amazonia

Oxisols	Ultisols
1. Low plant nutrient stores in soils	1. Low plant nutrient stores in soils
2. Low (often exceptionally low) cation exchange capacity	2. High levels of exchangeable aluminum
3. Weak retention of bases	3. Strong nitrogen losses in the surface layers through leaching
4. Strong fixation or deficiencies of phosphorus	4. Low structural strength in the surface layers (particularly as soil organic matter levels decline), resulting in increased erodibility
5. Strong nitrogen losses (particularly in the surface layers) through leaching	5. Decrease in permeability of the argillic B horizon
6. Strong acidity	6. Possibly restricted rooting due to both argillic B horizon and aluminum toxicity
7. High levels of exchangeable aluminum	
8. Very low calcium content	
9. Restricted rooting volume as a result of aluminum toxicity may result in moisture stress	

by kaolinite, although there may be some 2:1 minerals present. The soils
have a low cation exchange capacity (< 24 cmolc kg^{-1} clay) but the low pH ($<$
5.5) frequently results in exchangeable aluminum dominating the exchange
complex, particularly in subsurface horizons. The presence of higher levels
of organic matter in surface horizons may result in lower exchangeable aluminum levels. These high levels of aluminum saturation impose a major re-
striction for agricultural development. Table 8.3 summarizes the major limitations to agricultural development for Oxisols and Ultisols in Amazonia.

Forest Clearance in the Amazon Basin

Clearance of the natural vegetation within the Amazon Basin takes place
for a variety of reasons. The following are some of the major justifications
that have been given in recent decades:

- To exploit the timber resource
- To exploit other forest products
- To provide access to land for agricultural production (i.e., crops and grazing)
- For timber plantation
- For other plantation crops (e.g., cocoa, rubber)
- To provide land for urban and industrial development
- To provide access to below-ground mineral resources

In addition, of course, there is clearance by the indigenous population, using traditional slash-and-burn methods to sustain short-term exploitation of the land; this frequently takes the form of low levels of agricultural productivity followed by long fallow, or recovery, periods. This type of clearance, if employed extensively and with prolonged fallow periods, appears to be sustainable in the context of subsistence agriculture. Soil properties and their management will have little or no impact on the success or failure of activities such as exploitation of mineral resources or urban and industrial land uses. For other activities, however, the inherent soil properties and the degree to which the soil is changed during the clearance of the natural vegetation may largely influence sustainability of the postclearance land management. The nutrient status of the majority of the soils has been shown to be exceedingly poor, and the nutrient cycling through the organic matter is potentially the key to the sustainability of both the natural system and any managed system that replaces it. In low-input systems it seems likely that some form of agroforestry approach to land management, where the natural vegetation is replaced by a mix of trees and other harvestable materials, may be appropriate to achieve a degree of sustainability (Nortcliff 1998). In agroforestry practices the aim is to mix the positive benefits of a perennial tree component with shorter-term cropping systems. The trees supply organic litter to the surface and may provide a more efficient nutrient cycle; if trees capable of fixing atmospheric nitrogen are selected they may also provide nitrogen inputs to the system. The successful use of the soil in the postclearance phase is determined by the nature of the soil and the changes that take place during clearance.

The soil's nutrient status is important in determining its ability to sustain postclearance biomass production, and may be achieved through careful management of zero-external-input systems (e.g., agroforestry), through fertilizer applications, or through some combination of the two. Nicholaides et al. (1983) have clearly shown that, in the Amazon Basin, sustained relatively low output is possible with relatively low levels of external inputs. The changes in soil physical properties that occur during clearance may, however, be of considerably greater importance in determining the overall productive capacity of the soil in the long run, because these properties are far less readily remedied. Indeed, some changes may be permanent and others only slowly reversible. The changes in soil physical properties may take place during clearance or in the postclearance phase because of the changed environmental conditions. Nortcliff and Dias (1985) reviewed the changes in soil physical properties in the humid tropics resulting from different forest clearance activities. Dias and Nortcliff (1985) contrast three types of soils: those under forest that has been cleared using traditional slash-and-burn methods, those under forest cleared using bulldozers, and those under adjacent virgin forest. Although both clearance methods produced changes in soil bulk density, penetration resistance, infiltration capacity, macroporosity, and water-holding capacity, the major changes occurred only under the land cleared by bulldozers. The slash-and-burn site retained the good soil physical characteristics of the Oxisols found in the region. Observations of the cleared sites 12 months after the initial clearance noted that plant growth was vigorous on the slash-and-burn site but exceptionally poor on the bulldozed site (Dias 1983). It is obvious therefore that if mechanical clearance of the natural vegetation is to take place it must be undertaken with a considerable degree of care in order to avoid irreversible soil structural damage. If damage does occur it may be necessary to attempt remedial management in the postclearance phase to alleviate the structural damage.

In addition to damage caused during the process of forest clearance, the management of the postclearance phase may also have major significant effects on the long-term use of the soil. Studies in Roraima conducted as part of the RGS/SEMA/INPA Maracá Project (e.g., Milliken and Ratter 1998) on basic postclearance strategies, noted a marked reduction in the generation of runoff and loss of soil as a result of erosion along slope transects on an Ultisol (Nortcliff, Ross, and Thornes 1990). The postclearance strategies involved (1) total removal of all vegetation and litter from the site or clearance ("totally cleared") and (2) removal of all vegetation above breast height while retaining the litter and ground flora ("partially cleared"). These treatments were compared with a third transect in which the forest was left intact. Results showed both variation along the slope gradient and marked differences between the cleared plots—in comparison with each other and in comparison with the plots having intact forest (table 8.4).

TABLE 8.4 Relative Rates of Runoff and Sediment Yield
Across an Ultisol Landscape in Maracá, Roraima

	Top-Slope Runoff	Top-Slope Sediment Yield	Mid-Slope Runoff	Mid-Slope Sediment Yield	Bottom-Slope Runoff	Bottom-Slope Sediment Yield
Postclearance strategy:						
Intact forest	21.1	2.7	27.2	2.4	35.5	4.6
Partially cleared	24.9	5.2	36.3	9.9	18.0	4.5
Totally cleared	22.2	7.5	100	100	70.7	71.6

Note: Slope position and postclearance strategy were recorded during a 189-day period in 1988. The maximum recorded rates of runoff and sediment yield are scaled to 100.

Conclusions

It is apparent that development in Brazilian Amazonia must take account of the often finely tuned relationship that exists between the soil and its environment. An understanding of those relationships, and the changes in them that are likely to occur when the natural ecosystems are disturbed, holds the key to successful development. In many cases it is the failure to be aware of those relationships and to understand their nature that has resulted in some of the devastation associated with inappropriate development.

The soils of the Amazon are, on the whole, inherently infertile. The paradox of lush forest and poor soil must be highlighted. While the soils may be considered of poor quality in any sort of global comparison, there is variability: some soils are more suitable for development than others. If future development is to take place successfully and sustainably, identification of appropriate soils should be a priority. Although the soils are intrinsically of poor fertility with respect to the provision of nutrients, the physical properties are often relatively good and in some places exceedingly good. If development is to take place, care must be taken to ensure that the good physical properties are maintained or damaged only slightly.

Appropriate development in the context of soil constraints may require the adoption of traditional approaches, nontraditional approaches, or combinations of approaches. For many, a change from the low input:low output approach practiced in the present (and for many years in the past) by many indigenous peoples may be considered the most suitable form of land management; it also appears to be an approach that reflects most closely the best response to environmental conditions. There is very little evidence that land-use practices and cropping patterns adopted by indigenous land users result in the widespread land degradation seen in many places in the Amazon Basin. Indigenous practices do not produce high levels of output, but also do not totally destroy the ecosystems or permanently interrupt ecosystem processes and functions. There is, in contrast, very little evidence to suggest that widespread use of an agricultural strategy based on high input is likely to be successful or sustainable because of the constraints imposed by the soils and the environmental characteristics of the region. It would appear that an alternative approach covered by the broad term "agroforestry" offers a possibility for future sustainable development. Although heralded as a land development option of the latter part of the twentieth century, the broad principles and many of the practices known as agroforestry have their roots in strategies of land management long practiced by indigenous peoples in the Amazon and other tropical regions. A further option available to land developers is the use of plants that have been developed to be tolerant of the harsh soil and environmental conditions of the region but still capable of reasonable levels of productivity. There is evidence that programs from the International Centre for Tropical Agriculture (CIAT) in Cali, Colombia, have met with some degree of success in developing more productive and sustainable tropical pastures using acid-tolerant pasture species and grass mixtures with forage legumes.

If deforestation is to take place we must ensure that the replacement activity is productive and sustainable in the long term. Far too much clearance has been inappropriately undertaken and has often resulted in the production of cleared land that is considered useless and is immediately abandoned, necessitating the clearance of more forest! If development of the region requires that further natural forest must be cleared, there is a need to

- select forest more carefully, based upon the potential for development;
- ensure that the clearance does not drastically affect the potential of the soil for its postclearance production; and
- consider development options that are appropriate to the soil conditions as well as sustainable.

Clearing forest without knowledge of the soils and their response to development activities seems likely to lead to failed developments, in both the short and long term. Whatever the development option selected, it should be understood that there is a wealth of indigenous experience in soil and land management: future development efforts should recognize and incorporate this knowledge in the determination of appropriate management options.

References

Carmargo M. N., E. E. Klamt, and J. H. Kaufman. 1987. Sistema brasiliero de clasificaçao de solos. *Separata do B. Inf. Sociedade Brasiliero de Ciencia de Solo* 12(1):11–33.
Cassel, D. K and R. Lal. 1992. Soil physical properties of the tropics: Common beliefs and management constraints. In P. A. Sanchez and R. Lal, eds., *Myths and Science of Soils in the Tropics*, pp. 61–89. Madison, Wisconsin: Soil Science Society of America Special Publication 29.
Cochrane, T. T. and P. A. Sanchez. 1982. Land resources, soils and their management in the Amazon region: A state of knowledge report. In S. B. Hecht, ed., *Amazonia, Agriculture and Land Use Research*, pp.137–209. Cali, Colombia: Centro Internacional de Agricultura Tropical (CIAT).
Deckers, J.A., F. O. Nachtergaele, and O. C. Spaargaren, eds. 1998. *World Reference Base for Soils*. Leuven: Acco.
Dias, A. C. P. 1983. Effects of selected land clearing methods on the physical properties of an Oxisol in the Brazilian Amazon. Ph.D. dissertation, University of Reading, Reading, UK.
Dias, A. C. P. and S. Nortcliff. 1985. Effects of two land clearing methods on the physical properties of an Oxisol in the Brazilian Amazon. *Tropical Agriculture* 62:207–213.
FAO/UNESCO. 1984. *Soil Map of the World: Revised Legend*. Rome: Food and Agriculture Organization of the U.N. (FAO).
Hemming, J. 1978. *Red Gold: The Conquest of the Brazilian Indians*. Oxford: Macmillan.
Jordan, C. F. 1985. *Nutrient Cycling in Tropical Forest Ecosystems: Principles and Applications in Management and Conservation*. Chichester: Wiley.
Lanly, J. P. 1982. *Tropical Forest Resources*. Rome: FAO Forestry Paper, no. 30.
Milliken, W. and J. A. Ratter, eds. 1998. *Maracá: The Biodiversity and Environment of an Amazonian Rainforest*. Chichester: Wiley.
Nicholaides, J. J. III, P. A. Sanchez, D. E. Brandy, J. H. Villachia, A. J. Couto, and C. S. Valverde. 1983. Crop production systems in the Amazon Basin. In E. F. Moran, ed., *The Dilemma of Amazonian Development*, pp. 101–153. Boulder, Colorado: Westview Press.
Nortcliff, S. 1998. Human activity and the tropical rainforest: Are the soils the forgotten component of the ecosystem? In B. K. Maloney, ed., *Human Activities and the Tropical Rainforest*, pp. 49–64. Dordrecht: Kluwer Academic.

Nortcliff, S. and A. C. P. Dias. 1985. The change in soil physical conditions resulting from forest clearance in the humid tropics. *Journal of Biogeography* 15:61–66.

Nortcliff, S., S. M. Ross, and J. B. Thornes. 1990. Soil moisture, runoff and sediment yield from differentially cleared tropical rainforest plots. In J. B. Thornes, ed., *Vegetation and Erosion*, pp. 419–436. Chichester: Wiley.

Nortcliff, S. and J. B. Thornes. 1989. Variation in soil nutrients in relation to soil moisture status in a tropical forested ecosystem. In J. Proctor, ed., *Mineral Nutrients in Tropical Forest and Savanna Ecosystems*, pp. 43–54. Oxford: Blackwell Scientific.

Sanchez, P. A. and T. T. Cochrane. 1980. Soil constraints in relation to major farming systems of tropical America. In M. Drosdoff, H. Zandstra, and W. G. Rockwood, eds., *Priorities for Alleviating Soil-Related Constraints to Food Production in the Tropics*, pp. 107–139. Los Banos, Philippines: International Rice Research Institute (IRRI).

Soil Survey Staff. 1975. *Soil Taxonomy*. Agricultural Handbook 426. Washington, DC: USDA-NRCS.

Stark, N. 1971. Nutrient cycling. I. Nutrient distribution in some Amazonian soils. *Tropical Ecology* 12:24–50.

Went, F. and N. Stark. 1968. Mychorrhiza. *Bioscience* 18:1035–1038.

Wong, M. T. F., E. Akyeampong, S. Nortcliff, M. R. Rao, and R. S. Swift. 1995. Initial response of maize and beans to decreased concentrations of monomeric inorganic aluminium with application of manure or tree prunings to an Oxisol in Burundi. *Plant and Soil* 171:275–282.

9   Fragile Soils and Deforestation Impacts

The Rationale for Environmental Services of Standing Forest as a Development Paradigm in Amazonia

Philip M. Fearnside

Development in Brazil's 5×10^6 km² Legal Amazon region has, until now, been based mainly on removing and selling natural resources, such as timber and minerals, or on agriculture or ranching operations that derive their products from the soil. Sale of commodities such as minerals and timber often fails to benefit the local population. Conversion of forest to cattle pasture, the most widespread land-use change in Brazilian Amazonia, brings benefits that are extremely meager (although not quite zero). High priority must be given to redirection of development to activities with local returns that are greater and longer lasting. Tapping the value of environmental services offers such an opportunity. Keeping benefits of these services for the inhabitants of the Amazonian interior is the most important challenge in turning these services into development (Fearnside 1997a).

Amazonian Development

Soils in Amazonia are almost universally of low fertility and are "fragile," that is, subject to degradation. The fragile nature of the soil is one reason why the yields from agriculture and ranching are low and short lived. Fragile soil should provide part of the justification, from Brazil's perspective, for giving priority to a new paradigm for Amazonian development that would be based on tapping the value of the environmental services afforded by standing forest. Impacts of deforestation provide the other part of the justification from the Brazilian perspective, and also represent the main motivation from the perspective of countries from which monetary flows might one day be derived on the basis of forest services.

The predominant feature of development in Brazilian Amazonia has so far been conversion of forest to cattle pasture (Fearnside 1990a). Large and medium-sized ranches account for about 70 percent of the clearing activity, while small farms (with < 100 hectares of land) account for about 30 percent (Fearnside 1993a). Contrary to statements by the head of the Brazilian Institute for Environment and Renewable Natural Resources (IBAMA) (Traumann 1998), deforestation data for 1995 and 1996 released by Brazil's National Institute for Space Research (INPE) in January 1988 (INPE 1998) do not indicate that small-scale farmers are now the primary agents of deforestation. The fact that about half (59 percent in 1995 and 53 percent in 1996) of new clearings (as distinct from the properties in which the clearings were located) have areas under 100 hectares reinforces the conclusion that most deforestation is being done by ranchers, as no small-scale farmer can clear anywhere near 100 hectares in a single year. Twenty-one percent of the area of new clearings in 1995 and 18 percent in 1996 were less than 15 hectares. Small-scale farmer families are only capable of clearing about three hectares a year with family labor (Fearnside 1980), and this is reflected in deforestation behavior in settlement areas (Fearnside 1984).

Continuation of present land-use trends would result in the landscape approaching a stable equilibrium of 4.0 percent farmland, 43.8 percent productive pasture, 5.2 percent degraded pasture, 2.0 percent secondary forest derived from agriculture, and 44.9 percent secondary forest derived from pasture (Fearnside 1996a). Although the landscape would not reach this equilibrium for over a century, most of the adjustment would occur within the first few decades. Conversion to cattle pasture is the land-use choice that causes maximum impact on the forest while supporting only a very sparse human population (Fearnside 1983; Hecht 1983).

New initiatives may alter land-use patterns in significant ways. Soybeans are being promoted by the national and state governments (Fearnside 2001). Soybean cultivation has expanded into large areas of *cerrado* (central Brazilian savannas) and is now spreading to other vegetation types. The first major plantations are in natural grasslands near Humaitá, Amazonas. The Madeira River waterway, opened in March 1997, lowers the cost of transport from this part of the region to one-third of its former cost, thus radically altering the economic picture for more intensive agriculture there. A warehouse capable of storing 90,000 metric tons has been inaugurated in Itacoatiara, Amazonas, and a second such warehouse, adjacent to the first, is expected in a subsequent phase. Soybean cultivation is also expanding from the cerrado into forested areas in southern Amazonas, and already represents an important crop in northern Mato Grosso and eastern Rondônia. Soybean planting initiatives are under way in Roraima and near Santarém and Paragominas in Pará. Little employment results from soybean cultivation, which is done using mechanized agriculture. For example, in the state of Paraná in southern Brazil, when small farms growing coffee and annual crops were replaced by larger landholdings growing soybeans in the 1970s, for every person who found employment in the new system 11 people were expelled from the state (many of whom moved to Rondônia) (Zokun 1980).

The question of who is to blame for tropical deforestation has profound implications for the priorities of programs intended to reduce forest loss. The prominence of cattle ranchers in Brazil (as distinguished from many other parts of the tropics) means that the social cost of substantially reducing deforestation rates would be much less than is implied by frequent pronouncements that blame "poverty" for environmental problems in the region. It also means that measures aimed at containing deforestation by, for example, promoting agroforestry among small-scale farmers can never achieve this goal, although some of the same tools (such as agroforestry) have important reasons for being supported independent of efforts to combat deforestation (Fearnside 1995).

Fragile Soils and Development Options

The soils in Amazonia are notoriously infertile: indicators of soil fertility such as pH, cation exchange capacity, total exchangeable bases, and available phosphorus are low, while saturation of toxic aluminum ions is high (Fearnside and Leal Filho 2001). Under such circumstances, it is logical to maintain these areas under forest rather than to convert them to short-lived low-productivity land uses. But to what extent would the situation be different if the soils were more productive? What level of soil quality would make it worthwhile to sacrifice the forest? There are no simple answers to these questions. Rational decision making will require assessment of the value of both the agricultural production that can realistically be expected from the area and the environmental cost of sacrificing the forest.

One must consider the extent to which the relative economic attractiveness of forest over agriculture would change if technical advances were to occur that removed or relaxed the barriers presented by fragile soils. For example, recent progress has been made on removing aluminum saturation limitations through development of transgenic crop plants (Barinaga 1997; de la Fuente et al. 1997). It is not inconceivable that phosphorus limitations could be relaxed by developing crop plants having appropriate micorrhizal associations. Nitrogen limitations of various nonleguminous crops may be relaxed through pseudosymbiotic relationships with a variety of nitrogen-fixing bacteria; this is an area in which significant advances have been achieved in Brazil through the work of Johanna Döbereiner (e.g., Döbereiner 1992).

The temptation is strong to view Amazonia as a potential cornucopia, capable of solving its population and land distribution problems in other parts of the country and the world. However, the limitations to large-scale expansion of intensive agriculture make this a cruel illusion. These limits are best illustrated by the failure of any significant portion of Amazonia to adopt the "Yurimaguas technology"—a program for continuous cultivation developed in Amazonian Peru (Sánchez et al. 1982; Fearnside 1987, 1988; Walker, Lavelle, and Weischet 1987). Despite a long list of subsidies ranging from chemical inputs to free soil analyses and technical advice on a field-by-field basis, this high-input management package did not gain popular acceptance, even in the area surrounding the experiment station in Peru. The limits of physical resources such as phosphate deposits, as well as financial and institutional restraints, make widespread use of such systems unlikely (see Fearnside 1997b).

Land-use and zoning decisions are often based on permitting the maximum use intensity that physical conditions will allow, but when individual allocations are considered together, limits can quickly be reached in spheres other than the matching of crop-plant demands and tolerances to the physical characters of the site. One may examine each cell in a grid in a geographical information system (GIS), comparing components such as soil and rainfall with the demands of a given crop, and conclude that each individual hectare can be allocated to the use in question, and yet arrive at an overall conclusion that is patently unrealistic. This, for example, explains the conclusion that Brazil could support seven billion people—a conclusion reached in a study conducted by the Food and Agriculture Organization of the United Nations (FAO), in collaboration with the United Nations Fund for Population Activities (UNFPA) and the International Institute for Applied Systems Analysis (IIASA) (FAO 1980, 1981, 1984; Higgins et al. 1982). Unfortunately, the study's assumptions regarding Amazonian soil quality, land-use patterns, and availability of resources such as phosphates make this scenario totally unrealistic (see review in Fearnside 1990b).

In the early 1970s, when the fiscal incentives program for Amazonian pastures was rapidly expanding, the Brazilian Enterprise for Agriculture and

Cattle Ranching Research (EMBRAPA) maintained that pasture improved the soil. EMBRAPA's original publications did not indicate that phosphorus was one of the soil characters that improve as a result of pasture (Falesi 1974, 1976), but others added it to the list when the results were trumpeted to the world (e.g., Alvim 1981). EMBRAPA itself recognized that phosphorus was necessary, and in 1977 changed its position that pasture improves the soil, recommending instead that productivity be maintained by applying phosphate fertilizer at 200–300 kg/ha/dose (50 percent simple superphosphate, 50 percent hyperphosphate) (Serrão and Falesi 1977:55) to supply 50 kg/ha of P_2O_5 (Serrão et al. 1978:28). The dose of phosphorus needed to properly fertilize was subsequently modified to 25–50 kg/ha of P_2O_5 (Serrão et al. 1979:220), but more recent recommendations have returned to the original 50 kg/ha (Correa and Reichardt 1995).

The conditions that limit reliance on phosphate fertilizers are the cost of supplying phosphate and the absolute limits to minable stocks of this mineral. A report on Brazil's phosphate deposits published by the Ministry of Mines and Energy indicates that only one small deposit exists in Amazonia, located on the Atlantic coast near the border of Pará and Maranhão (de Lima 1976). In addition to the deposit's small size, it has the disadvantage of being made up of aluminum compounds that render its agricultural use suboptimal, but not impossible to use if new technologies are developed for fertilizer manufacture (dos Santos 1981:178). An additional deposit has been found on the Rio Maecuru, near Monte Alegre, Pará (Beisiegel and de Souza 1986), but information on its size is still incomplete. Almost all of Brazil's phosphates are in Minas Gerais, a site very distant from all but the easternmost states of Amazonia. Brazil as a whole is not blessed with a particularly large stock of phosphates—total deposits in the United States, for example, are about 20 times larger than all the deposits in Brazil (de Lima 1976:26), and Brazil's total reserves amount to only 1.6 percent of the global total (Sheldon 1982). On a global scale most phosphates are located in Africa (Sheldon 1982). Continuation of post–World War II trends in phosphate use would exhaust the world's stocks by the middle of the twenty-first century (Smith et al. 1972; U.S. Council on Environmental Quality and U.S. Department of State 1980). Although simple extrapolation of these trends is questionable because of limits to continued human population increase at past rates (Wells 1976), the conversion of a substantial portion of Amazonia to fertilized pasture would hasten the day when phosphate stocks become exhausted in Brazil and the world. Brazil would be wise to ponder carefully whether its remaining stocks of this limited resource should be allocated to Amazonian pastures.

A rough calculation can be made of the adequacy of Brazilian phosphate reserves to sustain pastures in Amazonia (Fearnside 1998). Brazilian reserves of phosphate rock total 780.6×10^6 metric tons, with an average P_2O_5 content of 12 percent (de Lima 1976:24), not counting the Maecuru deposit still being assessed. Discounting a loss of 8 percent of the P_2O_5 in the transformation of rock to phosphate fertilizer (de Lima 1976:10), the Brazilian reserves represent 86.2×10^6 metric tons of P_2O_5. The 53.0×10^6 hectares of forest cleared by 1997 in the Brazilian Legal Amazon (INPE 1998) would consume 1.06×10^6 metric tons of P_2O_5 annually if maintained in pasture. This assumes that pastures are fertilized once every two-and-a-half years (Serrão et al. 1979:220) at the 50 kg/ha dose of P_2O_5 per fertilization needed to attain a minimum critical level of 5 ppm P_2O_5 in the soil (rather than the traditional critical level of 10 ppm, which would require annual doses of fertilizer to maintain). If the entire 400×10^6 hectares of originally forested area in the Brazilian Legal Amazon were fertilized at the rate recommended for pasture, it would require 8.00×10^6 metric tons of P_2O_5 annually. Should all of Brazil's phosphate reserves be devoted to this purpose, they would last eighty-one years if the deforested area currently under pasture (an area the size of France) were maintained, and only eleven years if the remainder of the forested area were also converted to pasture. However Brazil's fertilizer deposits are already almost totally committed to maintaining agricultural production outside the Legal Amazon.

Nothing obliges Brazil or any other country to rely exclusively on domestic deposits of phosphates, as long as supplies continue to be available for import from more richly endowed countries. Some of the most richly endowed, such as Morocco, have little domestic agriculture to compete with an export trade. Brazil's reliance on imports implies disadvantages in terms of the price paid (including transport), the predictability of the price, and the security of availability.

The existence of limits such as phosphates to soil fertility leads to the inevitable conclusion that population and consumption cannot grow indefinitely. There is no such thing as sustainable development for an infinite number of people, or for a fixed population that is infinitely rapacious. There is also no way that development aimed at increasing the size of the pie can address problems that are rooted in highly unequal distribution of the pie. Many physical limits represent restrictions that need to be respected and lived with rather than an agenda of items to be attacked. Recognition of this fact forces one to face fundamental problems of development that many people would prefer not to think about. The result is a tendency to deny the existence of limits. Admitting to the finite potential for growth of the pie does not condemn the poor to poverty, but rather condemns the rich to dividing the pie (Fearnside 1993b).

Although a formidable array of limiting factors stands in the way of sustaining production in large areas of Amazonia if forests are converted to agriculture and ranching, this does not mean that the outlook need be gloomy for sustaining the region's current population, as long as the means of support is derived from the forest itself rather than from replacing it with nonforest land uses. This author believes that the best long-term strategy for providing a sustainable basis of development for the current residents of rural Amazonia and their descendants is to tap the potential monetary value of the environmental services supplied to the rest of the world by the natural forests in Amazonia (Fearnside 1997a).

Deforestation Impacts and Environmental Services

At least three classes of environmental services are provided by Amazonian forests: biodiversity maintenance, carbon storage, and water cycling. Amazonia's status as a "mega diversity" area is well known. The value of biodiversity has a variety of forms, some immediately utilitarian and others not. Regardless of the importance, if any, that a given individual may attribute to biodiversity, there is a substantial "willingness to pay" to avoid losing the values it provides (Cartwright 1985).

Emissions of greenhouse gases that contribute to global warming are a major consequence of deforestation. The 1990 rate of deforestation (13.8×10^6 km^2/year) resulted in at least 267×10^6 metric tons of CO_2-equivalent carbon as net committed emissions (the net long-term effect of one year's clearing activity). The equivalent value for the annual balance of net emissions in that year (emissions and uptakes over all of the originally forested area of Brazil's Legal Amazon) was 354×10^6 metric tons CO_2-equivalent carbon from deforestation and 62×10^6 metric tons CO_2-equivalent carbon from logging (Fearnside 2000a, updated from Fearnside 1996b, 1997c).

One of the impacts expected to result from the continued expansion of deforestation is reduction of rainfall during dry periods in Amazonia and in other parts of Brazil (e.g., Eagleson 1986; Fearnside 2004). In absolute terms, the reduction of rainfall would be approximately constant throughout the year in Amazonia, but in percentage terms the decline would increase substantially during the dry season. While the annual rainfall total would decrease by only 7 percent from conversion to pasture, the average rainfall in August alone would decrease from 2.2 mm/day with present levels of forest cover to 1.5 mm/day after conversion to pasture (a dropoff of 31.8 percent) (Lean et al. 1996:560–561).

Preliminary calculations of indicators of willingness to pay for the services lost from the 1990 deforestation in the Brazilian Legal Amazon total US$2.5 billion (assuming 5 percent annual discount); maintenance of the stock of forest, if regarded as producing 5 percent a year annuity, would be worth US$37 billion annually (Fearnside 2000b, updated from Fearnside 1997a). The magnitude and value of these services are poorly quantified, and the diplomatic and other steps through which such services might be compensated are also in their infancy. These facts do not diminish the importance of the services or of focusing effort on providing both the information and the political will needed to integrate them into the rest of the human economy in such a way that economic forces act to maintain rather than to destroy the forest.

What are the ingredients of a rational decision on the question of attempting or not attempting to overcome a limitation on development? The starting point must be a clear definition of the objectives of development. For example, if the objective of development is to provide a sustainable livelihood for the populations of the region, then little benefit will be achieved by augmenting the productivity or the life expectancy of cattle pastures on large ranches by supplying fertilizers and improving management. Many efforts to push back limits to agricultural crop production have as their rationale supporting an ever-larger population of farmers—for example, the immigrants who come to Amazonia from other parts of Brazil. This is not necessarily in the best interests of Amazonia's current inhabitants and their descendants. It would be better to recognize that the ability of Amazonia to support population is limited and to guide development in such ways that population size and environmental impacts are kept within those limits (Fearnside 1997b).

On many fronts, one of the major challenges to finding rational uses for Amazonian forest lies in gathering and interpreting relevant information. Making environmental services of the forest into a basis for sustainable development is, perhaps, the area where information is most critical. A better understanding of the dynamics of deforestation, and of deforestation's impacts on biodiversity, carbon storage, and water cycling, is a necessary starting point on the long road to turning environmental services into a basis for sustainable development in Amazonia.

It is essential that proposals for use of monetary flows from environmental services come from local communities themselves, rather than be packaged and imposed from outside. This is the most effective way to guarantee that poverty eradication is achieved, and that on-the-ground pressure is effective in avoiding invasion of the forest. A natural alliance of interests exists between those whose primary concern is environment and those whose primary concern is indigenous or other traditional peoples. It is important to realize, however, that it is an alliance, not an identity of interests. In other words, values for environmental services represent a payment for services rendered, not a human right. Adequate means of monitoring and accounting are needed (Fearnside 1997d).

Role of Traditional Peoples

Because the income stream that can be obtained from environmental services derives from services rendered, rather than simply as a human right, it is important to quantify and document the role that Amerindians, rubber tappers, and other traditional peoples play in providing the services. This is very much in the interests of the traditional peoples, since they could be major beneficiaries of making environmental services function as a basis for Amazonian development. Traditional peoples have so far had the best record of maintaining forest intact. In many heavily deforested parts of Amazonia, the only forest left standing is that in indigenous areas (Schwartzman, Moreira, and Nepstad 2000).

Traditional knowledge forms a part of the cultural system that makes traditional peoples effective guardians of the forest. If traditional communities lose their cultural identities, they are likely to be no more effective in defending the forest than are their nontraditional counterparts. Traditional populations must have an economic livelihood if they are to remain as viable communities. Income derived from traditional knowledge can make some contribution to this, particularly if appropriate mechanisms are created to guarantee that a portion of the value later derived from this knowledge is returned to the traditional group from which it came. However, it is important not to exaggerate the role of traditional knowledge (cf. controversies regarding the rosy periwinkle of Madagascar in Djerassi 1992). In addition, the dollar amounts that can be obtained from pharmaceutical products are not likely to be very large, contrary to the expectations of some. An indication of this is the well-known contract between Merck & Co., Inc., and Costa Rica's Institute of Biodiversity (Inbio), which provides approximately US$1 million annually in exchange for samples collected in that country's natural ecosystems by a network of parataxonomists. This flow of samples is sufficient to satisfy Merck's capacity to invest in searching for new compounds from tropical forests. The reality is, therefore, that Brazil is in competition with

Costa Rica and the rest of the tropics, and that the limiting factors to gaining income from biodiversity in this way are laboratories and taxonomists, not forests and tribal peoples (Fearnside 1999).

Traditional knowledge can be rewarded with a premium paid for information received, in a manner similar to that paid for certified forest products. The amounts available for this kind of subsidy depend on public perception of the importance of traditional peoples and of the environment: willingness to pay fluctuates widely depending on fashion and press interest in the subject. Although one may expect a long-term trend toward increasing willingness to pay for using and maintaining traditional knowledge and its products, much work will be required to build the appropriate institutional mechanisms needed to turn these assets into monetary flows, and to make those flows useful to the traditional peoples involved. This could be done in conjunction with the similar mechanisms that must be created to turn provision of environmental services into a form of sustainable development for rural Amazonia. The hurdles to be crossed in changing the development paradigm in Amazonia are many, but as the Chinese proverb explains: "A journey of a thousand leagues begins with a single step."

Acknowledgments

I thank the National Council of Scientific and Technological Development (CNPq AI 350230/97-98) and the National Institute for Research in the Amazon (INPA PPIs 5-3150 and 1-3160) for financial support. Portions of this paper have been updated from previously published material written by the author (Fearnside 1997a, 1997b, 1998, 1999; Fearnside and Leal Filho 2001). W. Capraro, D. A. Posey, S. V. Wilson, and one anonymous reviewer commented on the manuscript.

References

Alvim, P. de T. 1981. A perspective appraisal of perennial crops in the Amazon Basin. *Interciencia* 6(3):139-145.

Barinaga, M. 1997. Making plants aluminum tolerant. *Science* 276:1497.

Beisiegel, W. de R. and W. O. de Souza. 1986. Reservas de fosfatos—Panorama nacional e mundial. In Instituto Brasileiro de Fosfato (IBRAFOS) III. Encontro Nacional de Rocha Fosfática, Brasília, 16-18/06/86, pp. 55-67. Brasília, Brazil: IBRAFOS.

Cartwright, J. 1985. The politics of preserving natural areas in Third World states. *The Environmentalist* 5(3):179–186.

Correa, J. C. and K. Reichardt. 1995. Efeito do tempo de uso das pastagens sobre as propriedades de um latossolo amarelo da Amazônia Central. *Pesquisa Agropecuária Brasileira* 30(1):107–114.

De la Fuente, J. M., V. Ramírez-Rodríguez, J. L. Cabrera-Ponce, and L. Herrera-Estrella. 1997. Aluminum tolerance in transgenic plants by alteration of citrate synthesis. *Science* 276:1566–1568.

De Lima, J. M. G. 1976. *Perfil Analítico dos Fertilizantes Fosfatados*. Ministério das Minas e Energia, Departamento Nacional de Produção Mineral (DNPM) Boletim No. 39. Brasília, Brazil: DNPM.

Djerassi, C. 1992. Drugs from Third World plants: The future. *Science* 258:203–204.

Döbereiner, J. 1992. Recent changes in concepts of plant bacteria interactions: Endophytic N_2 fixing bacteria. *Ciência e Cultura* 44:310–313.

Dos Santos, B. A. 1981. *Amazônia: Potencial Mineral e Perspectivas de Desenvolvimento*. São Paulo, Brazil: Editora da Universidade de São Paulo (EDUSP).

Eagleson, P. S. 1986. The emergence of global-scale hydrology. *Water Resources Research* 22(9):6s–14s.

Falesi, I. C. 1974. O solo na Amazônia e sua relação com a definição de sistemas de produção agrícola. In *Empresa Brasileira de Pesquisa Agropecuária (EMBRAPA) Reunião do Grupo Interdisciplinar de Trabalho sobre Diretrizes de Pesquisa Agrícola para a Amazônia (Trópico Úmido), Brasília, May 6–10, 1974*, vol. 1, pp. 2.1–2.11. Brasília, Brazil: EMBRAPA.

Falesi, I. C. 1976. *Ecossistema de Pastagem Cultivada na Amazônia Brasileira*. Boletim Técnico, no. 1. Belém, Pará, Brazil: Centro de Pesquisa Agropecuária do Trópico Úmido (CPATU).

FAO. 1980. *Report of the Second FAO/UNFPA Expert Consultation on Land Resources for Populations of the Future*. Rome: Food and Agriculture Organization of the United Nations (FAO).

FAO. 1981. *Report on the Agro-Ecological Zones Project*. Vol. 3. World Soils Resources Report 48/3. Rome: Food and Agriculture Organization of the United Nations (FAO).

FAO. 1984. *Land, Food and People*. FAO Economic and Social Development Series No. 30. Rome: Food and Agriculture Organization of the United Nations (FAO).

Fearnside, P. M. 1980. Land use allocation of the Transamazon Highway colonists of Brazil and its relation to human carrying capacity. In F. Barbira-Scazzocchio, ed., *Land, People and Planning in Contemporary Amazonia*, pp. 114–138. Cambridge: University of Cambridge Centre of Latin American Studies Occasional Paper, no. 3.

Fearnside, P. M. 1983. Land-use trends in the Brazilian Amazon Region as factors in accelerating deforestation. *Environmental Conservation* 10(2):141–148.

Fearnside, P. M. 1984. Land clearing behaviour in small farmer settlement schemes in the Brazilian Amazon and its relation to human carrying capacity. In A. C.

Chadwick and S. L. Sutton, eds., *Tropical Rain Forest: The Leeds Symposium*, pp. 255–271. Leeds, UK: Leeds Philosophical and Literary Society.

Fearnside, P. M. 1987. Rethinking continuous cultivation in Amazonia. *BioScience* 37(3):209–214.

Fearnside, P. M. 1988. Yurimaguas reply. *BioScience* 38(8):525–527.

Fearnside, P. M. 1990a. Predominant land uses in the Brazilian Amazon. In A. B. Anderson, ed., *Alternatives to Deforestation: Steps Toward Sustainable Use of the Amazon Rain Forest*, pp. 235–251. New York: Columbia University Press.

Fearnside, P. M. 1990b. Human carrying capacity in rainforest areas. *Trends in Ecology and Evolution* 5(6):192–196.

Fearnside, P. M. 1993a. Deforestation in Brazilian Amazonia: The effect of population and land tenure. *Ambio* 22(8):537–545.

Fearnside, P. M. 1993b. Forests or fields: A response to the theory that tropical forest conservation poses a threat to the poor. *Land Use Policy* 10(2):108–121.

Fearnside, P. M. 1995. Agroforestry in Brazil's Amazonian development policy: The role and limits of a potential use for degraded lands. In M. Clüsener-Godt and I. Sachs, eds., *Brazilian Perspectives on Sustainable Development of the Amazon Region*, pp. 125–148. Paris: UNESCO; Carnforth, UK: Parthenon Publishing Group.

Fearnside, P. M. 1996a. Amazonian deforestation and global warming: Carbon stocks in vegetation replacing Brazil's Amazon forest. *Forest Ecology and Management* 80(1–3):21–34.

Fearnside, P. M. 1996b. Amazonia and global warming: Annual balance of greenhouse gas emissions from land-use change in Brazil's Amazon region. In J. Levine, ed., *Biomass Burning and Global Change*. Vol. 2, *Biomass Burning in South America, Southeast Asia and Temperate and Boreal Ecosystems and the Oil Fires of Kuwait*, pp. 606–617. Cambridge, Massachusetts: MIT Press.

Fearnside, P. M. 1997a. Environmental services as a strategy for sustainable development in rural Amazonia. *Ecological Economics* 20(1):53–70.

Fearnside, P. M. 1997b. Limiting factors for development of agriculture and ranching in Brazilian Amazonia. *Revista Brasileira de Biologia* 57(4):531–549.

Fearnside, P. M. 1997c. Greenhouse gases from deforestation in Brazilian Amazonia: Net committed emissions. *Climatic Change* 35(3):321–360.

Fearnside, P. M. 1997d. Monitoring needs to transform Amazonian forest maintenance into a global warming mitigation option. *Mitigation and Adaptation Strategies for Global Change* 2(2–3):285–302.

Fearnside, P. M. 1998. Phosphorus and human carrying capacity in Brazilian Amazonia. In J. P. Lynch and J. Deikman, eds., *Phosphorus in Plant Biology: Regulatory Roles in Molecular, Cellular, Organismic, and Ecosystem Processes*, pp. 94–108. Rockville, Maryland: American Society of Plant Physiologists.

Fearnside, P. M. 1999. Biodiversity as an environmental service in Brazil's Amazonian forests: Risks, value and conservation. *Environmental Conservation* 26(4):305–321.

Fearnside, P. M. 2000a. Greenhouse gas emissions from land use change in Brazil's Amazon region. In R. Lal, J. M. Kimble, and B. A. Stewart, eds., *Global Climate Change and Tropical Ecosystems*, pp. 231–249. Boca Raton, Florida: CRC Press, Advances in Soil Science.

Fearnside, P. M. 2000b. Environmental services as a strategy for sustainable development in rural Amazonia. In C. Cavalcanti, ed., *The Environment, Sustainable Development and Public Policies: Building Sustainability in Brazil*, pp. 154–185. Cheltenham, UK: Elgar.

Fearnside, P. M. 2001. Soybean cultivation as a threat to the environment in Brazil. *Environmental Conservation* 28:(1)23–38.

Fearnside, P. M. 2004. A água de São Paulo e a floresta amazônica. *Ciência Hoje* 34(203):63-65.

Fearnside, P. M. and N. Leal Filho. 2001. Soil and development in Amazonia: Lessons from the Biological Dynamics of Forest Fragments Project. In R. O. Bierregaard, C. Gascon, T. E. Lovejoy, and R. C. G. Mesquita, eds., *Lessons from Amazonia: The Ecology and Conservation of a Fragmented Forest*. pp. 291–312. New Haven: Yale University Press.

Hecht, S. B. 1983. Cattle ranching in the eastern Amazon: Environmental and social implications. In E. F. Moran, ed., *The Dilemma of Amazonian Development*, pp. 155–188. Boulder, Colorado: Westview Press.

Higgins, G. M., A. H. Kassam, L. Naiken, G. Fischer, and M. M. Shah. 1982. *Potential Population Supporting Capacities of Lands in the Developing World*. Technical Report of Project INT/75/P13, Land Resources for Populations of the Future. Rome: Food and Agriculture Organization of the United Nations (FAO).

INPE. 1998. *Amazonia: Deforestation, 1995–1997*. São José dos Campos, São Paulo, Brazil: Instituto Nacional de Pesquisas Espaciais.

Lean, J., C. B. Bunton, C. A. Nobre, and P. R. Rowntree. 1996. The simulated impact of Amazonian deforestation on climate using measured ABRACOS vegetation characteristics. In J. H. C. Gash, C. A. Nobre, J. M. Roberts, and R. L. Victoria, eds., *Amazonian Deforestation and Climate*, pp. 549–576. Chichester: Wiley.

Sánchez, P. A., D. E. Bandy, J. H. Villachica, and J. J. Nicholaides III. 1982. Amazon basin soils: Management for continuous crop production. *Science* 216:821–827.

Schwartzman, S., A. Moreira, and D. Nepstad. 2000. Rethinking tropical forest conservation: Perils in parks. *Conservation Biology* 14(5):1351–1357.

Serrão, E. A. S. and I. C. Falesi. 1977. *Pastagens do Trópico Úmido Brasileiro*. Belém, Pará, Brazil: Empresa Brasileira de Pesquisa Agropecuária-Centro de Pesquisa Agropecuária do Trópico Úmido (EMBRAPA-CPATU).

Serrão, E. A. S., I. C. Falesi, J. B. da Viega, and J. F. Teixeira Neto. 1978. *Produtividade de Pastagens Cultivadas em Solos de Baixa Fertilidade das Áreas de Floresta do Trópico Úmido Brasileiro*. Belém, Pará, Brazil: Empresa Brasileira de Pesquisa Agropecuária-Centro de Pesquisa Agropecuária do Trópico Úmido (EMBRAPA-CPATU).

Serrão, E. A. S., I. C. Falesi, J. B. da Viega, and J. F. Teixeira Neto. 1979. Productivity of cultivated pastures on low fertility soils in the Amazon of Brazil. In P. A. Sánchez and L. E. Tergas, eds., *Pasture Production in Acid Soils of the Tropics: Proceedings of a Seminar Held at CIAT, Cali, Colombia, April 17–21, 1978.* CIAT Series 03 EG-05, pp. 195–225. Cali, Colombia: Centro Internacional de Agricultura Tropical (CIAT).

Sheldon, R. P. 1982. Phosphate rock. *Scientific American* 246(6):31–37.

Smith, F., D. Fairbanks, R. Atlas, C. C. Delwiche, D. Gordon, W. Hazen, D. Hitchcock, D. Pramer, J. Skujins, and M. Stuiver. 1972. Cycles of elements. In *Man in the Living Environment*, pp. 41–89. Madison: University of Wisconsin Press.

Traumann, T. 1998. Os novos vilões: Ação dos sem-terra e de pequenos agricultores contribui para o desmatamento da Amazônia. *Veja* (São Paulo), February 4, pp. 34–35.

U. S. Council on Environmental Quality (CEQ) and U.S. Department of State. 1980. *The Global 2000 Report to the President.* 3 vols. New York: Pergamon Press.

Walker, B. H., P. Lavelle, and W. Weischet. 1987. Yurimaguas technology. *BioScience* 37(9):638–640.

Wells, F. J. 1976. *The Long-Run Availability of Phosphorus: A Case Study in Mineral Resource Analysis.* Baltimore, Maryland: Johns Hopkins University Press.

Zokun, M. H. G. P. 1980. *A Expansão da Soja no Brasil: Alguns Aspectos da Produçao.* São Paulo, Brazil: Instituto de Pesquisas Econômicas da Universidade de São Paulo.

10 Concurrent Activities and Invisible Technologies

An Example of Timber Management in Amazonia

Christine Padoch and Miguel Pinedo-Vásquez

As ever more research is done on smallholder resource management in the humid tropics and elsewhere, appreciation of the diversity and complexity of locally developed management practices has grown. By now a multitude of studies have pointed out that small-scale farmers tend to manage many species and varieties of plants and animals within their intricately structured fields and agroforests. (For a recent review of diversity in smallholder agricultural systems, see Brookfield 2002.) In these plots they often succeed in producing a wealth of useful products while protecting considerable biodiversity and maintaining crucial ecological functions.

Most researchers have recognized the diversity of plants, animals, and fields that characterize smallholder operations, but many still fail to appreciate adequately the complexity and diversity of effective, often subtle, and frequently overlapping management operations that take place in and around those fields. It is only in the last decade, for instance, that most field researchers have come to regard swiddens as stages in long-term, complex processes of land management, rather than as plots of land that are intensively farmed for a year or two and then abandoned (e.g., Denevan and Padoch 1988; Irvine 1989; Dufour 1990; and many others). The sequence of intensive cultivation of annuals phasing gradually into managed agroforestry production of fruits and other tree crops is now widely familiar. Emphasis on sequential changes in management technologies, intensities, and outputs may, however, be helping to obscure the concurrent complexity and diversity of locally developed management strategies. And the stress in recent works on agroforestry production of fruits and other nontimber products in swidden-fallow plots may be partly responsible for a continued lack of appreciation of smallholder patterns of timber management. (For a good overview of research on management of tropical forests by smallholders, see Wiersum 1997.)

Timber is economically important to numerous smallholders throughout the forested tropics. Nevertheless, local technologies of manipulation of landholdings for timber continue to be little explored. It is obvious that few field researchers have looked for or specifically asked about them. Formerly "invisible" processes of swidden-fallow management have now been widely described, evaluated, and discussed, but longer-term, often subtle and sophisticated timber-species-management processes have still to be "discovered" (Padoch and Pinedo-Vásquez 1996; Pinedo-Vásquez et al. 2001). The difficulty of detecting some timber management practices is doubtless due to their often being "concealed" by agricultural tasks. That is, while a plot of land may quite obviously be managed as an agricultural field by a smallholder, it may also be undergoing specific forest management operations aimed at enhancing the commercial timber value of the plot in a decade or two. Likewise, while one plant in the field is being managed for the grain or fruit it will yield within a month or two, a neighboring plant may be a seedling that is slated for timber harvesting in a quarter of a century or more.

Management of timber trees is often begun in agricultural fields concurrently with the management of many other resources. Simultaneous management and multipurpose management are among the most important characteristics that distinguish much smallholder management from the simpler patterns more characteristic of industrial agriculture and forestry. To illustrate and emphasize the complexity and concurrent diversity of smallholder management technologies—including long-term ones—we employ an example from our own research in Peruvian Amazonia. (For an example of concurrent management by smallholders in Brazilian Amazonia, see Pinedo-Vásquez et al. 2001.)

Fields in the Floodplain

For more than three years we studied local management of the tropical moist forests of the Napo-Amazon floodplain in lowland Peru. This floodplain lies at the confluence of the Napo and Amazon rivers. Situated between three degrees and four degrees south latitude and at a mere 109 meters above sea level, the climate of the site is warm and humid. (For a more detailed discussion of resource management in the Napo-Amazon floodplain, see Pinedo-Vásquez 1995.)

Amazonians have long used the Napo-Amazon floodplain for hunting, fishing, farming, and collecting various forest products. The study area—approximately 7,820 hectares in size—is inhabited by approximately 10,503 people. The villagers of the area usually refer to themselves as *ribereños* and are largely descendants of several indigenous groups (Chibnik 1991).

From the time of the Spanish and Portuguese colonization to the late 1960s, the extraction of animal and plant resources of the floodplain was the most important economic activity of local people living within the Napo-Amazon floodplain (San Roman 1975). However, from the mid-1970s to the present day, a mix of commercial and subsistence agriculture as well as forest management has become the dominant land use in this and other floodplains of the lowland Peruvian Amazon (Padoch and de Jong 1992).

Of the several types of forests that are under different degrees of management along the Napo-Amazon floodplain, *capinurales*—that is, forests dominated by the tree known locally as *capinuri* (*Maquira coriaceae*)—are among the most economically important. Residents of the floodplain extract several products from capinurales for both domestic use and occasional sale in regional markets. Products include wood (for plywood) and resin from the capinuri, many edible fruits, latex used as an antihelminthic from *oje* (*Ficus insipida*), and a variety of other construction materials, firewood, and medicinal plants.

Capinural forests mainly occupy natural levees along the floodplain. Much of the floodplain is also periodically farmed, with market-oriented production of rice, corn, cassava, and bananas predominating. Farmers exploiting the levees clear existing vegetation by using swidden techniques, or when vegetation is quite low and the annual flood very high, by slashing while the area is inundated.

Villagers then tend to follow a rather complex cropping sequence, planting a near-pure stand of rice as the first crop, followed by a near-monocultural planting of maize. Within a year the maize field tends to be succeeded by a rice-maize-cassava intercrop, a combination that may then be repeated. After one or two years, the grain crops are discontinued, and cassava tends to replace them. Cassava is often intercropped with bananas, and often with some fruit trees. After another two years or so, cassava is phased out and bananas and other fruits remain as the only important agricultural or agroforestry crops.

It should be noted that the above sequence of crops is an ideal. Since these agricultural operations are carried out on a natural levee of an extremely variable river, an unusually high flood occasionally interrupts the sequence outlined above. When the levee is inundated for a significant period, susceptible crops—especially cassava and most varieties of bananas—die, and a new or somewhat changed cropping sequence is initiated.

Agricultural management on levees of the Napo-Amazon floodplain involves tools and technologies that are familiar to researchers of smallholder agriculture. Clearing technologies were described above. Grain crops are dibbled in. Cassava and bananas are planted with the help of digging sticks. Weeding is done by hand or with bushknives. Harvesting is also done manually. Family labor predominates in all these tasks although communal work groups are employed for some larger chores. On very rare occasions Napo-Amazon villagers hire their neighbors to quickly complete agricultural tasks, and compensate them with wages.

While these agricultural tasks are being performed, the plots are readily identifiable and identified as swiddens by both researchers and villagers. When bananas and tree crops take over, the fields may be more accurately referred to as agroforestry plots or managed swidden-fallows. Cultivation then becomes largely confined to occasional slash weedings and periodic harvesting of the crops. Banana production rarely continues for more than eight years. Some fruit trees may be longer lived.

During the swidden and swidden-fallow phases there are other management activities being performed in the same sites and other important sequences of management operations taking place that tend to be less visible or recognizable. The plots that are being managed as swiddens for agricultural production are also being managed as capinural forests: the two processes are concurrent.

The Forest in the Field

Napo-Amazon villagers do not plant capinuri trees, but seedlings of capinuri tend to become established quickly in farm fields because smallholders encourage natural regeneration. In older fields or swidden fallows capinuri juveniles thrive, and in mature capinuri forests individuals are found in all layers. The species is highly adaptable, which is a great advantage in the management of capinural forests. Seedlings, juveniles, and adults respond well to human manipulation as well as to natural disturbances such as recurrent floods. Dense stands of capinuri, however, are formed only after a long process of management and the application of a complex of management techniques.

The management of capinurales can be divided into three principal stages. The precise number and kinds of techniques that are used varies from one stage to the other. People begin managing seedlings of capinuri and other valuable species during the first stages of agricultural production.

When levee plots are cleared for rice production, individual economically valuable trees and seedlings are often spared. The clearing of land for planting grain opens up the forest canopy for light-loving tree species, encouraging their development. Farmers using fire for clearing plots are careful not to burn any valuable tree seedlings. The burn, however, actually helps promote forest development by hastening the germination of tree seeds, a process that might be equated with seedbed preparation in a forestry context.

Since the late 1970s rice has commonly been the first crop planted. The planting of rice and maize was greatly encouraged in the 1970s and 1980s by the availability of credit programs for the production of these two grain crops (Chibnik 1994). On the other hand, the credit programs included a production package that the credit seeker had to promise to follow. Rice and maize were not only among the few crops for which credit was available, but planting of these crops in monoculture was also required. Although the production package specified that no other crops were to be planted among the rice or maize, nothing indicated whether concurrent management of trees was permitted. Farmers insisted that tree seedlings did not compete (*"no pelea"*) with the rice crop. By the end of the 1980s, the credit programs for rice and maize were no longer available, but production of the grains has continued.

While rice and maize—and later cassava and bananas—are being produced, seedlings of capinuri and other valuable timber species are also being managed. Weeding of the grain crops is done selectively. The process of cleaning out competitors of the planted crops coincides with the forestry operation of "brushing," and results in the selection of the most promising seedlings and their release for optimal growth.

As has been noted, many farming and forestry operations are conveniently combined. Among Napo-Amazon forest managers, however, the operations are carried out in a variety of ways and are often referred to by different terms. Two principal forest management techniques are used: *huactapeo* and *jaloneo*. Huactapeo is a technique that consists of three operations: selective weeding, cleaning (including the removal of the roots and stems of species that regenerate by sprouting), and controlled burning (including the elimination by burning of regeneration materials such as the roots and stems of sprouting species). Jaloneo is a thinning technique that includes the selection of healthy, well-formed seedlings, and the uprooting of seedlings with defects. By uprooting seedlings instead of merely cutting them, managers reduce the probability that the offending seedlings will resprout. These important forestry operations are easily missed by observers not focusing on timber management. The diverse (and disorderly) plots themselves are easily dismissed as abandoned fallows, and if noticed at all, the activities may be facilely identified as somewhat haphazard or incomplete weeding of old banana plots.

When yields of bananas decline, the fields can be better viewed as enriched fallows (*purmas*). During this second stage, local people manage the juveniles of capinuri and other useful timber species along with the planted and protected fruit and other valuable trees. While selective slash weeding is applied to enhance the growth of the many different types of resources in the fallows, three principal techniques—*huahuancheo*, *raleo*, and *mocheado*—are directed specifically toward the management of timber trees. The three techniques encompass selection, marking, pruning, liberation, and thinning activities that employ sophisticated procedures for removing undesirable individuals by felling, girdling, or burning. By removing unwanted individuals, space and light are made available, enabling the growth in height and diameter of stems of capinuri and other valuable species. Some of these operations are conducted mainly by village specialists.[1]

These three management techniques are applied in swidden-fallows every three to five years. During this management phase the majority of juveniles of capinuri and other economic species begin to form dense stands, and gradually the fallow becomes an established capinural forest.

The third phase begins when the fallow becomes a mature capinural forest. Two management techniques are commonly used during this phase in the Napo-Amazon: *anillado* and *desangrado*. Anillado involves the killing of selected stems of competitor species using girdling and fire. Its application causes the tree to die rapidly and deters resprouting from the roots or stem. Amazonians use it to kill stems of large species that are difficult to control. Desangrado is a commonly used girdling technique with which local people remove small stems of competitor tree species and individuals of stranglers and woody climbing vines. The desangrado and anillado management techniques are applied on an average of once every six to eight years.

The suite of traditional management techniques outlined above is employed to increase the value of capinural forests to local villagers. The application of these techniques also results in an increase in the commercial volume per hectare of timber in the capinurales. The mean commercial volume of managed capinurales in the Napo-Amazon floodplain region studied

---

1. Management of timber species is often a rather specialized activity. Villagers tend to vary greatly in their levels of expertise as timber managers (Padoch and Pinedo-Vásquez 1996).

was 81 m³/ha for areas that had been managed as mature (i.e., post-purma stages) capinural forests for eight years, 89 m³/ha for those managed for sixteen years, and 85 m³/ha for those managed for twenty-four years. All these values were significantly higher than the estimated mean of 54 m³/ha for the unmanaged capinural areas.

The commercial value of timber species to rural Amazonians living in many floodplain regions is significant (Pinedo-Vásquez and Padoch 1996; Pinedo-Vásquez et al. 2001). For instance, in the Amazon estuary near the city of Macapá, local sawmills are mainly supplied with timber from lands managed by small-scale farmers and forest managers (Pinedo-Vásquez et al. 2001). This pattern is becoming increasingly widespread throughout the Amazon floodplain and in its large cities. Locally produced floodplain timbers are common materials for house construction in the burgeoning shantytowns of the basin.

Invisible Practices and Threatened Knowledge

Much of the management process discussed here has doubtless gone unremarked and unrecognized by researchers because it involves subtle operations that do not easily conform to familiar categories of agricultural and forestry management. It is in the initial phases of the management process described above that the "simultaneity" or concurrence of agricultural and forestry management is most evident. As table 10.1 illustrates, the tasks of agriculture and forestry in Amazonian plots are often the very same, although the terminology and the specific purposes may differ. The "invisibility" of forestry activities in the midst of the performance of agricultural work also becomes obvious.

TABLE 10.1 Concurrent Agricultural and Forestry Activities

Agriculture	Forestry
Clearing	Opening forest canopy
Burning	Enhancing seed germination
Planting	Transplanting
Weeding	Brushing, seedling selection, and release
Harvesting	Liberation

The failure to recognize the complex nature of smallholder resource management has led to multiple misunderstandings, including critically inaccurate assessments of the sustainability and economic potential of locally developed production systems. The labor costs of the early stages of management of capinuri and other timber species in the Amazon floodplain are, for instance, impossible to disaggregate from agricultural labor costs. Smallholders get "two-for-one" benefits; such dividends are not realized by industrial timber plantations. This difference may radically alter the balance of economic costs and benefits of tropical timber production and management by large- and small-scale foresters.

Conclusions

The lack of attention given to long-term timber management by smallholders has obviously led to considerable misinterpretation about use of tropical forests by local peoples, and in turn to the formulation of some inappropriate policies. For instance, several authors have suggested that forest management for the production of nontimber plant products—assumed to be the province of smallholders—is ecologically sustainable, while forest management for timber production—assumed to be done exclusively on an industrial scale—is not sustainable in the tropics (Fearnside 1989; Anderson and Ioris 1992). Based on such generalizations, several countries are using the timber/nontimber distinction to implement policies and regulations for conservation and forest management (IBGE 1984; Allegretti 1990). For example, extraction of nontimber products by rubber tappers is promoted in Brazilian extractive reserves, while extraction of timber is prohibited (Allegretti 1990). Such inappropriate policies threaten not only the economic well-being of forest-dependent people, but also the continued existence and development of an important and still little understood domain of local technical knowledge.

References

Allegretti, M. H. 1990. Extractive reserves: An alternative reconciling development and environmental conservation in Amazonia. In A. B. Anderson, ed., *Alternatives to Deforestation: Steps Toward Sustainable Use of the Amazon Rain Forest.* New York: Columbia University Press.

Anderson, A. B. and E. M. Ioris. 1992. The logic of extraction: Resource management and income generation by extractive producers in the Amazon estuary. In K. Redford and C. Padoch, eds., *Conservation of Neotropical Forests:Working from Traditional Resource Use*, pp.179–199. New York: Columbia University Press.

Brookfield, H. 2002. *Exploring Agrodiversity*. New York: Columbia University Press.

Chibnik, M. 1991. Quasi-ethnic groups in Amazonia. *Ethnology* 30:167–182.

Chibnik, M. 1994. *Risky Rivers: The Economics and Politics of Floodplain Farming in Amazonia*. Tucson: University of Arizona Press.

Denevan, W. M. and C. Padoch, eds. 1988. *Swidden-Fallow Agroforestry in the Peruvian Amazon*. Advances in Economic Botany 5. Bronx: New York Botanical Garden.

Dufour, D. L. 1990. Use of tropical rainforest by native Amazonians. *BioScience* 40:652–659.

Fearnside, P. N. 1989. A prescription for slowing deforestation in Amazonia. *Environment* 32(4):16–40.

IBGE. 1984. *Producao Extrativa Vegetal 10*. Rio de Janeiro: Fundacao Instituto Brasileiro de Geografia e Estatistica (IBGE).

Irvine, D. 1989. Succession management and resource distribution in an Amazonian rain forest. In D. A. Posey and W. Balée, eds., *Resource Management in Amazonia: Indigenous and Folk Strategies*, pp. 223–237. Advances in Economic Botany 7. Bronx: New York Botanical Garden.

Padoch, C. and M. Pinedo-Vásquez. 1996. Smallholder forest management: Looking beyond non-timber forest products. In M. Ruiz Perez and J. E. M. Arnold, eds., *Current Issues in Non-Timber Forest Product Research*. Bogor, Indonesia: Centre for International Forest Research (CIFOR).

Padoch, C. and W. de Jong. 1992. Diversity, variation, and change in ribereño agriculture. In K. Redford and C. Padoch, eds., *Conservation of Neotropical Forests: Working from Traditional Resource Use*, pp. 158–174. New York: Columbia University Press.

Pinedo-Vásquez, M. 1995. Human impact on varzea ecosystems in the Napo-Amazon, Peru. Ph.D. dissertation, Yale School of Forestry and Environmental Studies.

Pinedo-Vásquez, M. and C. Padoch. 1996. Managing forest remnants and forest gardens in Peru and Indonesia. In J. Schelhas and R. Greenbers, eds., *Forest Patches in Tropical Landscapes*, pp. 327–341. Washington, DC: Island Press.

Pinedo-Vásquez, M., D. J. Zarin, K. Coffey, C. Padoch, and F. Rabelo. 2001. Post-boom logging in Amazonia. *Human Ecology* 29(2):219–239.

San Roman, J. 1975. *Perfiles Históricos de la Amazonía Peruana*. Lima: Ediciones Paulinas.

Wiersum, K. F. 1997. Indigenous exploitation and management of tropical forest resources: An evolutionary continuum in forest-people interactions. *Agriculture, Ecosystems and Environment* 63:1–16.

11 Institutional and Economic Issues in the Promotion of Commercial Forest Management in Amerindian Societies

Michael Richards

Following decades of frustration with state-managed approaches such as protected areas and industrial forest concessions, conservationists and others have seen the promotion of "natural-forest management" (NFM) as one of the main hopes for conserving the tropical rain forest (e.g., Clay 1988). The term NFM, as used here, refers to a more commercial management approach than is present in more subsistence-oriented swidden agriculture and extensive natural-resource management systems. In the case of timber production, NFM commonly implies using "classical" forest management concepts like a cutting cycle, division of the forest into coupes (areas where trees are felled), use of machinery, and animal traction. Above all, the emphasis is on supplying the market and generating enough revenue to ensure financial sustainability.

Because of their social cohesion and tendency to have a longer-term perspective, indigenous groups are seen by some donors and international nongovernmental organizations (NGOs) as offering particular promise for market-based NFM. For example, Anderson (1990) states that the potential for NFM in Brazilian Amazonia and elsewhere is highest in traditional communities, including among *caboclos* (mixed-blood descendants of Indians, Europeans, and Africans; often classed as "semi-indigenous"), where the technical understanding, experience, and common-pool regimes appear ideally suited to it.

The data presented in table 11.1, though based on broad estimates from a number of different sources, suggest that about a quarter of the Amazon forest region may lie within "indigenous territories," although only about 20 percent of these are legally demarcated and recognized. Thus the policies of

TABLE 11.1 Importance of Indigenous Territories in the Amazon Region

Country	Amazon Region (km²)[1]	Indigenous Territories (km²)[1]	Indigenous Population
Bolivia	824,000	350,000	175,000
Brazil	4,982,000	800,000	216,000
Colombia	406,000	220,000	80,000
Ecuador	123,000	> 30,000	95,000
Peru	957,000	500,000	250,000
Venezuela	530,000	80,000	39,000
Others[2]	400,000	140,000	70,000
Total	8,222,000	2,120,000	925,000

[1]Exact measurements do not exist for the indigenous territories, nor is there agreement on the exact boundaries of the Amazon region. The figures here should be treated as very approximate, and are based on a range of assumptions, including (in some cases) an estimated population density of 2–4 square kilometers per person, criteria established by the various governments, and estimates from anthropological and biological studies.
[2]Paraguay, Surinam, Guyana, and French Guiana.
Source: Greenpeace International (1993) drawing on various sources.

donors and international NGOs toward Amerindian natural-resource management are very important for environmental as well as social reasons.

Market-Based Natural-Forest Management and Indigenous Institutions

The Market Approach to Forest Conservation

For the global community, the least expensive approach to achieving environmental and biodiversity objectives is to rely on the market to support sustainable forest (or other natural resource) management. Reliance on the market appears to offer cost savings to the global community, which can avoid having to pay for the real costs and value of sustainable forest management, including the cost of providing environmental services. This "let-the-market-do-the-work" approach effectively places most of the onus and costs of conservation on local communities, except to the extent that donors and governments subsidize the process through institutional and technical support to the local communities. Experience has shown that the market approach, which ignores the need to compensate the provision of environmental services by forest managers, has not been particularly effective for either local communities or national and global interests (Southgate 1998).

The proponents of "green capitalism" also cite welfare arguments to strengthen their case. With indigenous societies increasingly being sucked into the market economy owing to globalization, and experiencing severe poverty that threatens to undermine their societies, environmental and social objectives appear to conveniently coincide in market-based "sustainable forest management."

The Clash of Incentives

Anthropologists and political scientists have long studied the effects of market integration on indigenous societies in Amazonia. For example, Murphy and Steward (1956) described how acculturation and economic dependency due to the late-nineteenth-century rubber boom caused the complete transformation of these societies. Among others, Davis and Wali (1993) point out that a large body of anthropological research shows there are close links between cultural conceptions and social institutions of lowland forest dwellers and their land-use practices. The overwhelming finding is that market integration and Western values are gradually eroding indigenous "cosmological" views of the universe, and that this has major implications for natural-resource management.

Box 11.1 presents a representative example of the role of traditional institutions in natural-resource management. The problem for NFM occurs when products from the forest are equated with manufactured goods in the "market economy," thereby removing the checks on extractive practices and eroding the reciprocal "gift economy" that maintains the incentive for group cooperation at the heart of the common-pool regime (Richards 1997). As the former Head of Indian Affairs in Colombia has pointed out: "If the market economy is characterized by the exchange of goods with a commercial value and the accumulation of surplus, then the local economic model of the Indians is anti-market, since the exchange of an object has value only in terms of the relationship of reciprocity that it establishes between people" (Martin von Hildebrand quoted in Bunyard, Maresch, and Renjifo 1993).

According to Chase Smith (1995), a long-time observer of Amerindian natural-resource management, "the most fundamental" problem between the two systems is the contradiction "between the gift economy's pressure on each person to give away his wealth generously in order to enhance his

BOX 11.1
TRADITIONAL EXCHANGE SYSTEMS IN THE COLOMBIAN AMAZON

The *Tanimuca* interpret living forms as the external manifestation of "thought." Each group of animals, plants, and people needs a certain amount of thought, which emanates chiefly from the sun. The "guardians" of each group (e.g., the tapir is the guardian of wild fruit) see to it that each group has enough thought and that no one group takes more than its fair share. The guardian of humans is the *shaman* or "jaguar man." When people hunt or collect plants they must do so under the direction of the shaman so that they obtain the correct amount of thought for their group.

On the basis of his negotiations with the other guardians, the shaman tells his people where, what, and how much they may hunt or collect. Permission varies with the seasons, the reproductive cycle of the animals and plants, and the use that the hunters or gatherers make of different areas of the forest. It is believed that the thoughts of people who consume too much become visible to the guardians who will then hunt the individuals down and punish them. In this system, therefore, it is regarded as antisocial to accumulate wealth, and there is strong pressure to maintain the traditional reciprocity-based economic system by exchanging gifts.

SOURCE: Bunyard, Maresch, and Renjifo 1993.

status, and the market economy's injunction to accumulate wealth privately as a means to heighten personal status." The evidence suggests that the latter incentive is more powerful when the two systems meet, especially among younger people.

Chase Smith (1995) mentions several cases in which tribal groups have ignored ritual taboos and spiritual limits as they have responded to commercial opportunities. His research included some well-known market-oriented NFM projects among Amerindian groups in Amazonia, such as the Yanesha Forestry Cooperative in the Palcazu Valley, Peru, and the Lomerio Project among the Chiquitano Indians of lowland Bolivia. Not one of these projects proved viable. In the case of the Palcazu project, soon after it finished the farmers reverted to their previous unsustainable logging practices and cleared additional forest areas for annual crops (Southgate 1998). According to Chase Smith, donors' insistence on collective action by communities has resulted in a confounding mixture of the influences of the market system and traditional gift economy institutions. As younger people, in particular, have moved away from tradition and embraced more individualistic market economy values, the result has been a "confusion and ambiguity among community members over access to resources, usufruct rights and property rights" (1995). Donations, where they have been made, have not carried with them the necessary reciprocal obligations and have further deepened the ambiguities and eroded the dignity of indigenous groups. Chapin (1991), another anthropologist, noted how market incentives caused younger Kuna in Panama to question— or simply to fail to learn—the old beliefs that made up the Kuna "conservation ethic."

The Lomerio Project, involving community NFM and timber processing among the Chiquitano in the Bolivian Amazon, provides a good example of the clash of institutions. This project was set up in the early 1980s and received considerable outside support, but there have always been problems with the administration of the sawmill (pers. comm. with project staff). At one point the project was in danger of closing because the indigenous sawmill administrator had made so many loans from the sawmill income to members of the community—a clear response to the reciprocal pressures of the gift economy. Finally, Corry (1993) claims that the Body Shop's initiative involving Brazil nut oil and the Kayapó has led to divisions that threaten to "rupture" the community, although this viewpoint has been contested by Roddick (1993).

The Economics of Market-Based Natural-Forest Management

A further problem for those keen to promote more market-oriented production is the economics of NFM. As discussed in box 11.2, there is no firm evidence, even in an industrial forest management context (whether on state or private forest land), that NFM is a commercially viable or attractive land use, except possibly in some of the higher value and less diverse oligarchic forests, as for example in some *várzea* (seasonally flooded forest) areas.

Apart from the biological, market, and policy factors that drive the incentives for NFM, a major determinant of viability and economic sustainability is the discount rate of forest-resource decision-makers. The discount rate is highly dependent on the level of risk associated with the market, that is, the institutional and tenurial uncertainties of any particular activity. Unfortunately, because they tend to lack experience with the market and new organizational and administrative procedures, as well as access to market and technical information (including information on new technologies), indigenous or other local users are usually at a higher risk than commercial operators. That understandably causes indigenous (or other) peoples to have less faith in future, as opposed to present, forest values, and thus reduces their belief in the economic viability of NFM. Traditional societies also suffer from the high transaction costs involved in endless meetings and negotiation with

BOX 11.2
IS "SUSTAINABLE" NATURAL-FOREST MANAGEMENT VIABLE?

Leslie (1987) pointed out over a decade ago that returns to forest management are likely to be low if only the forest product values are included, mainly because of the combined "cost" of time (resulting from slow natural growth) and high interest or discount rates (associated with high risk). This combination results in what Schneider (1993) calls *imediatismo* (literally, short-termism), and encourages forest mining as opposed to sustainable management. Another problem has been the slow growth in forest product prices. Southgate (1998) points out that timber prices are depressed due to the abundant supply of timber, much of it illegal, from unmanaged natural forests. Thus one study concludes that "taking these three factors together—tree growth, [forest product] price growth, and interest rates—most studies have found that there is no financial incentive for a logger to engage in [sustainable] natural forest management in the tropics" (Reid and Rice 1997). Repetto and Gillis (1988) also describe how market and policy failures, especially those associated with macroeconomic policies, have caused market distortions that favor alternative land uses.

Several studies—most famously that of Peters, Gentry, and Mendelsohn (1989)—purport to show the viability of sustainable NTFP extraction. Their calculations, which were desk-based or theoretical estimates rather than recorded production levels, gave rise to criticism (for example, by Pinedo-Vázquez, Zarin, and Jipp 1992) for various invalid assumptions about market size, transport and marketing efficiency, tenure security, and resource sustainability. There has also been confusion in various studies, including that of Peters, Gentry, and Mendelsohn (1989), between stock (what's in the forest) and flow (what comes out of it) values, and economic results have not been expressed in terms of the decision maker's limiting factor: production—which is usually family labor and/or capital. By contrast, studies based on observed or recorded results, like that of Pinedo-Vázquez, Zarin, and Jipp (1992), show that actual profitability is a fraction of that calculated in the theoretical estimates, and that alternative land uses are generally much more attractive to farmers or decision makers. It is not only economic conditions that have to be right for NFM to work. The necessary conditions for NFM include, among other things, land and tree tenure security and access to technical and economic information on NFM (Poore et al. 1989). Finally, Southgate (1998) points out that NTFP extraction is hardly a high-welfare option.

outsiders. Such costs have rarely been taken into account when assessing whether an NFM project should proceed (Byron 1991).

There has been considerable hope that timber certification would provide the necessary incentive for sustainable NFM. Several indigenous groups in Latin America, for example in Mexico and Guatemala, now have a "green seal" for their forest products, but so far they have found it difficult to meet the continuity and quality demands of timber importers. There is also much optimism about extractive reserves in Brazil, and there are strong social and environmental arguments for continuing to support them. However, here again there are some important economic problems (see box 11.3).

The Fragile Nature of Forest-Product Markets

Another problem for market-based NFM is vulnerability to market fluctuations, particularly for nontimber forest products (NTFPs). Homma (1992) points out the effect of markets that demand large quantities of uniform quality forest products and require a rapid initial supply response. The effects can be particularly severe when the product is destructively harvested, as in the case of such products as rosewood oil, and (most) palm-heart and fauna extraction. Many export-oriented NTFPs in Amazonia have been subject to boom-and-bust cycles. As the supplies of natural products have proved unable to meet longer-term market requirements following a phase of rapid and unsustainable harvesting, the tendency has been for synthetic or planted (domesticated) products to replace the NTFPs (Homma 1992; Richards 1993). For example, the Kaxinawa Rubber Tapper Association in the state of Acre (Brazil) used to produce some 55,000 pounds of rubber for sale to local tire companies. However, Brazilian companies switched to synthetic rubber, the natural rubber subsidy was removed, and the cooperative's future became uncertain (Irvine, Kosek, and Olson 1993). In addition, markets tend to encourage greater livelihood specialization. It is widely recognized that

---

BOX 11.3
EXTRACTIVE RESERVES

Browder (1992) doubted the economic viability of extractive reserves, pointing out that low prices for extractive NTFPs are forcing extractivists into less benign forms of land use. To this can be added the problems of how to organize efficient marketing, given the physical isolation of individual rubber tappers, for example, and a range of marketing problems for NTFPs (discussed in Richards 1993). Allegretti (1994) also found that "these reserves are not working as well as they could. Few have organized marketing cooperatives to provide economic support to their members. To survive, forest dwellers are either leaving their holdings or increasing their area of slash and burn plots." But recent legislation has provided a boost for rubber tappers in the state of Acre: the Chico Mendes Law, passed in 1999, subsidizes rubber production in recognition of the environmental services provided by the rubber tappers' role in forest conservation (Rosa, Kandel, and Dimas 2003).

the key to sustainable management is integrated multiple-product management and the broad livelihood base that maintains forest societies (Anderson
1990). Therefore any narrowing of the livelihood basis can have serious consequences.

Alternative Approaches to Indigenous Natural-Resource Management

The above arguments imply that more consideration needs to be given
to an "alternative incentives" approach to indigenous natural-resource management, thereby achieving conservation objectives. Von Hildebrand (in
Bunyard, Maresch, and Renjifo 1993) believes that conservation efforts in
indigenous areas can be better promoted by an exchange system more in
tune with indigenous reciprocal logic: commitment to conservation (for
international and national beneficiaries) in exchange for territorial rights,
effective defense of these rights against colonists and other vested interest
groups, and social and (some) financial assistance. Other commentators like
Davis and Wali (1993) and Redford and Stearman (1993) also support a new
relationship with indigenous peoples based on legal, scientific, and financial
support in exchange for a commitment to biodiversity conservation.

A second observation is that progress in the legal and policy areas is more
important than an approach to indigenous natural-resource management
that relies mainly on technical or management changes, although clearly the
two approaches are not mutually exclusive. There is considerable evidence
that policies such as those favoring the land claims of colonists and other
stakeholders have generally been inimical to indigenous natural-resource
management. An example of this was the creation of the Roraima National
Forest, which was followed by the erosion of Yanomami land rights and the
invasion of Yanomami territory by *garimpeiros* (mainly nonindigenous gold
extractors) (Hecht and Cockburn 1990). The main policy priorities are for
demarcation, including participatory delimitation, and protection of indigenous land rights following the legal recognition of the autonomy of indigenous political and social institutions (Colchester 1994). This is in line with
the "indigenous territory model" of land tenure promoted by the Coordinating Body of Indigenous Organizations of the Amazon Basin (COICA) and
others (Davis and Wali 1993). This general approach has been followed in a
number of cases in the Amazon region, including

- development of the Awa Ethnic Forest Reserve on the Ecuador-
  Colombia border, in which the Awa delineated their own boundaries;

- the proposal to establish a million-hectare communal reserve for the Ashanika peoples of the Pichis Valley in Peru, for which usufruct rights were to be granted in exchange for protecting the flora and fauna of the region; and
- the Colombian government's policies of (1) recognizing indigenous land rights, following successful collaboration between indigenous communities and scientists of the Puerto Rastrojo Foundation to protect endangered species; and (2) legitimization of the *resguardos* (the traditional Amerindian political authorities) with their associated community councils, effectively opening the door to self-government (Davis and Wali 1993).

In spite of the overall direction of this paper, it is recognized that, in situations where market integration is a *fait accompli*, there is an obvious need to respond to market forces. Although longer-term approaches are being developed, there is an immediate need for the involvement of stronger grassroots institutions in the processing and marketing of forest and farm products. May (1991) gives several examples where this has been done with some success in Amazonia, for example at the Center for Indigenous Research created by the Union of Indian Nations at the University of Goiás. Emerging alliances between grassroots organizations, universities, research institutes, and NGOs need to be fostered, as do initiatives by progressive elements of the private sector to establish more secure marketing channels.

Whatever the approach adopted, any intervention must be based on a possibly lengthy process of consultation with indigenous nations and on a careful and participatory analysis of the likely impacts of interventions on the underlying social institutions. There appears to be a strong argument in many circumstances for leaving traditional natural-resource management systems alone (or for helping to protect them from the effects of market forces), and for focusing market-based support on those parts of the livelihood system, like agricultural and handicraft production, which have become individualized.

Conclusions

This paper presents a number of fundamental flaws in the case for market-oriented NFM among indigenous societies in Amazonia. The first of these is that there is likely to be a clash of market and institutional incentives whenever there has been limited prior exposure to market forces. The second problem is the long-term economic viability of NFM caused particularly by high risk and discount rates. These two problems interact because the conflict of incentives tends to weaken the underlying institutional basis, which in turn increases risk and discount rates. Finally, market-based logic that rests on the perception that a forest will not be conserved unless it is given value (by the market) overlooks the possibility that nonmarket values provide the stronger conservation incentive for indigenous peoples.

It is argued here that "contract exchange approaches" more in tune with indigenous reciprocal logic should be promoted. These approaches are based on recognition of the environmental externalities to national and international stakeholders provided by sustainable natural-resource management, and represent a more just and equitable relationship between the various interested parties in rainforest conservation. Contract exchange approaches also have the advantage of not interfering with the traditional ecological knowledge that forms the basis of indigenous management systems. However North-based economists regard such ideas with great suspicion because they raise doubts about the market system, run counter to neoliberal demands for resource privatization, imply unpalatable international transfer payments, and raise the specter of increased socioeconomic dependency of indigenous peoples on the international community.

Policy and laws that support indigenous societies, land rights, and institutions—in particular, legitimization of the political autonomy of indigenous governing systems and decision-making structures—should also take precedence over more technically based approaches. As Marcus Colchester (1994) has argued, "Reconciling indigenous self-determination with conservation objectives is possible if conservation agencies cede power to those who are presently marginalized by current development and conservation models." In the final outcome, it is probably the case that a combination of market and nonmarket approaches is necessary to support indigenous natural-resource management, but the argument here is that any market-based interventions should only occur after considerable consultation and research into the likely impacts on the underlying institutions and incentives.

Acknowledgments

This paper stems from a research grant from the U.K. DFID (Department for International Development) to investigate institutional change in the forest sector in Latin America. It has benefited from comments or information received by Wendel Trio of the European Alliance with Indigenous Peoples,

David Brown of the Overseas Development Institute, and David Kaimowitz of CIFOR (Centre for International Forest Research). However the opinions expressed and responsibility for any errors are entirely the author's.

References

Allegretti, L. F. 1994. The unpaved road from the Rio Summit: Brazilian Amazon's extractive reserves. *Tropical Resources Institute News* (Spring). New Haven: Yale University.

Anderson, A., ed. 1990. *Alternatives to Deforestation: Steps Toward Sustainable Use of the Amazon Rain Forest.* New York: Columbia University Press.

Browder, J. 1992. The limits of extractivism: Tropical forest strategies beyond extractive reserves. *BioScience* 42(3):174–182.

Bunyard, P., S. Maresch, and J. Renjifo. 1993. *New Responsibilities: The Indigenous Peoples of the Colombian Amazon.* Withiel, UK: Ecological Press.

Byron, N. 1991. Cost benefit analysis and community forestry projects. In D. Gilmour and R. Fisher, eds., *Villagers, Forests and Foresters,* pp.163–180. Kathmandu: Sahayogi Press.

Chapin, M. 1991. Losing the way of the Great Father. *New Scientist* 10:40–44.

Chase Smith, R. 1995. The gift that wounds: Charity, the gift economy and social solidarity in indigenous Amazonia. Paper presented at Symposium on Community Forest Management and Sustainability in the Americas, University of Wisconsin, Madison, February 3–4.

Clay, J. 1988. Indigenous peoples and tropical forests: Models of land use and management from Latin America. *Cultural Survival Report,* no. 27.

Colchester, M. 1994. Salvaging nature, indigenous peoples, protected areas and biodiversity conservation. *UNRISD Discussion Paper,* no. 55.

Corry, S. 1993. The rainforest harvest: Who reaps the benefit? *The Ecologist* 23(4):148–153.

Davis, S. and A. Wali. 1993. Indigenous territories and tropical forest management in Latin America. *Policy Research Working Papers Series 1100.* Washington, DC: World Bank Environmental Assessments and Programs Division.

Greenpeace International. 1993. *Amazon Region and Indigenous Peoples' Territories: Basic Information.* Working document.

Hecht, S. and A. Cockburn. 1990. *The Fate of the Forest: Developers, Destroyers and Defenders of the Amazon.* London: Penguin.

Homma, A. K. O. 1992. The dynamics of extractivism in Amazonia: A historical perspective. In D. C. Nepstad and D. Schwartzman, eds., *Non-Timber Products from Tropical Forests: Evaluation of a Conservation and Development Strategy,* pp. 23–31. Advances in Economic Botany 9. Bronx: New York Botanical Garden.

Irvine, D., J. Kosek, and J. Olson. 1993. Indigenous forestry in the Americas: A roster. *Cultural Survival Quarterly* 17(1): 60–63.

Leslie, A. J. 1987. A second look at the economics of natural management systems in tropical moist forests. *Unasylva* 39:46–58.

May, P. 1991. Building institutions and markets for non-wood forest products from the Brazilian Amazon. *Unasylva* 42:9–16.

Murphy, R. and J. Steward. 1956. Tappers and trappers: Parallel processes in acculturation. *Economic Development and Culture Change* 4:335–353.

Peters, C., A. Gentry, and R. Mendelsohn. 1989. Valuation of an Amazon rainforest. *Nature* 339:655–656.

Pinedo-Vásquez, M., D. Zarin, and P. Jipp. 1992. Economic returns from forest conversion in the Peruvian Amazon. *Ecological Economics* 6:163–173.

Poore, D., P. Burgess, J. Palmer, S. Rietbergen, and T. Synott. 1989. *No Timber Without Trees: Sustainability in the Tropical Rainforest.* London: Earthscan Publications.

Redford, K. H. and A. Stearman. 1993. Forest dwelling native Amazonians and the conservation of biodiversity: Interests in common or in collision. *Conservation Biology* 7(2):248–255.

Reid, J. and R. Rice. 1997. Assessing natural forest management as a tool for tropical forest conservation. *Ambio* 26(6):382–386.

Repetto, R. and M. Gillis, eds. 1988. *Public Policies and the Misuse of Forest Resources.* Cambridge: Cambridge University Press.

Richards, M. 1993. The potential of non-timber forest products in sustainable natural forest management in Amazonia. *Commonwealth Forestry Review* 72(1):21–27.

Richards, M. 1997. Common property resource institutions and forest management in Latin America. *Development and Change* 28(1):95–117.

Roddick, T. G. 1993. The Body Shop replies. *The Ecologist* 23(5):198–200.

Rosa, H., S. Kandel, and L. Dimas. 2003. Compensation for environmental services and rural communities: Lessons from the Americas and key issues for strengthening community strategies. San Salvador: Programa Salvadoreño de Investigación sobre Desarrollo y Medio Ambiente (PRISMA). Available at PRISMA Web-site, http://www.prisma.org.sv/pubs/CES_RC_En.pdf (accessed on January 4, 2005).

Schneider, R. 1993. Land abandonment, property rights and agricultural sustainability in the Amazon. LATEN Dissemination Note 3. Washington, DC: World Bank Latin America Technical Department, Environmental Division.

Southgate, D. 1998. *Tropical Forest Conservation: An Economic Assessment of the Alternatives for Latin America.* Oxford: Oxford University Press.

12   Collect or Cultivate—A Conundrum

Comparative Population Ecology of Ipecac (*Carapichea ipecacuanha* (Brot.) L. Andersson), a Neotropical Understory Herb

Jan Salick

Domestication (e.g., Smith 1995) and the seed (e.g., Heiser 1973) have been considered cornerstones of "civilization." With the ascendancy of molecular biology, it seems absurd that a single gene change (i.e., the definition of domestication, Simpson and Ogorzaly 2000) should represent a watershed for plant/people interactions. For one thing, we know that single gene changes come about from all manner of plant/animal interactions, and for another, genetic change or no, people interact with plants in a multitude of very sophisticated ways. Root and tuber crops, not the seed, have been the staples in the Amazon since the beginning of agriculture. Vegetative reproduction (or cloning) rather than sexual reproduction by seed is disproportionately represented in crop plants (Salick 1992a). This may lead one to wonder: is civilization a product or a process?

Darrell Posey (1984, 1992; Posey and Balée 1989) encouraged ethnoecologists to focus more clearly on the multiple processes of plant/people interactions in the Amazon. Traditional peoples manage forests, fields, and water resources with practices that are being applied constructively to conservation and development (Redford and Padoch 1992; Redford and Mansour 1996). Cultivation and cloning are two traditional management techniques less well studied.

Cultivation—the tending of undomesticated crops—is a traditional management practice often neglected in our impatience to get on with the important business of domestication. What is the need to document a bit of scrabbling in the soil with "wild" plants, especially if no seed is planted? Vegetative reproduction likewise is ignored. A stick stuck in the soil seems a trivial business, not warranting the words "to plant."

Nontimber forest products (NTFPs), however, have brought cultivation into focus. Here arises the conundrum. As we try to develop NTFPs as a component of natural-forest management (Anderson 1990; Nepstad and Schwartzman 1992; Panayotou and Ashton 1992; Plotkin and Famolare 1992; Salick 1992b, 1992c, 1995a), again and again we find people taking these plants of value out of the forest and into cultivation rather than conserving the forest in which they grow (Browder 1992; Clement 1993). Why do people switch from collecting NTFPs to cultivating? Often the switch can be tied to the cycle of booms, busts, and rebirths of economic value. Ipecac (Rubiaceae: *Carapichea ipecacuanha* (Brot.) L. Andersson) is one example: cultivated for international markets (Gomes 1801; Rajkhowa 1969a, 1969b) that collapsed upon synthesizing of emetine (Fisher 1973), ipecac went "wild" again until recent demand for the more effective natural product launched recultivation in both the Amazon and Central America. To understand why people switch from collection to cultivation, I apply ecological ethnobotany (Salick 1995b) to the conundrum, and experimentally and quantitatively compare cultivation with collection. I also compare reproduction from seed with vegetative reproduction to understand why people propagate crops vegetatively.

Population ecology has been used to investigate sustainable harvests of NTFPs (Peters 1990; Charron and Gagnon 1991; Nault and Gagnon 1993). Using basic experimental design from population ecology, I tested the sustainability of ipecac harvest, and compared growth and reproduction under cultivation with that of wild harvest, and cultivation by seed with that of vegetative material. All these management options were the same as those commonly used by local producers.

I began this research by looking for a nontimber forest product that is cultivated but not domesticated, and that is reproduced both sexually and vegetatively. I also wanted a plant that would provide research results that farmers and foresters could apply to their resource management. Ipecac fit all these requirements admirably. Ipecac is promoted for both conservation and development (Jimenez 1964; Castrillo and Querol 1990; Akerele, Heywood, and Singe 1991; Centro Nacional de Medicina Popular Tradicional 1992; Nepstad and Schwartzman 1992; Balick, Elisabetsky, and Laird 1996). Its collection as an NTFP in extractive reserves and other conservation areas will allow people to participate in conservation efforts and will give value to conserving the tropical rain forest (Peters, Gentry, and Mendelsohn 1989; Allegretti 1990; Anderson, May, and Balick 1991; Balick and Mendelsohn 1992; Plotkin and Famolare 1992; Redford and Padoch 1992; Mendelsohn and Balick 1995; Redford and Mansour 1996). Integrated with natural-forest management or cultivated in agroforestry, ipecac can give extra value by increasing forest or farm income and by encouraging these types of natural-resource management (UCA/CATIE/SAREC 1991; Panayotou and Ashton 1992; Salick 1992b, 1992c, 1995a). The experiments discussed here are designed to evaluate levels of sustainable harvest and techniques for best cultivation, and incorporate variables requested by local producers (Chambers, Pacey, and Thrupp 1989). I also set out to determine if there is an inherent discrepancy between developing nontimber forest products for income generation (for which cultivation is a real option) and for conservation (for which collection from natural forests best maintains biodiversity and habitat).

Thus the objectives of this study are multiple: to address theoretical questions about the importance of cultivation and cloning, to test methods and analyses for ecological ethnobotany, to answer applied questions on production and sustainable harvest, and to plumb the potentials for natural-resource management of the collection and cultivation of a nontimber forest product. All of these objectives were approached simultaneously with a field experimental design comparing cultivated with forest-grown ipecac, and seed with vegetative propagation of ipecac. This integration of ethnobotanical objectives through multivariate ecological experimentation is, to me, the essence and power of ecological ethnobotany (Salick 1995b).

Methods

Ipecac is an understory herb native to neotropical lowland rain forests, including those in Amazonia and Central America (figure 12.1a and 12.1b). All planting stock for these experiments was collected from natural forests in the Atlantic coastal lowlands of Costa Rica and transplanted to holding beds for one year. Then seeds from the transplants and vegetative material were transported to Turrialba (Costa Rica). There was no genetic change or domestication to confound the simple effects of cultivation.

The experimental portion of this research took place at the Tropical Agricultural Center for Research and Teaching (CATIE) in Turrialba; field interviews took place in various parts of the Amazon Basin and Central America. At CATIE the vegetation is tropical wet forest that receives 2,600 millimeters of precipitation per year, lies at 600 meters elevation, and grows in Inceptisols of volcanic origin. It was in these conditions that the experimental blocks of forest-grown ipecac were planted. The plots used for "cultivation" were prepared so as to simulate the conditions used by local producers: the soils were well mixed with sand and organic matter with no fertilizers added; shade was

FIGURE 12.1 Ipecac (Rubiaceae: *Carapichea ipecacuanha* (Brot.) L.Andersson): (a) habit: a neotropical understory herb; (b) rhizomes from which the powerful emetic is extracted; (c) forest-grown plants were consistently smaller with less rhizome production; (d) cultivated and vegetatively reproduced ipecac were larger, with larger rhizomes, flowers, and fruits.

provided artificially with shade cloth (50 percent); and mist irrigation was administered twice daily for 15 minutes at a time. The blocks were weeded by hand and sprayed as needed with fungicides (Benlate and copper; Laguna, Gómez, and Gómez 1992)—the same fungicides commonly used for coffee, which, like ipecac, is in the Rubiaceae family; however, leafcutter ant attacks could not be controlled in either the experimental plots or in farmers' fields. In contrast to the cultivated plots, the forest plots never experienced drought (because of the moist understory conditions), weeds (because of the heavy leaf litter), or the pathogens or insects that usually accompany agriculture, and they received more than 50 percent shade. Thus, cultivation is very multivariate.

The experimental design—to test the differences between cultivated and forest-grown ipecac, between seed and vegetatively reproduced ipecac, and among types of vegetatively reproduced ipecac—was as follows: For the vegetative sample, three independent blocks of plantings were cultivated and three identical independent blocks were planted and grown without cultivation in natural forest. Each block held nine rows of eight plants surrounded by a border of unmeasured plants. The nine rows consisted of three repetitions of three randomly placed vegetative treatments: ipecac grown from roots, stems, or growing tips (three leaves dibbled directly into the soil). After one, one-and-a-half, and two years, the plants were measured and one block was harvested from each site (cultivation and forest).

Seeds could not be planted within the same blocks as vegetative material because they grew much more slowly and could not compete. Therefore seed germination was tested separately in trays containing porous, well-watered sand. One hundred seeds were planted in each of four treatments (three under cultivation and the fourth in the forest), with each treatment having four repetitions. The three treatments under cultivation included: (1) soaked in water for 24 hours, (2) seed coat minutely clipped (as with coffee), and (3) seeds not treated (control). The fourth treatment was another control group (seeds not treated) planted directly into forest soil. Seeds germinated slowly, beginning after six months, and were monitored monthly to 11 months when the final percentages of germination were determined. At one year, seedlings were transplanted into their own fourth experimental blocks. The seedlings were planted in two blocks of 9×8 plants plus borders, in methods identical to the vegetatively reproduced seedlings; one block was kept under cultivation and one was planted in the forest.

Variables for each plant reproduced from seed and measured after one, one-and-a-half, and two years included mortality, plant height, leaf area, biomass (dry weight), and flowering. Because of high mortality and low growth rates, seedlings were not harvested for biomass measurements until the end of the two-year experiment.

Statistical testing was done by analyses of variance with orthogonal contrasts where data were normal or where transformations (log or arcsine) provided normal data. For data with too many zeros to normalize (i.e., treatments with plants that grew very poorly) nonparametric Kruskal-Wallis analyses of variance (one-way) or Mann-Whitney U statistics (two-way) were used.

Results

Results are shown for seed germination trials; height, leaf area, and mortality; flowering; and biomass and yield.

• *Seed Germination Trials.* Simulated "wild" seeds (W) germinated only at 11 months, whereas cultivated treatments began germinating at 6 months and continued through 11 months (figure 12.2a). At 11 months there were no statistically different germination rates among treatments except for the clipping treatment ("cut"), which showed lower germination than the other treatments ($p < 0.026$); although cutting is used to increase coffee germination, it does not help ipecac (figure 12.2b). Forest germination, although delayed, was not significantly different.

• *Height, Leaf Area, and Mortality.* Over time, these variables showed many similar responses across treatments (figure 12.3). Under cultivation, ipecac grown from seeds grew significantly slower than vegetative material ($p < 0.0001$) in both height and leaf area. Mortality of ipecac from seed was initially high (1 year, $p < 0.02$), then none (1.5 years, $p < 0.0001$), and then equivalent to vegetative material (2 years, $p < 0.38$). Mortality was lowest among ipecac vegetatively reproduced under cultivation ($p < 0.0001$). Among the different types of vegetative reproduction under cultivation, ipecac grown from stems were significantly taller (2 years, $p < 0.0001$) and had more leaf area (2 years, $p < 0.0001$) than ipecac grown from roots or growth tips.

Forest-grown ipecac (figure 12.1c) could not be statistically compared with cultivated ipecac, but were consistently smaller (shorter and with less leaf area), and had higher mortality (figure 12.3). In the forest, only ipecac grown vegetatively from roots showed significantly lower mortality (all stages, $p < 0.0001$) than ipecac under other forest-grown treatments. In the forest, ipecac from seed did not survive beyond germination.

• *Flowering.* Flowers were found only among the vegetatively reproduced ipecac under cultivation (figure 12.1d). Among these (figure 12.4), ipecac flowered significantly more ($p < 0.0001$) when propagated from stems and roots than from growth tips. Ipecac grown from seed never flowered after two years under any treatment.

• *Biomass and Yield.* For all plant parts (including roots), biomass and yield were greatest for vegetative material under cultivation (figures 12.1b and 12.5; leaf $p < 0.0001$, woody stem $p < 0.003$, green stem $p < 0.0001$, root $p < 0.001$). There was at least an order of magnitude difference in overall plant biomass and in root yield between cultivated and forest-grown ipecac

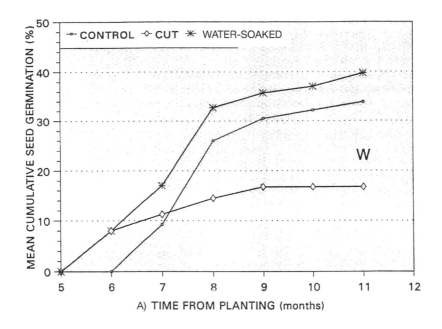

A) TIME FROM PLANTING (months)

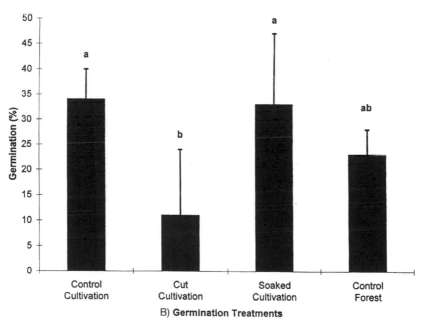

B) Germination Treatments

FIGURE 12.2 Seed germination trials comparing ipecac seed treatments: control, cut, and water-soaked under cultivation, and control in forest or "wild" (W) simulated. (a) Cumulative germination (%) over time to 11 months. (b) Statistical comparison of treatments at 11 months: means (bars), standard deviations (whiskers), and results of orthogonal contrasts (letters) following significant ANOVA (p < 0.026).

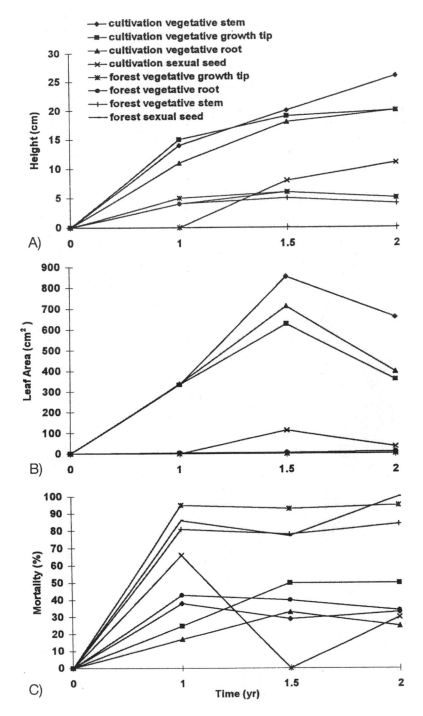

FIGURE 12.3 (a) Height, (b) leaf area, and (c) mortality showed many similar treatment responses over time.

TABLE 12.1 Growth and Life-History Characteristics of Ipecac

| | Management | | Reproduction | |
	Cultivation	Forest-Grown	Vegetative	Seed
Seed Germination	+	–	+	–
Height	+	–	+	–
Leaf Area	+	–	+	–
Flowering	+	–	+	–
Biomass	+	–	+	–
Root Yield	+	–	+	–
Mortality	Steady	High	Steady	Initially high

(figures 12.5 and 12.6; note differences in scales). For cultivated ipecac, root yield leveled off after one-and-a-half years for plants grown from stems ($p < 0.001$), and continued to increase up through two years for plants grown from roots ($p < 0.002$).

These experimental results on ipecac population ecology can be summarized most simply with positive and negative comparisons among the four main treatments: cultivation versus forest management, and vegetative versus sexual reproduction (table 12.1).

Discussion

Small wonder that people begin to cultivate nontimber forest products when it is worth their while. No surprise that vegetative reproduction is selected for among agricultural plants. These techniques—cultivation and cloning—readily increase production and reproduction of crops. The conundrum is solved. The results, at least for ipecac, are so startlingly signifi-

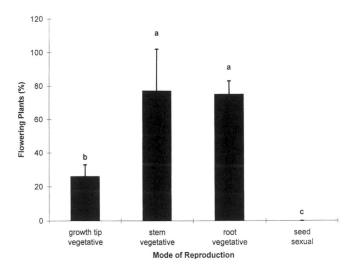

FIGURE12.4 Flowering after two years was only found among the vegetatively reproduced ipecac under cultivation. Statistical comparisons: means (bars), standard deviations (whiskers), and results of orthogonal contrasts (letters) following significant ANOVA (p < 0.0001).

cant that one wonders again: why is so little attention paid to cultivation and vegetative reproduction with regard to natural-resource management and the evolution of agriculture? On the basis of this research, let me propose that cultivation of vegetatively propagated roots and stems — rather than the seed and domestication — catalyzed agricultural evolution and led to civilization. This was most likely the case in the Neotropics (Piperno et al. 2000) but also in other tropical regions (e.g., Denham et al. 2003).

Today, no less than before the advent of domestication, cultivation and cloning seem to be the garden path along which we continue with modern agriculture. However we might note the warning signals provided by experiences with ipecac, as with other crops. Immediately upon cultivation in these experiments, ipecac — only one generation removed from the forest — experienced pest, pathogen (Laguna, Gómez, and Gómez 1992), and weed pressures. That there was no extended period of pest buildup may be due to ipecac's close relationship to ubiquitous coffee, from which pests easily transfer to ipecac, and to ipecac's susceptibility to generalist pests such as leafcutter ants and weeds. Nonetheless, intensive crop cultivation and cloning foster concomitant pest pressures.

Sustainable harvests of NTFPs sometimes can be estimated using population ecology (Peters 1990) and sometimes cannot (Nault and Gagnon 1993). If growth and reproduction rates are very low or if mortality and collection rates are very high, harvesting NTFPs may adversely affect plant populations. With ipecac, there is little indication of sustainable harvest from natural forest. These experiments demonstrate the negative factors affecting ipecac populations: forest-grown ipecac had very low growth rates and reproduction along with high mortality. In addition, the collection of wild-harvested ipecac that I have observed is pure plunder, where every plant encountered is taken without replanting. Collectors have told me about methods by which only large roots are harvested, small roots are left intact, and stems and tops are replanted; but I have never personally observed these practices. At present the only hope for sustainable harvests from wild populations of ipecac is in the cycling of market prices, which have previously had long periods of depression that allow ipecac populations to recover from periodic overharvesting. On the other hand, data show that sustainable harvest of cultivated ipecac could occur regularly at approximately two-year intervals, adding another reason for cultivation over wild-harvesting.

Management of ipecac teaches us a lesson for conservation as well. If NTFPs are to be used for development, then cultivation may be an option; however, if NTFPs are to be used for conservation of natural forests or other habitats, then buyers beware. Giving value to single products in a natural forest (e.g., Clay 1992) will more likely encourage producers to cultivate the product rather than to conserve forest (Browder 1992; Clement 1993). For conservation purposes, our strategies must give value to the forest as a whole. Management of the whole vegetation community of timber and NTFPs—integrated natural-forest management—must be our focus in order to avoid cultivation as the obvious solution. The same is true for timber: when markets for individual timbers are profitable, then tree plantations become the obvious choice over natural-forest management. For conservation purposes, integrated natural-forest management must profit from the biodiversity it provides rather than from individual products.

Thus traditional human management skills in cultivation and cloning have affected the Amazon and continue to have important roles in both development and conservation. Simple cultivation and cloning techniques should no longer be disregarded in our expanding appreciation of astute processes of traditional natural-resource management.

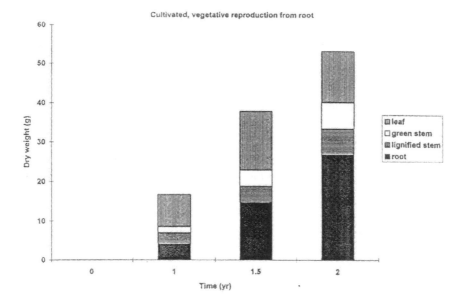

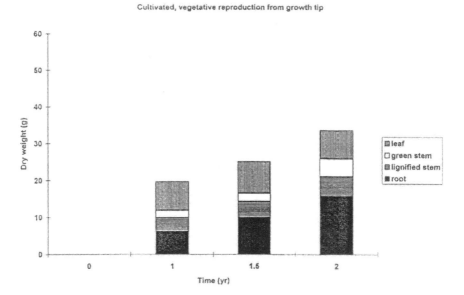

FIGURE 12.5 Biomass and yield of cultivated ipecac increased over time for all plant parts. Biomass is compartmentalized into leaf (vertical hatching), green stem (white), lignified stem (horizontal hatching), and root (black).

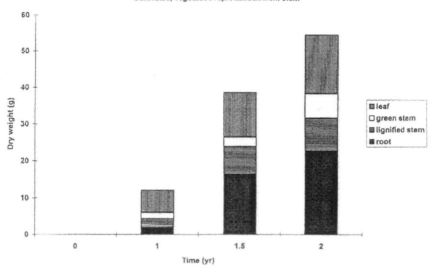

Cultivated, vegetative reproduction from stem

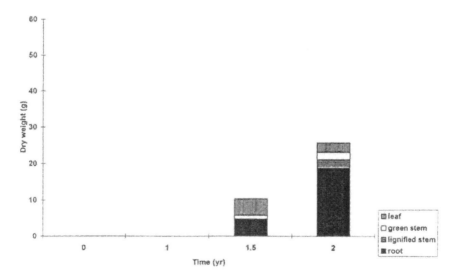

Cultivated, sexual reproduction from seed

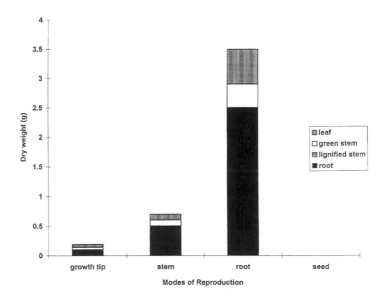

FIGURE 12.6 Biomass and yield of forest-grown ipecac after two years was an order of magnitude less than of cultivated ipecac (compare figures 12.5 and 12.6; note difference in scales). Biomass is compartmentalized into leaf (vertical hatching), green stem (white), lignified stem (horizontal hatching), and root (black).

Acknowledgments

I gratefully acknowledge funding and support from the Swedish Government (SAREC), CATIE, and Centro International de Agricultura Tropical (CIAT).

References

Akerele, O., V. Heywood, H. Singe, eds. 1991. *Conservation of Medicinal Plants.* Cambridge: Cambridge University Press.

Allegretti, M. H. 1990. Extractive reserves: An alternative for reconciling development and environmental conservation in Amazonia. In A. B. Anderson, ed., *Alternatives to Deforestation: Steps Toward Alternative Use of the Amazon Rain Forest,* pp. 252–264. New York: Columbia University Press.

Anderson, A. B., ed. 1990. *Alternatives to Deforestation: Steps Toward Alternative Use of the Amazon Rain Forest.* New York: Columbia University Press.

Anderson, A. B., P. H. May, and M. J. Balick. 1991. *The Subsidy from Nature: Palm Forests, Peasantry, and Development on an Amazon Frontier*. New York: Columbia University Press.

Balick, M. J., E. Elisabetsky, and S. A. Laird, eds. 1996. *Medicinal Resources of the Tropical Forest: Biodiversity and Its Importance to Human Health*. New York: Columbia University Press.

Balick, M. J. and R. Mendelsohn. 1992. The economic value of traditional medicine from tropical rain forests. *Conservation Biology* 6:128–139.

Browder, J. O. 1992. The limits of extractivism: Tropical forest strategies beyond extractive reserves. *Bioscience* 42:174–182.

Castrillo, S. and D. Querol. 1990. *La Raicilla*. Buena Vista, Rio Sabalo y Managua: La Cooperativa de Raicilla.

Centro Nacional de Medicina Popular Tradicional. 1992. *Isnaya: Manual de plantas medicinales para el promotor de medicina preventiva y salud comunitaia*. Esteli, Nicaragua: RESCATE.

Chambers, R., A. Pacey, and L. Thrupp. 1989. *Farmer First: Farmer Innovation and Agricultural Research*. London: Intermediate Technology Publications.

Charron, C. and D. Gagnon. 1991. The demography of northern populations of *Panax quinquefolium* (American ginseng). *Journal of Ecology* 79:431–445.

Clay, J. 1992. Some general principles and strategies for developing markets in North America and Europe for non-timber forest products. In M. Plotkin and L. Famolare, eds., *Sustainable Harvest and Marketing of Rain Forest Products*, pp. 302–309. Washington, DC: Island Press.

Clement, C. R. 1993. Review of *Non-Timber Products from Tropical Forests: Evaluation of a Conservation and Development Strategy*, D. C. Nepstad and S. Schwartzman, eds. *BioScience* 43(9):644–646.

Denham, T. P., S. G. Haberle, C. Lentfer, R. Fullagar, J. Field, M. Therin, N. Porch, and B. Winsborough. 2003. Origins of agriculture at Kuk Swamp in the highlands of New Guinea. *Science* 301:189–193.

Fisher, H. H. 1973. Origin and uses of Ipecac. *Economic Botany* 27:231–234.

Gomes, B. A. 1801. *Memoria sobre Ipecacuanha fusca do Brasil*. Lisboa: Litteraria do Arco do Cego.

Heiser, C. B. 1973. *Seed to Civilization*. San Francisco: Freeman.

Jimenez, O. 1964. La Ipecacuana o "Raicilla." *Revista de Agricultura* (San José, Costa Rica), Setiembre:228–232.

Laguna M., R. Gómez, and J. R. Gómez R. 1992. Diagnostico de las principales enfermedades de la raicilla (*Cephaelis acuminata* K.) bajo regimen de cultivo. *Germoplasma* (UNA, Nicaragua) 1:49–55.

Mendelsohn, R. and M. J. Balick. 1995. The value of undiscovered pharmaceuticals in tropical forests. *Economic Botany* 49:223–228.

Nault, A. and D. Gagnon. 1993. Ramet demography of *Allium tricoccum*, a spring ephemeral, perennial forest herb. *Journal of Ecology* 81:101–119.

Nepstad, D. C. and S. Schwartzman, eds. 1992. *Non-Timber Products from Tropical Forests: Evaluation of a Conservation and Development Strategy.* Advances in Economic Botany 9. Bronx: New York Botanical Garden.

Panayotou T. and P. S. Ashton. 1992. *Not by Timber Alone: Economics and Ecology for Sustaining Tropical Forests.* Washington, DC: Island Press.

Peters, C. M. 1990. Population ecology and management of forest fruit trees in Peruvian Amazonia. In A. B. Anderson, ed., *Alternatives to Deforestation: Steps Toward Alternative Use of the Amazon Rain Forest,* pp. 86–98. New York: Columbia University Press.

Peters, C. M., A. Gentry, and R. Mendelsohn. 1989. Valuation of an Amazonian rainforest. *Nature* 339:655–656.

Piperno, D. R., A. J. Ranere, I. Holst, and P. Hansell. 2000. Starch grains reveal early root crop horticulture in the Panamanian tropical forest. *Nature* 407:894–897.

Plotkin, M. and L. Famolare, eds. 1992. *Sustainable Harvest and Marketing of Rain Forest Products.* Washington, DC: Island Press.

Posey, D. A. 1984. A preliminary report on diversified management of tropical forest by the Kayapó Indians of the Brazilian Amazon. In G. T. Prance and J. A. Kallunki, eds., *Ethnobotany in the Neotropics,* pp. 112–126. Advances in Economic Botany 1. Bronx: New York Botanical Garden.

Posey, D. A. 1992. Traditional knowledge, conservation and "the rainforest harvest." In M. Plotkin and L. Famolare, eds., *Sustainable Harvest and Marketing of Rain Forest Products,* pp. 46–50. Washington, DC: Island Press.

Posey, D. A. and W. Balée. 1989. *Resource Management in Amazonia: Indigenous and Folk Strategies.* Advances in Economic Botany 7. Bronx: New York Botanical Garden.

Rajkhowa, S. 1969a. The cultivation of Ipecac roots in Assam—I. *Indian Forester* (April):246–252.

Rajkhowa, S. 1969b. The cultivation of Ipecac roots in Assam—II. *Indian Forester* (August):568–575.

Redford, K. and C. Padoch, eds. 1992. *Conservation of Neotropical Forests: Working from Traditional Resource Use.* New York: Columbia University Press.

Redford, K. and J. A. Mansour, eds. 1996. *Traditional Peoples and Biodiversity Conservation in Large Tropical Landscapes.* Arlington, Virginia: The Nature Conservancy.

Salick, J. 1992a. Crop domestication and the evolutionary ecology of cocona (*Solanum sessiliflorum* Dunal). *Evolutionary Biology* 26: 247–285.

Salick, J. 1992b. Forest products and natural forest management. In F. R. Miller and K. L. Adam, eds., *Wise Management of Tropical Forests,* pp. 118–124. UK: Oxford Forestry Institute.

Salick, J. 1992c. The sustainable management of non-timber rain forest products in the Si-a-Paz Peace Park, Nicaragua. In M. Plotkin and L. Famolare, eds., *Sustainable Harvest and Marketing of Rain Forest Products,* pp. 118–124. Washington, DC: Island Press.

Salick, J. 1995a. Non-timber forest products integrated with natural forest management. *Ecological Applications* 5:922–954.

Salick, J. 1995b. Toward an integration of evolutionary ecology and economic botany: Personal perspectives on plant/people interactions. *Annals of the Missouri Botanical Garden* 82:68–85.

Simpson, B. B. and M. C. Ogorzaly. 2000. *Economic Botany: Plants of Our World*, 3rd ed. New York: McGraw-Hill.

Smith, B. D. 1995. *The Emergence of Agriculture*. New York: Scientific American Library.

UCA/CATIE/SAREC. 1991. *Plan operativo para el desarrollo de sistemas de manejo sostanible para el aprovechamiento de los bosques humedos tropicales de Nicaragua.* Turrialba, CR: Tropical Agriculture Center for Research and Teaching (CATIE).

13 Extractivism, Domestication, and Privatization of a Native Plant Resource

The Case of Jaborandi (*Pilocarpus microphyllus* Stapf ex Holmes) in Maranhão, Brazil

Claudio Urbano B. Pinheiro

One of the most important drugs in ophthalmology is derived from the leaves of a plant in the citrus family collected by Indians and peasants in the forests of Brazil: *jaborandi* (*Pilocarpus* species; Rutaceae–Pilocarpinae). This genus is a group of shrubs or treelets 3–7.5 meters tall, widely distributed in Brazil, and ranging from the northern state of Pará to the southern state of Rio Grande do Sul (Joseph 1967).

Jaborandi has been, for the past three decades, one of the most important commercial species of the native Brazilian flora. It is the only source of the drug *pilocarpine*, a cholinomimetic alkaloid used in ophthalmology for contracting the pupil in certain optical surgical procedures. It is also used to reduce intraocular pressure in the treatment of particular types of glaucoma. And it is a powerful stimulant of salivation and perspiration (Merck 1989); in 1994 it was approved by the U.S. Food and Drug Administration for treating postirradiation xerostomia (dry mouth) in patients with head and neck cancer (Joensuu, Bostrom, and Makkonen 1993; Miller 1993; Valdez et al. 1993; Rieke et al. 1995; Wynn 1996).

Jaborandi was first introduced to Western medicine in 1873, when Symphronio Coutinho, a Brazilian doctor born in the northeastern Brazilian state of Pernambuco, took leaf samples to France. Coutinho routinely employed jaborandi leaves in his practice of medicine as an active diaphoretic and sialagogue (Holmstedt, Wassén, and Schultes 1979); its value in ophthalmology was actually a secondary discovery. What first piqued the interest of the French physicians was its use among Brazilian Indians to promote profuse sweating and salivation. Ironically, this long-abandoned effect on the parasympathetic nervous system only recently returned to clinical use.

Plants of several families, mainly those known to have similar physiological properties, have been classed under the name jaborandi—for example, some species of *Piper* (Piperaceae), *Verbena* (Verbenaceae), and *Herpestis* (= *Bacopa*, Scrophulariaceae), and even some other rutaceous species in the genera *Esenbeckia*, *Zanthoxylum*, and *Monnieria* (Joseph 1967; Holmstedt, Wássen, and Schultes 1979). However the true jaborandi belongs to the genus *Pilocarpus* of the Rutaceae, with about 18 species described for Brazil according to Joseph (1967), or only 10 species according to Kaastra (1982).

The center of production of jaborandi leaves in Brazil is the mid-northern state of Maranhão, which accounts for approximately 95 percent of the entire national production (IBGE 1975–1992, 1993–2000). Three species are reported to occur in Maranhão—*Pilocarpus jaborandi* Holmes, *P. trachyllophus* Holmes, and *P. microphyllus* Stapf ex Holmes (Joseph 1967)—although only the last is cited by Kaastra (1982) as occurring in the state. *Pilocarpus microphyllus* is considered the "legitimate jaborandi" (Corrêa 1969)—or "Maranhão jaborandi" because of its more intensive occurrence in Maranhão—and its leaves are reputed (and proven in the laboratory) to have the highest pilocarpine content (up to 1 percent) of the three *Pilocarpus* species in Maranhão. The other two species contain pilocarpine in varying concentrations (an average of 0.5 percent), plus about 0.5 percent of the other alkaloids (isopilocarpine, jaborine, jaboridine, and pilocarpidine) combined. The leaves of *P. microphyllus* also contain 0.24 percent to 0.38 percent of an essential oil with a balsam-like smell (Merck 1983).

The German pharmaceutical company Merck & Co., Inc., has retained a three-decade monopoly on the purchase of jaborandi leaves and the production of pilocarpine in Brazil, particularly in Maranhão. The exploitation of jaborandi in Maranhão was initiated in 1968 with the installation of Merck's factory in São Luís, capital of the state. From the very beginning, it was clear to Merck that the wild supply of *Pilocarpus* leaves could be exhausted, and that the only solution was the domestication of the wild source. In 1989 the pharmaceutical company acquired a 2,250-hectare farm in the municipality of Barra do Corda, in the Pre-Amazon region of Maranhão. This was the final step in the process toward the domestication, and ultimately the privatization, of jaborandi. On this farm, experiments were carried out in order to select the best plants to provide genetic material for the cultivation of jaborandi.

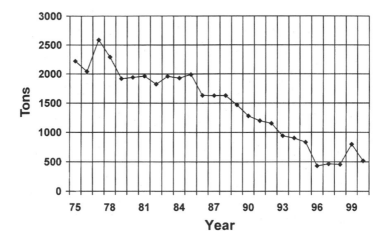

FIGURE 13.1 Production of jaborandi leaves in Maranhão (1975–2000).

The Jaborandi Research Project

The jaborandi research project was designed to study the current system of production in Maranhão, from the initial phases of leaf collection to the final industrialization. Fieldwork resulted in the botanical mapping of the population ranges of *Pilocarpus microphyllus* as well as in the characterization and description of the current systems of production practiced in Maranhão. Ecological information was also obtained, with a focus on the natural habitats of jaborandi plants in the different ecological regions where they occur. Data on production of pilocarpine was also collected (figure 13.1). Six trips to the production areas were carried out by researchers in a period of two years: three trips to the Pre-Amazon region, two to Cerrado, and one to Litoral. In total, 36 informants were interviewed (informal and semistructured interviews), including collectors, middlemen, and employees of the Merck farm.

The results from the study provided:

1. An understanding and description of the systems of production practiced by *caboclos* (i.e., the traditional inhabitants of the study regions) and indigenous groups.
2. The location and mapping of the main areas of production in the major ecological regions of Maranhão.

3. The collection of data on production for the last twenty-five years for each municipality in the state, in order to study the dynamics of the process of production and to evaluate the changes that could have occurred as a result of the exploitation of jaborandi.
4. An elucidation of many aspects related to the system of production, areas of production, and amounts produced. This was done by comparing the information gathered during the fieldwork with data obtained from secondary sources.

Production Systems

The Traditional System

No significant differences among the extraction systems practiced in the different regions of the state were observed; the process in all regions follows essentially the same steps. During the dry season (July–December), leaf collectors go into the forest in search of jaborandi plants. Once they find a group of plants, they strip the leaves from the branches by hand. Collectors assert that the plants tolerate this procedure and that new leaves sprout after the harvest, when the rainy season starts. However the excessive and frequent collection of leaves from the same plant or group of plants does not allow full development of new leaves, and according to local people, it causes a high percentage of plant mortality, considerable reduction in the height and vigor of the plants, and reduction in the size of the new leaves that sprout (Pinheiro 1997).

Merck practiced virtually the same system as extractors. In the company's system, an organization of leaf suppliers recruited men to go into the forests and to collect the leaves. Like extractors, leaf collectors recruited by the company's suppliers used the same method of manually stripping the leaves. The company asserted that harvesting was conducted only during the dry season because in the rainy season (1) penetration of the forest is more difficult; (2) drying of leaves is slower, resulting in loss of alkaloid content and higher costs; (3) labor is hard to recruit because pickers are involved in agriculture; and (4) for conservation purposes, the buying of leaves is suspended in order to allow the plants a recovery period.

In the dry season, the moisture content in the leaves is approximately 40 percent. The leaves are dried to bring the moisture content down to about 10–12 percent because higher moisture content can cause fermentation and spontaneous combustion during storage (Frazão and Pereira 1979). Most leaf collectors simply air-dry the leaves by spreading them on plastic sheets on the ground and turning them periodically. The leaves are dried under direct sunlight for about two days or until they turn a greenish cream color. At night, the leaves are kept in bags under shelters in order to avoid reabsorption of moisture from the air.

The dried leaves are cleaned by manually removing twigs and other non-leaf material, and are then packed in jute or polypropylene bags. The bags are transported, usually by animals, from the collection areas in the forests to established points of trade at the edges of the forests or within the villages. Jaborandi leaves are purchased based on dry weight. Between 1998 and 2000, buyers were paying in the range of US$2.50 to $4.00 per kilogram of leaves, depending on grading. In the final step, the harvested, dried, and packed leaves are transported by trucks to Merck's factory in Parnaíba, a small city in the neighboring state of Piauí.

Jaborandi in Cultivation

The Fazenda Chapada, the farm on which Merck's jaborandi plantations have been established, is located in the municipality of Barra do Corda, approximately 350 kilometers from São Luis. The pharmaceutical company originally (in 1990) planted 3 million shrubs on 300 hectares, and carried out its first harvest of leaves in 1993. Since then, many investments and improvements have been made in order to increase the number of plantations and maximize the means of production as well as productivity.

The area planted with jaborandi practically doubled in five years (1993–1998). During the same period, the number of cultivated jaborandi plants on the farm increased to five times the original number as the result of advances made in Merck's agricultural technology. Compared to the beginning of the project, the areas were more densely planted (from 23,000 to 48,000 plants per hectare), chemical fertilization was applied more efficiently and systematically through irrigation, and pests and nematodes were kept to reasonable levels with the use of biological controls.

By the end of 1998, roughly 15 million plants of jaborandi were planted on about 500 hectares, producing approximately 10,000 kilograms of leaves per year. The productivity on the plantations reached about 1,800 kilograms per hectare for each of 5–8 annual harvests. With an average content of 1 percent of pilocarpine in the jaborandi leaves, the farm's annual production of this alkaloid reached approximately 1,000 kilograms, which had an equivalent value of US$2 million (at a wholesale price of US$2.00 per kilogram). A local newspaper (*Jornal Estado Maranhão* 1999) published data on the production of pilocarpine by Vegetex (Merck's subsidiary, based in Parnaíba, Piauí) showing that the company's total production of pilocarpine (including the production of all leaves from extractivism in Maranhão and elsewhere) reached 9 tons annually, at a value of about US$15 million on the world market.

The System of Leaf Classification

For the purpose of setting standard prices for the jaborandi leaves harvested in Maranhão, Merck established four classes of leaves (A, B, C, and D). The most valued type (A) presents a number of desirable characteristics, such as a greenish cream color, which indicates the optimal level of dehydration (approximately 10–12 percent moisture content); a large size; and a high level of pilocarpine (as noted earlier, this was most evident in *Pilocarpus microphyllus* leaves). The establishment and use of these classes of leaves were the result of experience accumulated over a long period by Merck; they allowed the company to divide the regions of production according to the four leaf classes and to set the price paid for the leaves on the basis of geographic origin. Most of the areas of production in the Pre-Amazon region contained the A and B types of leaves; the Litoral and Cerrado regions produced mostly C and D types.

The Cerrado and part of the Litoral region are probably the oldest areas of exploitation of the jaborandi plant and the production of its leaves. That has been confirmed by leaf collectors and buyers, who have harvested leaves in these regions for almost three decades. In these areas of exploitation, jaborandi leaves now receive a C or D classification. Once, however, they produced at least the B type of leaves. Thus the Merck classification reflects both intensity and antiquity of the jaborandi exploitation in these regions. The extensive period of exploitation and the overharvesting in these areas caused a decrease in both the size of the leaves (produced by smaller plants) and the level of pilocarpine.

The long period of intensive harvesting of leaves in Maranhão has caused a drastic decrease in the level of production over the last ten years. The current volume of leaves produced (fewer than 500 tons per year from 1996 to 2000) is significantly smaller than that recorded for the period 1975–1985 (approximately 2,000 tons per year), a decrease of 75 percent. Since then leaf production has been in a consistent decline (figure 13.1).

At the level of the macroregions, some production changes were observed. The main areas of the Cerrado region produced approximately 200 tons of leaves per year in the 1970s; by the period 1998–2000, production was only about 30 tons or less per year. In the Pre-Amazon region, the decrease in production occurred at the same order of magnitude: in the 1970s, the region's leaf production was around 1,300–1,500 tons per year; in the period 1998–2000, it was always below 200 tons per year.

At the municipality level, the changes have been even more dramatic. The intensity of leaf collection has caused the complete disappearance of jaborandi in some areas: harvesting of jaborandi leaves has become just a memory in the older people's minds.

In the Pre-Amazon region the major producers of jaborandi leaves—and of the most valued types of leaves (A and B)—have historically been the municipalities of Santa Luzia (600 tons of leaves annually in the 1970s), Barra do Corda (700 tons), and Grajaú (300–500 tons). Santa Luzia, Barra do Corda, and Arame (recently created from the division of the Grajaú territory) are the only Pre-Amazon municipalities that maintained, until 1995, yearly production at around 100 tons each. However in the period from 1996 to 2000, only Arame maintained a consistent level of production (about 150 tons per year). The production of leaves from extractivism in the municipality of Barra do Corda, Merck's plantation area, declined to zero.

During the 1970s in the Cerrado region, the municipality of São Benedito do Rio Preto was always the major producer of leaves, averaging 200 tons a year; however reports of annual production show declines since then: about 30 tons a year were produced until 1995, none in 1996, and about 60 tons a year from 1997 to 2000. Leaves harvested in this region received C and D classifications during the last ten years, yet leaf collectors say they used to be in class B in the 1970s. Overharvesting was blamed for the decrease in the volume of production and for the decline in the quality of leaves from the region.

Discussion

In the last five years, the market for pilocarpine has reflected the economic problems in several parts of the world, especially in Russia, one of the most important buyers of this alkaloid. The levels of production of natural pilocarpine were also affected by the manufacture of the synthetic pilocarpine, although use of the synthetic form was reduced for some years until the industry could finally overcome some of its undesirable side effects in ophthalmology.

Although there has been no growth in the pilocarpine market for the past five years, no significant reduction in the demand for the alkaloid has been reported in the same period. High corporate investments from Merck (more than US$25 million in the last ten years) and technological improvements to increase the productivity and production of leaves are indicators that jaborandi continues to be a good business investment—at least for Merck. For local populations that were introduced to this activity by Merck, and were somehow induced to take part in the business of pilocarpine, no benefits could be clearly discerned.

Although most of the local populations involved in the leaf harvesting over the past three decades did not rely completely on that activity for income, many of them engaged in it very intensively as a profitable replacement for other daily forms of work. In spite of its regularity as a local activity in the areas where jaborandi is extracted, leaf harvesting did not seem to be culturally incorporated into the local traditional way of life, or perceived as the main supporting economic activity. However in some areas the activity had been around for so long that it could have developed social value to local communities, and might have been absorbed into the local culture because of its economic advantages.

The original use of pilocarpine by indigenous groups—as a stimulant of salivation and perspiration—required only sporadic and nondestructive leaf collecting. The nonsustainable exploitation of jaborandi was in fact initiated by Merck thirty years ago. In many areas of Maranhão, indigenous groups and peasants were stimulated by the payments they received from the pharmaceutical company to search for jaborandi plants. As a result, a consistent process of degradation of natural populations of the plant began. In several areas of Maranhão jaborandi is now either extinct or rare.

Merck representatives have stated that since 1993 approximately 90 percent of the company's demand for leaves has been satisfied by the plantations. The remaining 10 percent of leaves needed was obtained from wild harvests that Merck also controlled. The company continued to pay attractive prices in order to maintain the interest of the leaf collectors and retain its monopoly on jaborandi and pilocarpine: if Merck had left the extractivism to other companies, it would have created competition where there had been none (Pinheiro 1997). The 10 percent of leaves from extractivism actually reflected the percentage of natural populations of jaborandi left in Maranhão after the continued, intensive, and destructive harvest of leaves over the years.

Is there still a role for jaborandi and traditional knowledge about this plant in the future of the regions where jaborandi still occurs, considering the disappearance of natural populations in Maranhão, and the domestication and privatization of the resource by Merck? Traditional knowledge alone does not seem destined to play an important role in the future of the few areas where natural populations of jaborandi still persist. There are three reasons for this:

1. Traditional knowledge about jaborandi and its uses has been replaced by applications that require industrial processing.
2. The sources of raw material (jaborandi plants) have mostly become private, beginning with Merck's monopoly on the extractive activity, and consolidated by the establishment of the company's plantations.
3. With the apparent success of the synthetic pilocarpine, the future of the market for the natural alkaloid has become uncertain.

Developing a technology for the cultivation of jaborandi in small areas could have been a strategy for diminishing the pressure on wild populations. It could also have contributed to a more organized and sustainable way to help local people change from leaf collectors to producers of raw material to the industry by creating information and other conditions (e.g., credit, technical assistance, infrastructure) to help establish small plantations. However the difficulties in defining and establishing a system of production that would accommodate the interests of both small-scale farmers and industry were (and still are) many, as illustrated by the following two strategies that I have been considering.

1. *Planting of jaborandi in the understory of the forests in an attempt to reproduce their natural growing conditions in the wild.* This would be a way for indigenous peoples to grow jaborandi plants within the limits of their reserves. However the industry seems uninterested in this strategy owing to a number of technical problems, including:

- Harvest time becomes very long (10 years) compared with that obtained on the plantation (3 years under direct sunlight).
- Irrigation is not possible.
- The levels of pilocarpine in the leaves cannot be controlled as they can on the plantation with the use of sophisticated equipment.

2. *Cultivation of jaborandi in small areas associated (or not) with subsistence crops.* The first tests carried out by Pinheiro (unpublished) revealed

that this strategy could be viable, but at the level of small-scale farmers some problems are limiting to the success of this system:

- Pests: a number of insects have been identified as pests to jaborandi in cultivation, and the damage caused by them can be severe. Control can be difficult for small-scale growers because chemical products cannot be used owing to their effects on ophthalmologic patients.
- Nematodes: Jaborandi is susceptible to nematodes, and chemical nematicides cannot be used because they would accumulate in the leaves. Biological solutions (e.g., minimal weeding, use of nematode-repellent plants) have been studied with unsatisfactory results. Although applying fertilization through irrigation water minimizes the problem by producing stronger and healthier plants, the best results have been achieved by inoculating jaborandi plants with the bacteria *Pasteuria penetrans*, an expensive method for small-scale farmers.
- The control of the level of pilocarpine in the leaves cannot be maintained in small-scale farming conditions.
- Irrigation—because of its costs and logistics—is also a limiting factor in this system.
- Finally, the current market limitations for natural pilocarpine have to be carefully considered. Merck did not seem to be interested in a system that could increase the production of leaves if the market did not demand more pilocarpine. Under such conditions, the company would have to reduce its planted areas in order to stimulate a system of small plantations. That option did not appeal to Merck once it developed the technology to control the levels of pilocarpine in jaborandi leaves.

Conclusions

The relationship between Merck and the communities that were induced to collect jaborandi leaves did not offer bilateral benefits. On the contrary, whole jaborandi populations were drastically reduced or eliminated, and the exploitation ceased in many parts of Maranhão. No progress, no improvements for communities can be discerned in these areas. In Merck's farm region one would naturally expect to see benefits accrue to local communities, at least for the generation employed at the farm. Employment, however, was a factor only in the early years of the plantations; mechanization of most activities replaced labor on the farm by 1998, when little more than 200 people were employed there because fertilization, irrigation, and the harvest and drying of leaves had become totally automated.

Economic benefits to the state of Maranhão were minimal and limited to ICMS (a state tax on the circulation of goods and services), while the IPI (a tax on industrialized products) was collected by the bordering state of Piauí, where jaborandi leaves were processed and from which pilocarpine was exported. Thus neither Maranhão nor its people received the profits and benefits of having this important plant resource; within thirty years whole populations of jaborandi were devastated to the point of being among the species of the Brazilian flora in danger of extinction (IBAMA 1992). Merck monopolized the profits and the benefits that should have been more evenly distributed among thousands of people in a large number of communities that were (and still are) involved in the exploitation of jaborandi in Maranhão. But Maranhão has suffered more than the lack of benefits: it has been left with irreversible damages to its environment.

Acknowledgments

I am grateful to the following people and institutions: Darrell A. Posey (in memoriam) for reading the manuscript and for his valuable comments; Michael J. Balick for supporting part of the fieldwork through The New York Botanical Garden, Institute of Economic Botany; and the Charles A. and Anne Morrow Lindbergh Foundation for the grant awarded to the author.

References

Corrêa, M. P. 1969. *Dicionário das plantas úteis do Brasil*. Vol. 4. Rio de Janeiro: Ministério da Agricultura.

Frazão, J. M. F. and R. L. S. Pereira. 1979. *Diagnóstico preliminar do* jaborandi *no Maranhão*. São Luis, Maranhão: Secretaria de Recursos Naturais, Tecnologia e Meio Ambiente, ITEMA.

Holmstedt, B., S. H. Wassén, and R. E. Schultes. 1979. Jaborandi: An interdisciplinary appraisal. *Journal of Ethnopharmacology* 1(1):3–21.

IBAMA (Instituto Brasileiro do Meio Ambiente e dos Recursos Naturais Renováveis). 1992. Portaria. *Diário Oficial* (Brasília), no. 06N (January 15):870–872.

IBGE (Instituto Brasileiro de Geografia e Estatistica). 1975–1992. *Produção Extrativa Vegetal*. 18 vols. Rio de Janeiro: Fundação Brasileira de Geografia e Estatística.

IBGE. 1993–2000. *SIDRA—Extração vegetal*. www.ibge.gov.br (accessed in January 2001).

Joensuu, H., P. Bostrom, and T. Makkonen. 1993. Pilocarpine and carbacholine in treatment of radiation-induced xerostomia. *Radiotherapy and Oncology* 26(1): 33–37.

Jornal Estado Maranhão. 1999. Plantio de Jaborandi no Maranhão. São Luís (MA), 12 de fevereiro de 1999, No 12.356, página 5.

Joseph, C. J. 1967. Revisão sistemática do gênero *Pilocarpus* (ssp. *brasileiras*). *Mecânica Popular* (Rio de Janeiro) October:1–9.

Kaastra, R. C. 1982. *Pilocarpinae (Rutaceae)*. Flora Neotropica 33. Bronx: New York Botanical Garden.

Merck. 1983. *The Merck Index: An Encyclopedia of Chemicals, Drugs, and Biologicals*. Rahway, New Jersey: Merck.

Merck. 1989. *The Merck Index: An Encyclopedia of Chemicals, Drugs, and Biologicals*. Rahway, New Jersey: Merck.

Miller, L. J. 1993. Oral pilocarpine for radiation-induced xerostomia. *Cancer Bulletin* (Houston) 45(6):549–550.

Pinheiro, C. U. B. 1997. Jaborandi (*Pilocarpus* spp., Rutaceae): A wild species and its rapid transformation into a crop. *Economic Botany* 51(1):49–58.

Rieke, J. W., M. D. Hafermann, J. T. Johnson, F. G. Leveque, R. Iwamoto, B. W. Steiger, C. Muscoplat, and S. C. Gallagher. 1995. Oral pilocarpine for radiation-induced xerostomia: Integrated efficacy and safety results from two prospective randomized clinical trials. *International Journal of Radiation Oncology Biology Physics* 31(3):661–669.

Valdez. I. H., A. Wolff, J. C. Atkinson, A. A. Macynski, and P. C. Fox. 1993. Use of pilocarpine during head and neck radiation therapy to reduce xerostomia and salivary dysfunction. *Cancer* (Philadelphia) 71(5):1848–1851.

Wynn, R. L. 1996. Oral pilocarpine (Salagen): A recently approved salivary stimulant. *General Dentistry* 44(1):26, 29–30.

14 Peasant Riverine Economies and Their Impact in the Lower Amazon

Mark Harris

A senior woman from a floodplain village in the Lower Amazon once told me that when she was young her father used to ask her and her siblings what kind of fish they would like to eat for lunch. They would tell him, he would go out, and usually he was able to obtain the desired species. Recently, some forty years later, she complained that this is no longer possible, the diversity had gone. "We have often to make do with fatty catfish," she added. She concluded that "it is not the times that change, but the people." I asked her to explain. She said that fish stocks were declining because outsiders were using predatory methods of fishing and were invading the lakes. These fishing enterprises used large nets and the seine technique because they did not care about the future. This forced her local villagers into one of two positions to protect the immediate future of their livelihoods: either they could use those same predatory techniques, but on a smaller scale, or they could prevent outsiders from entering the lake. One or the other of these actions was necessary in order to keep up with the changing times. In her view, people drive the engine of social change to avoid being stuck in tradition or the past. Floodplain dwellers, she was saying, reinvent themselves in order to survive each new historical challenge.

That woman's comments not only give an important insight into the self-knowledge of floodplain inhabitants — in particular as it reflects their perceptions of environmental and economic changes in the Lower Amazon — but also, I believe, provide a social-scientific understanding of the floodplain peasantry. This paper will elaborate on her comments with regard to the historical impact of local economic and social activities on the floodplain environment. A necessary part of this discussion is a consideration of the wider socioeconomic transformations that have occurred in Amazonia.

Riverine peasant communities in the Lower Amazon have been remarkably resilient at making a living in a market economy for at least two hundred years. In addition, the kinds of economic activities found in riverine areas have always been diverse, ranging from the large-scale and isolated dry-season cattle ranches to the small-scale and semi-independent patterns of peasant economic activities and dense family networks. Regardless of the ways people earn their livelihoods, the conditions for maintaining them lie outside local control. Consequently a portrait of floodplain peasants' ecological knowledge, which is my concern here, should demonstrate how local economic activities are framed by the wider system that influences change.

Part of the problem in understanding the peasant economy in the Lower Amazon, as well as elsewhere in the Brazilian Amazonia, is that so much of what takes place is invisible (Nugent 1993). Much of the activity of the peasant economy exists outside the purview of the state government and its accounting bodies (for tax purposes and statistical calculation). So it does not get officially recorded and is hidden by the larger and better remunerated forms of production, those that are associated with the market. Some of the commercial activity is illegal, such as selling turtles or the skins of animals. But some is part of the vast and inchoate sphere of noncommoditized relations, such as labor exchange, saints' festivals, and family "help." These various spheres can be labeled the informal economy—the shadowy zone, to use Braudel's memorable phrase (1981:23)—which underlies the market economy.

Peasants' practices change with market demands. These practices are acutely in tune with new opportunities, which arise out of seasonal variations or historical dynamics. Thus for the livelihoods of riverine peasants to be "sustainable" there must be political action that recognizes the interconnectedness between the local and the wider economies, while allowing the local economy to keep the traditional adaptability and flexibility that comes from access to local resources. It is this ability to meet changing conditions with changing activities that needs to be appreciated in order to have a meaningful discussion about the future of riverine livelihoods.

The Informal Economy and the Resilience of Peasant Livelihoods

Goulding, Smith, and Mahar (1996), in their book on the Amazon, claimed that the "floodplain forests are one of the most threatened of all habitats in

Amazonia" because deforestation of the floodplain in the lower and middle reaches of the Amazon has eroded a vital source of fish foods—namely fruits, nuts, and berries—during the wet season. As a result, fish stocks have declined, provoking peasant migration to the towns. In addition, biodiversity has been reduced by large-scale economic activities such as cattle ranching. Their claims overlooked the central importance of the informal economy in the development of Amazonia (see McGrath's 1989 work on traveling river traders [*regatões*], and Nugent's 1993 analysis of merchant capital). Semi-independent peasants have played a major role in contributing the products and distribution networks that make up an informal economy.

Goulding, Smith, and Mahar (1996) contend that deforestation and the degradation of land and resources from cattle ranching and overfishing arise from events within the last thirty years or so, beginning with the military government's attempts to integrate the Amazon into the Brazilian nation. The floodplain, and more generally the riverine environment, is one of the oldest sites to show the effects of colonization on Amazonia. This is so in both human and environmental terms. The aboriginal population was almost wiped out and a series of new mixed-blood peasantries emerged in its place. The export-dominated economy—in fulfilling demands for floodplain products such as fish, turtles, turtle eggs and oil, caiman skins, jute, cattle, and cocoa—has had successive negative impacts on resource levels. These short cycles of exploitation based on external demands have led to the near extinction of some animals and products in some areas (e.g., caimans, pirarucu, and turtles are quite rare nowadays).

There are three production activities that are rarely given much prominence in discussions of the economic history of riverine areas: cattle raising, cocoa production, and firewood extraction. Each of these activities had its own effect on both the social and environmental landscape before the mid-twentieth century. Furthermore, each activity was extremely important to the establishment of a peasantry in the nineteenth century (e.g., Alden 1973; Acevedo and Castro 1993; Cleary 1998). All three gave people an opportunity to use local resources for exchange with a merchant elite (McGrath 1989).

A Floodplain Community

Many social and economic changes are evident in the following brief account of a floodplain community in which I lived for one-and-a-half years in 1992 and 1994, and to which I made return visits in 2000, 2003, and 2004.

The village is situated a half-day's journey upstream from the city of Óbidos in the Brazilian state of Pará. Though my account relates to one small portion of reality in Amazonia, my relatively long-term experience of this community provides a useful window for viewing more general transformations.

The history of the colonization of Parú—the name of the floodplain region of lakes, rivers, and seasonally inundated land and forest—is quite heterogeneous. Ceramic pieces found by local folk indicate a preconquest Amerindian presence and most likely belonged to the Conduri (Nimuendaju 1981; Gomes 2002). There is no record as to precisely what happened to that Amerindian society, but by end of the seventeenth century there could have been little left of its size and distribution (Porro 1992). Beginning around the end of the seventeenth century Franciscan and Jesuit missions became established in the Lower Amazon, and Portuguese soldiers and officials settled nearby, using locally available Amerindian slave labor. On the basis of reconstruction, we can say that Parú, like other riverine areas near the towns along the trunk of the Amazon, became an important satellite for Óbidos in the late eighteenth century owing to the allocation of land grants (*sesmarias*) and to the rise of cocoa plantations and cattle raising. From the second half of the eighteenth century, there would have been some African slaves and their families, slave owners and their families, and smallholders (comprising mixed-blood Brazilians and acculturated Indians). In the third quarter of the nineteenth century many northeasterners came to the region, and then in the early twentieth century there was a small but significant arrival of Italians, who were escaping the poverty of southern Italy. According to oral histories, the population increase has been continual and even, though areas closest to towns became settled the earliest, particularly north of Santarém, between Alenquer and Óbidos (see Biblioteca e Archivo Público do Pará [1904] for a register of sesmarias up to the 1820s, and Le Cointe's wonderful map of the Lower Amazon [1911], which includes major landholders); it is only toward the mid-twentieth century that all the available land became fully occupied and used. This is in contrast to urban centers in the Lower Amazon, which in the last thirty years have received a large influx of rural migrants in search of jobs and schools for their children, disrupting the previously balanced rural-urban continuum.

Oral histories also reveal much about the flexibility and diversity of riverine livelihoods. The items discussed below were probably the most important for exchanging with either bosses in town or passing river traders. The bulk of trade was, however, done with the *regatões*, because the journey to town was made by canoe and involved two days of travel there and back. Many older people retell their grandparents' accounts of slavery in the area.

According to these stories, slaves used to work in Parú preparing cocoa for export and firewood for passing steamboats. When slavery was abolished the slaves just left, and as with the Indians, no one knows where they went (see Acevedo and Castro 1993 and Funes 1996 for ethnographic histories of slavery and escaped-slave communities in the Lower Amazon).

In the nineteenth century, Parú was covered in cocoa trees and the largest commodity crop in the village was cocoa. Those cocoa trees are now mostly gone, having either been drowned in a series of high floods or cut down for jute cultivation. There were also a few rubber stands; one man I spoke with remembered exactly how the rubber was prepared for sale to the passing trading boats. Firewood was another crucial resource; it was gathered for the steamboats that went up and down the river until diesel boats were introduced after World War II. The wood was exchanged for goods that were not locally produced. According to a number of Parúaros, there were four *depósitos de lenha* (firewood deposits) in Parú. Manioc flour, caiman skins, vegetables (e.g., squash), salted pirarucu, turtles and their eggs, kapok cotton, wild rice, sarsaparilla, and feathers were the products most important to the trade repertoire of the floodplain. A range of seasonal subsistence activities revolving around fishing, hunting, and gardening complemented the production of these exchangeable goods.

Cattle have been raised on the floodplain areas of Óbidos since the eighteenth century (Reis 1979). Inglés de Sousa, in his 1876 novel *O Cacualista* (2004), describes a floodplain area opposite Parú where cattle roamed freely under the cocoa trees. Older Parúaros recount how, in their youth, they worked occasionally for a large cattle owner. Toward the end of the dry season they would be employed to slaughter a number of bulls, then salt the meat and put it into large wooden boxes for export. Nowadays, cattle have become the single most important sign of wealth. Any large amount of money that is not spent on necessities and house building or maintenance is used to buy cattle. "Cattle are," one man told me, "like a bank," safe and secure investments that grow in size. And given that the eighties and early nineties in Brazil were characterized by high inflation with beef keeping pace with price increases, cattle became a very important source of income.

From the 1940s until the 1980s there was a boom in the production of jute on the floodplain. In Parú this meant that most, if not all, people were involved in jute production: clearing ground and planting, reaping, and bundling jute for sale. Jute was a relatively lucrative crop; many people in Parú told me that they had built their good-looking houses and had bought fishing nets and diesel-powered boats with jute money. The decline of jute coincided with the growth of the fishing industry, which brought improved technology, including ice-making plants and fish-exporting factories; and so Parúaros turned to the sale of fish as their primary source of revenue. This led to a rise in overall income for the area. From a very good month's sale of fish (about 8 tons) in the early and mid-1990s, a boat owner could have earned up to ten times the minimum salary (a value set by the Brazilian government which was the equivalent of about US$70 at the time).

Thus in the 1980s another set of petty commodity activities emerged on the floodplain of Parú, and they have continued into the early twenty-first century. They revolve around fishing for sale and cattle raising. Agriculture plays a small, but important, part in the repertoire. Access to technology and resources is a central factor governing the patterns of a household's economic activities. A wealthy household is one that owns a boat having a container suitable for storing fish and ice. A senior couple typically employs junior kin and neighbors on the boat and probably owns a score or two of cattle. A poorer household often consists of a much younger couple who has children, but very few resources and no ability to hire labor. This differentiation between households is an important communal dynamic. However disparities between households do not last for more than a generation because the resources accumulated by parents are equally divided among their children: the splitting up of the inheritance prevents longer-term inequality.

This is a brief summary of the economic and social history of one village as recounted to me by Parúaros. It is crucial to note the flexibility of cycles of exploitation from one product to another, although some commodities (e.g., turtles, firewood) were driven almost to the point of extinction in the process. Each cycle has had its own impact on the environment, as well. The ease with which people have moved from one economic activity to another is remarkable. Despite the historic upheavals in Amazonia, the inhabitants of this village have been able to maintain themselves without compromising their position in any way (for example, by migrating to cities, converting to Protestantism, relying more on credit).

This section gives a sense of the continuity of livelihoods in Parú. It indicates that the peasants living in floodplain areas are not recent arrivals; nor are they haphazardly making a living or simply accommodating external demands. On the contrary, the resilience of their livelihoods over two centuries suggests an ability to adapt not only to prevailing uncertainties in the regional markets, but also to internal needs.

Much of the resilience is dependent on an ability to continuously occupy plots of land, and therefore to have access to nonvalorized resources. Central to this process is kinship relations: keeping kin together heightens the sense of community. Kinship is crucial to the continuity of livelihoods, for sources of labor, and for access to resources. It is through kinship that ownership of local resources is maintained and that continuity is achieved.

The continuity of peasants' lives has depended on the ability to manage a diverse economy and maintain access to land and water resources. The identity that develops from this is not an ethnic one. Instead it is one that arises from people's continual engagement with a particular kind of environment. It is a concrete identity that is characterized by a way of doing things, rather than a conceptualized identity that is dependent upon selective representation (Harris 2000).

What Peasants Do: Ecological Knowledge and Its Future

A consequence of the previous argument is my claim that the ecological knowledge of floodplain dwellers in the Lower Amazon is neither "traditional," in the generally accepted sense, nor the result of an accumulation of inherited rules and representations. As Ingold argues: "Knowledge is not a matter of being in possession of information handed down from the past, but is rather indistinguishable from the life-activity of a person in an environment which has itself been, and continues to be, fashioned through the activities of predecessors and contemporaries" (Ingold 2004). Knowledge is thus continually generated and reevaluated in the context of what people "do," in the concrete practice of their life activities. It is flexible and up-to-date, and able to accommodate wide changes. It is not applied from some impersonal source, such as culture or tradition, which exists independently of a person's current relations with the environment and other people. All of this has the implication that ecological knowledge cannot easily be articulated in language.

This theory is in contrast to that of other scholars who have written on the ecological knowledge of riverine, or *caboclo* (the mixed-race offspring of Amerindian and Portuguese parents), peasantries. Wagley, for example, asserted that "Amazon folk culture is a regional variety of Brazilian national culture" (1967:226). He went on to add, "It might be said that all Indian residues in contemporary Amazon culture relate directly or indirectly to the environment" (1967:227). Moran has a very similar conception: "Europeans came to rely on the Amerindian to teach them what to eat, how to hunt, fish, row, and cultivate the basic staples from a riverine forest environment. In almost all aspects which deal with man's relation to the environment, the indigenous ways are still practised today through the cultural adaptations of the *caboclo amazonense*" (1974:139). Further on in the article, Moran details some of these indigenous ways, which form part of the folklore of the peasants. He points to *panema*, an Amazonian concept of bad luck (see also Matta 1973 and Harris 2000); the enchanting dolphin, or *o boto* (see Slater 1994), and other forest and river spirits (*encantados*); taboos on eating various foods (Motta-Maués and Maués 1980); and the practice of shamanism, or *pajelança* (see Maués 1995) as examples of the Amerindian heritage present today. Of interest, Moran does not mention land use as part of the indigenous legacy.

Wagley's and Moran's approach exemplifies the idea that knowledge can be possessed and stored in brains and in books. Yet paradoxically for them knowledge is also intrinsically determined by the natural environment in which it is put to use. The dilemma that this sets up is beyond the bounds of this paper. In my following remarks I would prefer to give more emphasis to the "presentness" of knowledge, that is, to the way it changes with context and necessity. The rest of the paper offers a consideration of the implication of this approach for the study of the floodplain and its inhabitants.

Seasonal Rhythms

I present here an analysis of seasonal changes on the floodplain and of the nature of land and water access and ownership. These phenomena are intimately bound to each other in the way they organize people's livelihoods and as social realities that need to be negotiated.

The floodplain of the Amazon has both aquatic and terrestrial habitats. It is an environment that is particularly challenging to human survival. It requires a special attitude to endure the periods of flooding, when water enters the house, mosquitoes abound, food is scarce, there is little to do, and a number of close kin live in cramped quarters. Elsewhere (Harris 1998) I have analyzed seasonality in terms of an ongoing rhythm that is intrinsic to the flow of economic and social activities, rather than external to or constraining of them.

Many people have written excellent descriptions and analyses of floodplain life. There is no need to repeat previous portraits here (e.g., Higbee 1954; Meggers 1971; McGrath et al. 1993; Goulding, Smith, and Mahar 1996; Sternberg 2000). Some of these authors have assumed that the marked seasonal variations have a negative impact on the development of society and economy and therefore need to be overcome (Barrow 1985; Townsend 1985).These writers claim that technological improvements are needed to overcome the limitations of the environment. Perhaps the most extreme of these suggestions was the proposal to build a dam that would flood much of the basin and remove much of its human population (Goulding, Smith, and Mahar 1996). Again I argue that these authors have overemphasized the nature-culture dichotomy. This has led them to ignore the existing ways of living with seasonal changes, such as the building of raised platforms to store cattle during times of flood, changing fishing technology and work relations, and the great efforts to organize successful crop processing in the buildup to the floods. There is also a more compelling example from the estuary of the Amazon.[1]

It is precisely the rhythm of economic life arising out of the coordination of these seasonal activities that I think is so significant. Instead of developing complex technical solutions to constrain the floodplain environment, I am arguing that, to use a cliche, people need to go with the flow. It follows that economic and social activities relate to and resonate with changes in the environment. People do not say they have to plant now, to give one example, because it is September. Instead they decide to plant when certain conditions (other seasonally variant phenomena) have been fulfilled, such as when the land has hardened, or a certain fish has migrated past the village, or a particular bird has appeared. In other words, the activities have a rhythmical order, a sequence that arises from their intrinsic and apprehensible relationships to each other. These relationships are produced in the continuous and practical engagement between persons and environment. This is the kind of knowledge and perception that needs to be analyzed. That may sound obvious, but it is quite different from claiming that seasonal changes have to be overcome or they will limit the full potential of society. Peasant floodplain society and economy are inherently seasonal, suggesting that the nature of the peasantry's integration into the global economy is dependent on an ability to manage a diversity of economic activities. The range of resources is influenced by the seasonality of the environment.

From this basis we can look at variation in incomes, labor organization over the year, fish catches, and so on. From there, we can examine longer time spans and the ways in which economic activities change from season to season in accordance with wider economic demands (see Guyer 1988 for the innovation of this analysis in West Africa).

_____

1. Scott Anderson (1993) has written on rum production in areas around Abaetetuba in the first two-thirds of this century, before it was undercut by cheaper imports from southern Brazil. In his analysis he shows how the daily tides were used to power the sugarcane mills and thus played an important role in the making of rum.

Perception of Land and Water on the Floodplain

The seasonal nature of the floodplain, with its aquatic and terrestrial synthesis, leads to a more direct appreciation of people's claims to the land.
Historically, floodplain-dwelling peasants have been denied land tenure.
Many authors who have written about the floodplain have analyzed local
notions of ownership of land and water (e.g., Lima 1992; McGrath et al.
1993; Goulding, Smith, and Mahar 1996). There are virtually no de jure
land titles on the floodplain. Most land is held de facto, that is, by continuous occupation, which confers usufruct rights. In order to secure land rights,
a complex and yet flexible system of family and marriage has developed.
Social continuity is ensured through horizontal kinship strategies such as
marriage between cousins and assertions of cousinhood between men. Let
us not forget that plots of land on the floodplain are quite small: a few hectares is the average in my fieldwork area, unless the plot belongs to a large
cattle ranch, which could cover many hundreds of hectares. In the perception of riverine peasants, areas outside individually controlled plots are considered communal; in other words, residential proximity confers rights of
access, use, and crucially, stewardship. Each community has an informal
understanding of the boundaries of its territory. And this is the point: all of
this is part of the informal nature, oral tradition if you like, of the history of
floodplain occupation.

These traditional notions of ownership are based on a landscape that is in
a constant process of redefinition and becoming. There are no fixed boundaries to demarcate discrete special domains of land, forest, house, garden,
lake, and stream. From one year to the next people do not know what will
be land or water. In the local view, the river creates the land and imparts
power (força) to it during the flood. Thus land and water are inseparable in
the local perception of the landscape, and their merging is an integral part
of "land" use. As one person told me: "Este rio é nossa terra" [This river is
our land].

The traditional peasant perception is that land and its resources are the
property of its occupant and worker. Water and its resources are for communal stewardship. During the flood there is open access; what is beneath
the water is irrelevant. Central to this model is the idea that labor creates
ownership. Land can be owned because it is worked, either for agriculture
or in preparation for cattle raising. Water, and in particular fish, cannot be
owned by anybody because, for example, fish cannot be controlled. As one
person told me: "Fish are looked after by nature and swim where they want."

A related part of this model is that agriculture is regarded as the only real type of work. Fishing is not considered work because it does not involve creating a product; rather it is the act of putting a net or line out and waiting. Needless to say, there is much discussion about this proposition with the rise of fishing as a dominant economic activity among men. In addition, any understanding of floodplain land and water use should take into account the possible depositions and erosions resulting from each year's flood.

The effect of the fluidity of the floodplain on people's claims to property needs to be fully understood from a human or social perspective. The problems of devising a legal framework of land and water tenure have come to be discussed only recently (Benatti 2003). Clearly any community-based integrated-development program in the floodplain must accord land titles to its dwellers (see, for example, the two nongovernmental organizations' research and development programs being conducted on the floodplain: Projeto Mamirauá, near Tefé in the state of Amazonas, and IPAM's [Instituto de Pesquisa Ambiental da Amazônia] projects around Santarém in the state of Pará). At the center of such a program should lie the seasonal nature of water/land use and the problems posed by erosion and deposition. This is nothing less than taking into account current reality and formalizing it. It would amount to a technical solution, and would highlight floodplain peasants' participation in the wider economy and society. Land titles throughout Amazonia could, in addition, prevent rural migration to towns.[2]

There are many complicating factors, of course: the heterogeneity of the floodplain population and the informal and pragmatic nature of peasant perceptions of the environment are two examples. Goulding, Smith, and Mahar (1996) identify a more difficult problem. The nature of fish movements and the interconnections in the ecosystem in general complicate the future success of small-scale projects such as Mamirauá and Iara. It is no good preserving the floodplain in isolated areas, the argument goes, if in the majority of other places it is being degraded. Nevertheless such local projects offer a

---

2. It is important to note that migration patterns differ along the trunk of the Amazon, that is there is no uniform pattern. Permanent rural-urban migration appears to depend on proximity to Manaus, and urban-urban migration upon class and wealth. Belém traditionally attracts the elites, who follow their family ties in many cases, while Manaus attracts the rest. There is of course much temporary lifecycle migration, for example, as part of a young person's desire "to get to know the world." In the village in which I did fieldwork there had been virtually no permanent migration of whole families in the preceding ten years.

tremendous incentive to move forward. The next steps could be recognition and then a legal conceptualization of floodplain lands according to peasant perceptions. Such undertakings are realistic within the wider demands for land reform in Brazil.

It is the flood time that probably presents the biggest obstacle to legalizing the peasant conceptualization of land. During the flood, fisherpeople can ideally move about with unlimited access to the whole floodplain.[3] I know of many instances when, during the flood, large fishing vessels enter lakes, often fishing extremely near peasant communities and extracting large amounts of migrating catfish. In effect these large boats are fishing in water that is over land considered privately owned by the peasants. I also know of peasants who want to fish in the flooded forests using only a canoe and hand-line, but who are prohibited from doing so by large-scale cattle ranchers who consider the water their own. One rancher's only justification was that it was his land and therefore anything over it was his. This is clearly in conflict with both the peasant view and the legal basis of water access.

Finally, the call for recognizing land ownership on the basis of local knowledge has been taken up in another important way since the end of the 1990s. The lower parts of the Tapajós and Arapiuns rivers, which are part of the Lower Amazon, have seen the resurgence of Amerindian identities (*índios resurgidos*). This is the process by which groups reclaim an indigenous identity by showing they are descended from Indians who used to live on the land the group presently occupies (Pacheco de Oliveira 1999). Previously the only recorded large indigenous group was the Mundurucu who lived in an indigenous reserve in the Tapajós valley. Many communities are now saying they are Indians, using ethnic names that have not been used in official documents for at least two hundred years and were unknown to the National Indian Foundation (Fundação Nacional do Índio, FUNAI). It is a moot point whether the names have been used in oral traditions among communities, and clearly this needs more research. These communities, through a Franciscan monk, have started a legal process with FUNAI to obtain recognition of their status. That will allow them to create an indigenous reserve, thereby giving them access to federal resources and complete con-

---

3. Notions of communal or private ownership do not apply in the wet season. In the past, the flood was a fallow period, when people fished on a small scale in the flooded forest. The vast mass of water increases manifold the effort needed to make small-scale fishing attractive. However with the recent increase in demand for catfish, and with improved technology, it is now possible to make large catches of migrating catfish throughout the flood period.

trol of their land, as well as protecting them from loggers, planters of soya, and gold miners, the three biggest predatory interests in the Lower Amazon. I hasten to add that these developments are quite localized to an area that has been subject to conflicts over access to land and that in 2000 was designated an extractive reserve, which gave locals basic but not full control over the area. Their intention is now to secure total ownership by forming an indigenous reserve. The fascinating story here is that these people were, perhaps two-hundred-and-fifty years ago, acculturated Indians (*índios* rather than *gentios*, or wild Indians) living in missions; then in the late-eighteenth and nineteenth centuries they were identified as caboclos; and now, in the early twenty-first century, they have adopted again the label *índio*, having identified the names of tribes who used to live in the region.

Conclusions

Bruce Albert (1997) argues that the processes of "local indigenous empowerment and politico-symbolic globalisation of ethnicity define the context in which the conditions and stakes of anthropological research" (56) are being carried out. He concludes that the opening up of anthropological fieldwork means the subversion of the notion that cultural change equals degeneration. I would like to add that knowledge changes, too, with the new conditions in which people find themselves. Change now equals continuity. The challenge, as Albert indicates, is to be able to show consideration for the demands of indigenous or peasant peoples and to appreciate how the context of political recognition shapes those needs.

This article has presented a brief portrait of a peasant floodplain group that is sensitive to both the wider historical dynamics and the environmental setting. It is only with such a perspective that it makes sense to consider ecological knowledge. I have argued that such knowledge cannot be abstracted from its context. This brings me back to my opening quote: "It is not the times that change but the people." The resilience of floodplain people in the Lower Amazon has come from their keeping abreast of environmental and economic changes while modifying their own practices in direct response to these changes — without being forced to do so.

This raises the questions of what is tradition and what maintains a way of life. The danger in asserting that something is traditional is that it focuses attention on abstract categories. In the development talk of modernist theorists, which assumes the superiority of the West, this is not far from dismissing people altogether. It seems obvious that it is people who need conserving, not traditions. I have suggested that the way to achieve this is to provide socioeconomic conditions for continuity of the peasantry, that is, to ensure land ownership. For the environment to remain able to support these people there needs to be political protection from large commercial exploitation that would force a change in local people's subsistence strategies. The extent to which "traditional ecological knowledge" may have a role in the future of the floodplains depends almost entirely on the ability of its inhabitants to secure legal ownership of land and water areas in their vicinity. In fact, I believe this is the only condition under which traditional ecological knowledge can help sustain the floodplain.

Acknowledgments

I am most grateful to Darrell Posey for editorial suggestions and to James Leach for his comments on an earlier draft. I also acknowledge and thank the British Academy for awarding me a postdoctoral fellowship (1996–1999), which allowed me to work on the ideas in this paper.

References

Acevedo, R. and E. Castro. 1993. *Negros de Trombetas: Guardiães de matas e rios.* Belém: Universidade Federal do Pará/Núcleo de Altos Estudos Amazônicos (UFPa/ NAEA).

Albert, B. 1997. "Ethnographic situation" and ethnic movements. *Critique of Anthropology* 17(1):53–65.

Alden, D. 1973. *O significado da produção de cacau na região Amazonica.* Belém: Universidade Federal do Pará / Núcleo de Altos Estudos Amazônicos (UFPa/ NAEA).

Anderson, S. 1993. Sugar cane on the floodplain: A systems approach to the study of change in traditional Amazonia. Ph.D. dissertation, University of Chicago, Illinois.

Barrow, C. 1985. The development of the várzeas (floodlands) of Brazilian Amazonia. In J. Hemming, ed., *Change in the Amazon Basin: Man's Impact on Forests and Rivers,* vol. 1, pp. 108-128. Manchester: Manchester University Press.

Benatti, J. H. 2003. Direito de propriedade e proteção ambiental no Brasil: Apropriação e o uso de recursos naturais no ímovel rural. Ph.D. dissertation, Universidade Federal do Pará/Núcleo de Altos Estudos Amazônicos (UFPa/ NAEA), Belém.

Biblioteca e Archivo Público do Pará. 1904. *Catálogo nominal dos posseiros de sesmarias*. In Annaes da Biblioteca e Arquivo Público do Pará, vol. 3, pp. 5–159. Belém: Typografia do Instituto do Lauro Sodré.

Braudel, F. 1981. *Civilisation and Capitalism, 15th–18th Century*. Vol. 1. *The Structures of Everyday Life: The Limits of the Possible*. London: Collins.

Cleary, D. 1998. "Lost altogether to the civilized world": Race and the Cabanagem in Northern Brazil, 1750 to 1850. *Comparative Studies in Society and History* 40(1):109–135.

Funes, E. 1996. Nasci nas matas, nunca tive senhor. In J. J. Reis and F. dos Santos Gomes, eds., *Liberdade por um Fio: História dos Quilombos no Brasil*, pp.467–498. São Paulo: Companhia das Letras.

Gomes, D. 2002. *Cerâmica Arqueológica da Amazônia: Vasilhas da Coleção Tapajônica MAE-USP*. São Paulo: Editora da Universidade de São Paulo; Imprensa Oficial do Estado.

Goulding, M., N. J. H. Smith, and D. J. Mahar. 1996. *Floods of Fortune: Ecology and Economy Along the Amazon River*. New York: Columbia University Press.

Guyer, J. 1988. The multiplication of labour: Historical methods in the study of gender and agricultural change in modern Africa. *Current Anthropology* 29(2):247–272.

Harris, M. 1998. The rhythm of life on the Amazon floodplain: Seasonality and sociality in a caboclo village. *Journal of the Royal Anthropological Institute* 4:65–78.

Harris, M. 2000. *Life on the Amazon: An Anthropology of a Brazilian Peasant Village*. Oxford: British Academy/Oxford University Press.

Higbee. E. 1954. The river is the plow. *Scientific Monthly* 60:405–416.

Ingold, T. 2004. Two reflections on ecological knowledge. In G. Sanga and G. Ortalli, eds., *Nature Knowledge: Ethnoscience, Cognition and Utility*, pp. 301–311. Oxford: Berghahn Books.

Le Cointe, P. 1911. *Carte du Bas Amazone de Santarém a Parintins: Municipe de Óbidos et partie des Municipes limitrophes—Etat du Pará*. Paris: Armand Colin.

Lima, D. 1992. The social category caboclo: History, social organisation, identity and outsider's classification of the rural population of an Amazonian region. Ph.D. dissertation, University of Cambridge, Cambridge.

Matta, R. da. 1973. "Panema: uma tentativa de analise estrutural." In *Ensiaos de Antropologia Estrutura*, pp. 63–92. Petrópolis: Vozes.

Maués, R. H. 1995. *Padres, pajés, santos e festas: Catolicismo popular e controle eclesiástico*. Belém: Editora Cejup.

McGrath, D. G. 1989. The Paraense traders: Small-scale, long distance trade in the Brazilian Amazon. Ph.D. dissertation, University of Wisconsin, Madison.

McGrath, D. G., F. de Castro, C. Futemma, B. D. do Amaral, and J. Calabria. 1993. Fisheries and the evolution of resource management on the Lower Amazon floodplain. *Human Ecology* 21(3):167–196.

Meggers, B. 1971. *Amazonia: Man in Counterfeit Paradise*. Chicago: Aldine.

Moran, E. 1974. The adaptive system of the Amazonian caboclo. In C. Wagley, ed., *Man in the Amazon*, pp. 136–159. Gainesville: University of Florida Press.

Motta-Maués, M. A. and Maués, R. H. 1980. *O Folklore da Alimentação: Tabus Alimentares da Amazônia*. Belém: Falangola.

Nimuendaju, C. 1981. *Mapa etno-história de Curt Nimuendaju*. Rio de Janeiro: Instituto Brasileiro de Geográfia e Estastica (IBGE).

Nugent, S. 1993. *Amazon Caboclo Society: An Essay on Invisibility and Peasant Economy*. Oxford: Berg.

Pacheco de Oliveira, J. 1999. *Ensaios em Antropologia Histórica*. Rio de Janeiro: Editora Universidade Federal do Rio de Janeiro (UFRJ).

Porro, A. 1992. *As crônicas do Rio Amazonas: Tradução, Introdução e notas etnohistoricas sobre as Antigas Populações Indígenas da Amazônia*. Petrópolis: Vozes

Reis, A. C..F. 1979. *História de Óbidos*. Rio de Janeiro: Civilisação Brasileira.

Slater, C. 1994. *Dance of the Dolphin: Transformation and Disenchantment in the Amazonian Imagination*. Chicago: University of Chicago Press.

Sousa, I. de. 2004. *O Cacualista: Cenas da Vida do Amazonas*. Belém, Brazil: Editora Universitária, UFPa. (Orig. pub. 1876.)

Sternberg, H. O. 2000. *A Agua e o Homen na Várzea do Careiro*. 2 Vols. Belém, Brazil: Museu Paraense do Emilio Goeldi (MPEG).

Townsend, J. 1985. Seasonality and capitalist penetration in the Amazon Basin. In J. Hemming, ed., *Change in the Amazon Basin: The Frontier After a Decade of Colonisation*, vol. 2, pp. 140-157. Manchester: Manchester University Press.

Wagley, C. 1967. The folk culture of the Brazilian Amazon. In S. Tax, ed., *Acculturation in the Americas*, pp. 224–230. New York: Cooper Square Publishers.

15 Conservation, Economics, Traditional Knowledge, and the Yanomami

Implications and Benefits for Whom?

William Milliken

There are three principal ways in which native peoples of the Amazon may be seen to contribute toward regional forest conservation: by broadening and strengthening national and international concern about Amazon destruction, by managing the forests in a sustainable manner, and by providing the knowledge necessary for sustainable forest management by others. The first of these (i.e., the capacity of indigenous peoples to focus international interest on rainforest destruction—either actively or passively) was effectively demonstrated during the 1998 droughts and fires in the northern Brazilian state of Roraima (Allen, chapter 4 this volume).

By the time the Roraima fires were at their height, the world's press had tired of stories of burning rain forests, and few would have taken notice of what was happening had a small corner of forest in the Yanomami territory not been set alight. But, thanks partly to the efforts of their spokesmen, partly to the support they received from nongovernmental organizations (NGOs), partly to their exposure to the media during the devastating Roraima gold rush of the late 1980s and early 1990s, partly to the polemic that has surrounded the controversial and somewhat ingenuous interpretations of some of the earliest anthropologists to work among them, and partly to their close resemblance to the popular Rousseauesque image of what a native Amazonian ought to be, the Yanomami are one of the few peoples of the Amazon whose name is internationally known. Media attention at the end of the twentieth century remained focused on these people, and when their forest caught fire it became a "burning issue." After months of combustion and very little action, the Brazilian government finally mobilized what firefighting forces were available, and then it rained.

The Yanomami maintain an intimate relationship with their natural environment and resources (their forests), and have apparently developed a sustainable system of managing them. Their continued occupation of their land and their maintenance of these management systems have gone a long way toward preventing a substantial tract of the Amazon (their Brazilian territory covers some 192,000 square kilometers) from being destroyed or degraded by informal- and formal-sector miners, settlers, loggers, and ranchers, all of whom represent a very real threat to the biodiversity of the region. Thus, while struggling to maintain their livelihoods and their forests, the Yanomami have acted as important proponents of conservation. Had they not tenure to their land, this might be of limited significance in the long term. Fortunately, however, and thanks to their own efforts and those of their supporters, their territory was officially recognized by the government in 1991.

Traditional Resource Management

It can be, and indeed has been, argued that the only reasons why native Amazonians have maintained their environment in a relatively pristine state are that their populations were, at least until recently, significantly limited by disease and warfare, and that they have hitherto lacked the technology for wholesale resource exploitation. It seems clear from historical evidence that since the introduction of steel tools, for example, the Yanomami have been felling more trees and planting larger gardens than they used to. This technological advancement appears also to have allowed them to expand their territory beyond the Parima highlands, into areas that were formerly unsuitable for traditional cultivation techniques (Colchester 1984; Lizot 1984; Milliken, Albert, and Goodwin Gomez 1999).

Even if one accepts that the traditional resource management practices of the Yanomami are intrinsically sustainable, it is important to observe that these practices are dynamic, and that changes may have significant implications for conservation. One of the most important contemporary trends is the tendency toward larger and geographically more settled communities (i.e., a shift from the traditional seminomadic state), placing a heavier and more sustained burden on local natural resources. This is a common phenomenon among Amazonian peoples exposed to significant levels of contact with outside society and its goods and diseases. For logistical reasons, health organizations, missionaries, and government agencies generally prefer to work with sedentary communities, thereby encouraging the settlement process. One of the common consequences of this trend is degradation of the surrounding land, sometimes to the point where it can no longer sustain the population (Furley, chapter 7 this volume; Nortcliff, chapter 8 this volume). At Toototobi, for example, where Yanomami communities became permanently established around a mission in the early 1950s (the mission was subsequently replaced by a health clinic), there is virtually no land left for productive gardens. The Yanomami, of course, were already influencing their environment prior to these changes. In the highlands of their territory there are expanding patches of fern savanna that Huber et al. (1984) attribute to anthropogenic origins, primarily as a result of deliberate burning practices. Moreover, as Balée (1989) has pointed out, a significant proportion of what we used to regard as the pristine Amazonian environment now appears, at least to some degree, to have been influenced by man.

The village of Watoriki in the lowlands of the Yanomami area (state of Amazonas) is an example of an expanding and increasingly sedentary Yanomami community. This group has made a conscious decision to remain in the same place, close to the health clinic and airstrip where it settled in 1993. The effects of these changes are already making themselves felt on forest resources. Long-term settlement, for example, has contributed to a growing scarcity of *Geonoma* palm fronds for rethatching their roundhouses, a problem also afflicting long-settled Ka'apor communities in the eastern Amazon (Balée 1994). Game has also become scarcer, and this has been exacerbated by the abandonment of traditional trekking systems that previously took community members away from their village for as much as half the year. In the past, a serious shortage of game would have served as a stimulus for translocation. Instead, the people of Watoriki have extended their hunting range by constructing subsidiary camps for seasonal use.

Conservation, Traditional Knowledge, and Economic Development

In the long term, management of Amazonian environments by indigenous peoples will contribute most to the region's conservation if equilibrium can be found between the use of natural resources and the growing requirements for manufactured goods. More important, the establishment of such equilibrium is vital for the welfare of indigenous groups. Unsustainable practices, such as the logging activities that some indigenous Amazonians have been encouraged to adopt, are likely to prove detrimental not only to the environment but eventually to the people themselves.

It is in this context that we examine the relationship between traditional knowledge and economic development—in this case not at the national or regional level, but at the community level. Taking the Yanomami as an example again, several communities now find themselves at the transitional phase between an almost entirely self-contained and sustainable material culture and a developing economic relationship with the outside world. Can their traditional knowledge help them develop economies sufficient to satisfy their growing material requirements, and if not, then what are the implications for conservation?

Many Amazonian peoples, thanks to the accessibility of markets via the river system, are able to make a living from "classical" nontimber forest products such as Brazil nuts, rubber, and guaraná. For the Yanomami, however, products like these are proving unrealistic to transport from their isolated communities, and excessively demanding of labor. At Watoriki, for example, a significant demand has now developed for manufactured goods. The community has for some time been seeking a sustainable source of income, but so far without visible success. For a time the villagers collected Brazil nuts for sale in Manaus, but gathering these resources proved time consuming and relatively unrewarding. Shipping them downstream was slow and costly, and the price of the nuts was low. More recently a large manioc plantation was established on the bank of the nearest large river for *farinha* (flour) production, but this enterprise encountered similar obstacles. These difficulties are not unique to the indigenous peoples in the Amazon, but are also experienced by *caboclos* (mixed-race Amazonians) living subsistence lifestyles in remote regions.

In 2000 the Comissão Pró-Yanomami (CCPY), a Brazilian NGO working with Yanomami health-care and education projects, initiated an agroforestry (fruit-tree cultivation) project with selected Yanomami villages. Similar projects are already running (or are in the process of being established), with varying degrees of success, in a number of other indigenous communities in the Amazon. In the case of CCPY, the initiative was set up in the hope that it might contribute toward mitigating the effects of the communities' increasingly sedentary lifestyle. Initially, however, Yanomami interest in the scheme stemmed from the misplaced idea that fruit production might provide a source of capital. However the economics and logistics of exporting fresh rainforest fruits from a village such as Watoriki, where it takes too long to get out by boat and costs too much to get out by air, don't add up. Instead it seems that the Yanomami, like others in similarly isolated situations, need to find lightweight and high-value products if they wish to market their forest resources.

Medicinal Plants

Some very high figures have been quoted for the potential value of tropical forest medicinal plants—Mendelsohn and Balick (1995) estimated US$147 billion worldwide—and it is repeatedly pointed out that only 1 percent of the world's plants has ever been screened for pharmacological activity. Yet so far there has been little evidence of a bonanza, and as Posey, Dutfield, and Plenderleith (1995) have pointed out, less than 0.001 percent of the profits made on plant-based medicines has ever been returned to the source communities. Marketing their traditional knowledge of medicinal plants seems therefore to offer little chance of an economic solution for the Yanomami.

Nevertheless, the need for greater respect and protection of intellectual property rights, and for the establishment of viable legislation to enforce them, is incontestable. It is clearly right, for example, that unscrupulous drug companies should be prevented from unregulated exploitation of other peoples' knowledge, and that agribusiness should not be given free rein with other peoples' crop cultivars. Our approach, however, needs to be tempered with realism and perhaps an element of pragmatism. Expectations of financial benefits appear to have been raised unrealistically high, at both national and local levels, resulting in paranoia and in some cases paralysis of research efforts. Institutions and nations have progressively lost sight of the fact that in most cases neither their collective knowledge nor their genetic resources are unique to them, and that historically their greatest rewards have often been derived from the knowledge and genetic resources of *other* peoples and *other* countries. In practice, therefore, there may be a great deal more to be gained from an ethos of sharing than from one of protectionism.

Examining the plant resources upon which the Yanomami rely, for instance, one can see that many of their most important species and cultivars originated elsewhere. Trade and exchange of germplasm and knowledge tend to occur whenever one Yanomami group meets another, and likewise when they meet outsiders. It was through such trade routes that their bananas originally arrived from Southeast Asia, their sugarcane from New Guinea, their peach palms from the western Amazon, and their bottle gourds from Africa. Their calabashes were introduced relatively recently by missionaries, and several of their cassava varieties, including the increasingly important high-producing varieties, were likewise acquired from outsiders (either other indigenous groups or white people).

Continuing the example, studies over the last few years have shown that contrary to earlier reports, Yanomami knowledge of medicinal plants is probably as great as—if not greater than—that of any other Amazonian indigenous people (Milliken and Albert 1996, 1997). So far, at least 200 species have been recorded, and there are doubtless a great many more. It is clear from a detailed examination of this knowledge that a large proportion of it is by no means unique to the Yanomami. On the contrary, many of the medicinal uses are widely known across Latin America and have been published several times in the scientific literature (see table 15.1). Furthermore, a part of that knowledge was almost certainly borrowed from the Ye'kuana and other neighboring peoples (extinct or extant). In some cases the medicinal plants themselves, such as ginger, have been introduced from elsewhere. The Yanomami are not unique in this respect: in their review of the indigenous pharmacopoeia of northern South America, Bennett and Prance (2000) listed 216 introduced species in use among *mestizo* (mixed European and Amerindian ancestry) and indigenous peoples.

In other words, it would be unrealistic to attempt to copyright most of this information and equally unrealistic to expect significant financial gain from doing so. This is no argument for ignoring intellectual property rights, nor does it suggest that the Yanomami have nothing to gain from their traditional medicine. There may indeed be scope for deriving economic benefits from these resources, but perhaps from marketing the plants themselves rather than the knowledge of how to use them. The greatest benefits that medicinal plants offer the Yanomami, however, probably lie with their continued role in health care, helping them to resist the growing dependence on outside medical support and thus to avoid the health risks and loss of autonomy that such dependence engenders.

Conclusions

What knowledge or resources, then, traditional or otherwise, might help the Yanomami meet their economic needs and remain the long-term autonomous guardians of their forests? For them, as for other peoples in similar situations, the answers may lie in looking beyond their own traditional resource use and finding what can be learned from others. As with the potential agroforestry project already discussed, it is quite likely that the most appropriate species will prove to be ones that *other* people have discovered and domesticated, either in the Amazon or elsewhere.

For effective conservation and management of the Amazon region, we would benefit from a better understanding of its biology and ecology—of the complex interrelationships between its flora, its fauna, and its human

TABLE 15.1 Selected Medicinal Plants and Their Uses by the Yanomami and Others

Species	Family	Uses by Yanomami	Parallel Uses in the Neotropics	
Aspidosperma nitidum Benth.	Apocynaceae	Malaria	Brazil (widespread in Amazonia)	Malaria
Bauhinia guianensis Aubl.	Leguminosae	Diarrhea and stomachaches	Brazil (Waimiri Atroari); French Guiana (Wayãpi)	Diarrhea and dysentery
Clusia spp.	Guttiferae	Infected wounds	Colombia (Makuna, Karijona, and Makú); Mexico; Costa Rica	Sores, wounds, and leprosy
Cyperus articulatus L.	Cyperaceae	Fevers	Ecuador (Secoya); Brazil (Tiriyó)	Fevers
Geophila repens (L.) I. M. Johnst.	Rubiaceae	Eye infections	Ecuador (Ketchwa); French Guiana (Palikur)	Fungal infections
Peperomia macrostachya (Vahl) A. Dietr.	Piperaceae	Fevers	Colombia (Taiwano); French Guiana (Wayãpi)	Fevers
Peperomia rotundifolia (L.) Kunth	Piperaceae	Coughs	French Guiana (Wayãpi)	Coughs
Philodendron spp.	Araceae	Scorpion stings	Peru; Brazil (Waimiri Atroari); Colombia	Insect stings/bites
Phlebodium decumanum Willd.	Polypodiaceae	Coughs and congestion	Colombia	Pulmonary disorders
Phytolacca rivinoides Kunth and Bouché	Phytolaccaceae	Chigger holes in feet	Colombia (Andoke); Ecuador	Inflamed and infected wounds (disinfectant)

Protium unifoliolatum Spruce ex Engl.	Burseraceae	Congestion and respiratory infections	Colombia (Tikuna); Brazil (Ka'apor)	Colds and congestion
Siparuna guianensis Aubl.	Monimiaceae	Dizziness	Brazil; Colombia (Tikuna)	Headaches; nausea
Spondias mombin L.	Anacardiaceae	Fevers	Mexico; Brazil	Fevers
Uncaria guianensis (Aubl.) Gaud.	Rubiaceae	Diarrhea and stomachaches	Bolivia (Chácobo); Colombia	Diarrhea and stomachaches
Urera baccifera (L.) Gaud.	Urticaceae	Body aches/pains (including headaches)	Ecuador (Siona); Colombia; Brazil	Muscular pains; paralysis
Vismia angusta Miq.	Guttiferae	Fungal skin infections	Colombia (Tikuna); French Guiana (Wayãpi); Brazil	Fungal infections
Zingiber officinale Roscoe	Zingiberaceae	Toothaches	Colombia	Toothaches

inhabitants. This is one of the aspects of traditional indigenous knowledge that has been most neglected by researchers, at least in terms of systematic documentation. Tropical ecologists have tended to set about their own observations and experiments without bothering to discuss the subject with local people, who in many cases could tell them in a matter of hours what it would take scientists years to find out by themselves.

This brings us to the third means by which indigenous peoples contribute to conservation—through their knowledge per se. In many cases indigenous peoples are in a position to provide conservationists and forest managers with information that, if properly handled, could make an important contribution to sustainable development and conservation initiatives in the Amazon. Indigenous peoples might secure a livelihood from this knowledge, either as compensation for knowledge transferred or for their role as forest "managers." It has been suggested that the Yanomami could market themselves as teachers of Amazon ecology, and that as guardians of an increasingly valued commodity they might charge researchers for access to the forest itself (J. Boyle, pers. comm.). The future of the Yanomami, like that of other indigenous peoples in Brazil, is uncertain, but their traditional knowledge has the potential to play a significant role in determining it.

References

Balée, W. L. 1989. The culture of Amazonian forests. In D. A. Posey and W. Balée, eds., *Resource Management in Amazonia: Indigenous and Folk Strategies*, pp. 1–21. Advances in Economic Botany 7. Bronx: New York Botanical Garden.

Balée, W. L. 1994. *Footprints of the Forest: Ka'apor Ethnobotany—the Historical Ecology of Plant Utilization by an Amazon People*. New York: Columbia University Press.

Bennett, B. C. and G. T. Prance. 2000. Introduced plants in the indigenous pharmacopoeia of northern South America. *Economic Botany* 54(1):90–102.

Colchester, M. 1984. Rethinking stone age economics: Some speculations concerning the Pre-Colombian Yanoama economy. *Human Ecology* 12(3):291–314.

Huber, O., J. A. Steyermark, G. T. Prance, and C. Alès. 1984. The vegetation of the Sierra Parima, Venezuela-Brazil: Some results of recent exploration. *Brittonia* 36(2):104–139.

Lizot, J. 1984. *Les Yanõmami centraux*. Paris: Cahiers de l'Homme, Éditions de L'EHESS.

Mendelsohn, R. and M. J. Balick. 1995. The value of undiscovered pharmaceuticals in tropical forests. *Economic Botany* 49(2):223–228.

Milliken, W. and B. Albert. 1996. The use of medicinal plants by the Yanomami Indians of Brazil. *Economic Botany* 50(1):10–25.

Milliken, W. and B. Albert. 1997. The use of medicinal plants by the Yanomami Indians of Brazil. Part II. *Economic Botany* 51(3):264–278.

Milliken, W., B. Albert, and G. Goodwin Gomez. 1999. *Yanomami—A Forest People.* Kew: Royal Botanic Gardens.

Posey, D. A., G. Dutfield, and K. Plenderleith. 1995. Collaborative research and intellectual property rights. *Biology and Conservation* 4(8):892–901.

16   The Commodification of the Indian

Alcida Rita Ramos

Frontiers Old and New

The advance of the economic frontier has been a recurring theme in the Amazon since the first decades of the sixteenth century. In the nearly five-hundred-year history of Western rapacity, the region has witnessed a number of ways in which non-Indians appropriated its wealth, first, at the expense of indigenous peoples, and later, of regional populations as well. In their attempts to secure quick profits, colonizers promoted massive indigenous slave labor, spread devastating epidemics, looted forest products (*drogas do sertão*), and grabbed territories, water, and subsoil resources as if Amazonia were an immense no-man's-land, literally up for grabs (Pinto 1980; Ramos 1991). The infernal paradise of both tropical fevers and tropical bounty served well the dominant vision of Amazonia as an empty expanse to be exploited for future gain. But this vision has not been limited to the Amazon. It has appeared whenever the "Whiteman" (Basso 1979) has ventured into the "wilderness." In Africa, for example, "The European improving eye produces subsistence habitats as 'empty landscapes,' meaningful only in terms of a capitalist future and of their potential for producing a marketable surplus" (Pratt 1992:61).

The twentieth century has seen acceleration in the swell of the economic frontier, particularly in the 1960s and 1970s when South American governments entered the era of megaprojects financed by international capital. As devastation of the rain forest reached an unprecedented speed, numerous indigenous peoples found themselves severely depopulated by both new and old infectious diseases, and without an adequate subsistence base, were forced to join the multitude of citizens who live well below the poverty line (Davis 1977; Ramos 1994).

After decades of struggle for the guarantee of their land rights, some indigenous peoples in Brazil are still desperately fighting to have their territories legally secured. Although most other ethnic groups have had their territories officially demarcated, invasions and other kinds of abuses continue to occur throughout the region. Nevertheless, as the legal battle for securing indigenous lands slowly but steadily approaches its goal, the demand for land rights is being replaced by another type of call. Economic interest in the labor force and in what was traditionally designated as natural resources is decreasing compared with an interest in other sources of profit. The Indians are now being bombarded with new threats. Not only do commercial interests continue to exploit natural products (e.g., soil, latex, gold, and Brazil nuts) as integral entities, but increasingly these interests covet the dissected fragments of the Amazonian gene pool. In short, the economic frontier has entered its microscopic phase. Associated with this new demand is the attempt to tap the indigenous cultural pool as a shortcut to reaching the final scientific-commercial product more quickly—for native guidance helps save researchers the often excruciatingly time-consuming task of random sampling. Thus turned into marketable items, Amazonian Indians now face the herculean task of having to inform themselves, however minimally, about the baffling fields of genetics and world economics in order to better defend their persons and their resources from scientists and other highly sophisticated predators. In other words, in addition to enduring the habitual modes of spoliation, the Indians are now being attacked at the very heart of their cultural and physical integrity. Indigenous knowledge and selected cultural features have served as appealing bait in advertising campaigns, generating profits—the dimensions of which the best-informed Indians were until recently unable to fathom—for non-Indians. These new forays into the innermost spaces of indigenous minds and bodies may not directly endanger lives, but they pose a threat to the Indians' social integrity nonetheless. Pressure is mounting on the Indians to either conform to market forces, or to surrender their precious territories and resources to the "rational" demands of "development" (read Western market). In order to justify the protection of their lands, many indigenous peoples are obliged to prove themselves productive in market terms. To put it in a nutshell, the expansion of the knowledge and genetic frontiers is continuing the plunder begun by the old-style natural resources frontier.

Three fronts have lately troubled indigenous peoples, particularly in the
Amazon. I am referring to the commercial exploitation of their knowledge,
their genes, and their image. Although closely related, these aspects deserve
to be examined separately, since each represents a distinct mercantile province.

Knowledge

The virile vision of a European on horseback, surrounded by excited natives, triumphantly reaching a promontory in the act of "discovering" a geographical sensation—be it the source of the Niger or the Nile, or any other
Western fetish—became a praised icon in the eighteenth and nineteenth
centuries (Pratt 1992:202). A mixture of arrogance and heroism, the white
explorer, usually male, entered Western history as the one who, in the name
of knowledge (read Western knowledge), endured unbelievable hardships
in indomitable jungles. He can be exemplified by a number of travelers in
South America and, especially, Africa, like Mungo Park in search of the
sources of the Niger: "He traverses desert wildernesses, suffering the trials of
thirst, beasts, and banditry" (Pratt 1992:75). Yet the promontory syndrome
was cultivated to "render momentously significant what is ... practically a
non-event" (202), for all the explorer needed to do—and actually did—was
to ask the natives about lakes, watersheds, and the like. As Pratt goes on to
remark, "Discovery in this context consisted of a gesture of converting local
knowledges (discourses) into European national and continental knowledges
associated with European forms and relations of power" (202).

Well, the figure of the intrepid explorer or naturalist roughing it in the
jungle is being replaced by that of the pragmatic scientist who wants to get
to the point as quickly as possible, using Indian knowledge as a shortcut.
The promontory has flattened considerably. In centuries past, the natives
were acknowledged for being carriers, not of useful knowledge but merely of
the Whiteman's gear, if not of the Whiteman himself (Buarque de Holanda
1986; Taussig 1987). (Who was whose burden, we may well ask.) Nowadays
it is proper to praise indigenous ingeniousness. For instance, according to a
statement put out by the World Bank in 1981, "Tribal groups can make valuable contributions to the wider society, especially to the national society's
knowledge of socioeconomic adaptations to fragile ecosystems" (World Bank
1981:3). Some actually regret that there isn't more indigenous knowledge
available to non-Indians; for example, the Brazilian Eletrobrás technocrats
lament the Indians' illiteracy because it precludes the preservation of the

"valuable collection of knowledge accumulated by indigenous tribes, especially with regard to the thousands of native biological species that exist in
their ecosystems and with which they have lived for centuries" (quoted in
Viveiros de Castro and Andrade 1988:19).[1] Lack of writing, however, does
not seem too serious a deterrent:

> Tapping this reservoir of knowledge has already proven effective. Three-
> quarters of the plants that provide active ingredients for prescription drugs
> originally came to the attention of researchers because of their uses in tra-
> ditional medicine. Accordingly, the NCI [U.S. National Cancer Institute]
> collection strategy involves close attention to indigenous medical prac-
> tice and especially to the expertise of traditional healers and *curanderos*.
> Similarly, the USDA's [U.S. Department of Agriculture] crop germplasm
> acquisition policy now gives priority to obtaining samples for which the
> ethnic source of the cultivar is described.   (Kloppenburg 1991:15)

Repeated failures to turn quick profits from the Amazon into long-term
returns, plus the emergence of the ideology of environmentalism, particularly in industrialized countries, have steered the market's attention to indigenous systems of knowledge. There is now a demand for cooperation
between economic agents and anthropologists, ethnobiologists, and other
experts who are expected to know how to "bridge cultures." Comparing traditional knowledge with industrial knowledge, Cunningham considers that
"ethnobiologists and anthropologists are in a position to act as brokers to
facilitate a partnership agreement for the benefit of both rural communities
and urban-industrial society" (1991:4). Anthropologists in particular are put
in the disturbing role of indigenous surrogates, the next best choice to the
real thing, namely real Indians (Ramos 1994).

Even among Indian sympathizers, respect for cultural wealth is not always accompanied by respect for ethnic agency. Two common assumptions
seem to underlie their positions. One is that it is preferable to deal with specialists in indigenous cultures than with the Indians themselves. The other
is the tacit disregard for the Indians' thoughts and feelings: would they want
to put their knowledge at the service of those directly responsible for their
centuries-old plight? Like fauna, flora, or stones, the Indians seem to be just
there, passively accessible to Western science and markets. When the Indi-

---

1. All translations from Portuguese into English have been rendered by the author of
this chapter.

ans and their allies protest the poaching of biological materials—removed from Indian territories as if they were free for the taking—entrepreneurs and scientists engage in elaborate justifications for denying indigenous property rights to the resulting research products. The casuistic filigree of patent laws is worth examining.

As the argument goes, there is a basic difference between the genetic management of species in their natural habitat (in situ), and the genetic processing of samples in laboratories or botanical gardens (ex situ). Although the Indians may be entitled to compensation for in situ species, these resources are considered to be the "common heritage of humankind" (Kloppenburg 1991:14). As to ex situ products, which are comparable to quotations taken out of context, Indians are denied royalties because these products are classified as human manufactures and hence the object of private property. The imbalance of world power imposes U.S. patent laws upon the rest of the world, especially southern tropical countries. According to U.S. law, what is in nature cannot be the object of private property, but the moment you manipulate elements extracted from nature, you have the right of patent so long as your product is new, not common sense, and clearly useful (RAFI 1995:8–10). Furthermore, "you can patent the invention of an inventor in another country—if it hasn't been patented at home or if an article hasn't been written about it. (American inventions, however, are always protected—even if not unpatented or unpublished.)" (McGou'an 1991:20).

The result has been a predictable tendency for resources in nature to be transformed into objects made by humans, or as Latour (1993) would put it, into quasi objects or hybrids, a composite of natural and fabricated materials. Whenever the natural "impurity" of an entity or substance (for example, different DNA cells mixed together in a living organism) is isolated, purified, and altered, it is no longer a product found in nature. It becomes a man-made artifact, and as such, it is patentable and the source of private revenue. Hence the protection of nature's products from privatization "has been converted into something hollow" (RAFI 1995:8). Far from being a spontaneous, chance phenomenon, this state of affairs is the result of concerted action on the part of drug corporations. "Business interests in the developed nations have worked very hard over the past ten years to assemble a legal framework that ensures that genetically engineered materials—whole organisms, tissue cultures, cells, DNA sequences—can be owned" (Kloppenburg 1991:16). In other words, the distance between impurity in nature and purity in the lab is the distance between natives deprived of fair compensation and the business groups and individuals likely to make millions.

Whether actual millions are made is not the real issue. From the perspective of the indigenous peoples involved, what matters is, on the one hand, the often clandestine manner in which prospecting is undertaken, and on the other hand, the high expectations, unrealistic as they may be, that the product of research will bring big gains to the communities. "We are not opposed to having our knowledge favor non-Indians.... But we do not accept that our knowledge be used without our due permission. Researchers and industries cannot get rich at our expense without the proper compensation due us" (Wapichana 1999:42).

As genetic research progresses, more indigenous peoples are affected by Western exploitation of their resources. Amazonia is one of the richest regions in the world in biodiversity. In the Brazilian Amazon alone one finds 30 percent of all tropical forests, concentrating more than 50 percent of the Amazon's species. It is in the Amazon region that 98 percent of indigenous territories are located. "Therefore, the correlation between indigenous peoples, tropical forest, and biodiversity is obvious" (Santilli 1996:19). Consequently researchers are constantly probing Amazonia for marketable forest products, be they the anticoagulant found in Uru-eu-Wau-Wau territory, the pain killer from Marubo villages, the ubiquitous *urucum* (*Bixa orellana*), or the equally common *ayahuasca* (*Banisteriopsis caapi*).

I shall now briefly refer to the case of the Wapichana Indians, based on work by Ávila (2001). This is one of scores of examples of the unethical patenting of indigenous resources occurring the world over, and specifically in the Amazon.

- *1840.* "I have endeavoured to make the reader acquainted with some of the trees which, with reference to their timber, are of importance to commerce. Of equal, if not greater value, are the trees and plants from which medicinal substances may be obtained" (Schomburgh quoted in Ávila 2001:64).
- *1996.* "It has been known for some time that Amerindian peoples of the Rupununi area of Guyana, South America, chew the nuts of the greenheart (*Ocotea rodiaei*) as a crude form of contraception. Also, infusions of the bark of the greenheart tree have been used as a febrifuge and as an antiperiodic in fevers" (United States Patent No. 5,569,456, Oct. 29, 1996, quoted in Ávila 2001, Anex 3:101).
- *1998.* "The term 'cunani' has long been used by Amerindians for a group of fast-acting fish poisons. Such fish poisons are generally derived from plants, and especially from the leaves thereof. South America probably pos-

sesses greater numbers of recorded fish-poison plants than any other continent" (United States Patent No. 5,786,385, Jul. 28, 1998, quoted in Ávila
2001, Anex 3:106).

• *1999*. During an international seminar on rights over biodiversity, Wapichana leader Clóvis Ambrósio expressed his irritation at the news that U.S.
patents for plants grown in Wapichana lands straddling the Brazil-Guyana
border had been registered in the name of Gorinsky:

> My people, the Wapichana, live in the Brazilian grasslands and in
> British Guiana.... With our common knowledge of the vegetation ...
> we use a plant named *cunani* for fishing. We also produce medicines
> extracted from a tree known as *tibiru*, or greenheart.... Many of our
> kinsmen don't even imagine what our knowledge can represent to the
> [Western] industry. That is why chemist Conrad Gorinsk[y], the son of
> a Wapichana woman and a German man ... researched the *cunani* and
> the *tibiru* while promising to help our communities with medicines.
> He never did.... Mr. Conrad Gorinsk[y] has patented the cunaniol
> and the rupununi in the United States, Europe, and Great Britain. He
> has contacted multinationals for the exploitation of his "discoveries."
>
> (Wapichana 1999:42)

Two years earlier, the National Indian Agency (FUNAI) tried, apparently
without success, to guarantee compensation for, or the cancellation of, a
patent registered in Great Britain to the chemist Conrad Gorinsky for the
substance extracted from the seeds of the *tibiru* tree—the substance that
the Wapichana used as a contraceptive (Braga 1997). During the 1990s, two
other indigenous substances were also patented in Europe: *cunani (Clyba-
dium sylvestre)* was patented in 1998; *tibiru (Octotea rodiaei)*, also called *tipir*
or greenheart, was patented in 1994 and given the name Rupununines, after
Rupununi, the Wapichana homeland in Guyana (Ávila 2001:67).

Conrad Gorinsky belongs to one of the powerful families that for more
than a century have occupied indigenous lands, primarily for cattle ranching, on both sides of the international border (Rivière 1972, 1995; Farage
1991). Although the son of a Wapichana woman, Gorinsky is not recognized
either in Brazil or in Guyana as a member of the Wapichana group because
ethnic recognition is based on social rather than on genetic criteria. The
Wapichana are angered that his family holds large portions of land at their
expense. That he is now patenting indigenous products adds insult to injury.
Decades of struggle to eradicate illegal ranching have sharpened indigenous
consciousness about the extent of both external abuses and indigenous con-

stitutional rights. No wonder Indians make a close association between biopiracy and the issue of land demarcation. Although these matters involve two countries, as far as the indigenous peoples of the Guiana region are concerned lack of credibility and private rapacity do not stop at national borders. For this reason, cross-boundary indigenous organizations join forces when it comes to expressing their grievances (Ávila 2001:80).

Again Clóvis Ambrósio: "If the government has not resolved the problems of land demarcation, if the perpetrators of crimes against our communities go unpunished, if predatory fishing and illegal exploitation of minerals and timber happen every day with no concrete measures to curb them, then how is the Brazilian state going to protect biodiversity and our traditional knowledge?" (Ávila 2001:74). Negligence on the part of the federal government along with the explicit anti-indigenous policies of all Roraima governors to date make up an explosive combination that has cost the indigenous peoples of east Roraima great losses in resources and lives, a situation that is far from being resolved. The Web site of the Conselho Indígena de Roraima (CIR), available at www.cir.org.br, gives an idea of the social unrest that deeply disturbs the savanna communities practically on a daily basis.

I would like to call attention to the role of the Brazilian press as detonator in the explosive field of biopiracy. Ambrósio himself first heard about the Gorinsky patents from the daily *Folha de São Paulo* while he worked at the CIR headquarters in the capital town of Boa Vista (Ávila 2001:71). In the 1990s Brazilian newspapers reported numerous cases of scientists, filmmakers, and other professionals, especially from abroad, to whom FUNAI granted official authorization to carry out their declared activities; once in the field, however, those groups engaged in illegal collecting of samples of genetic materials. "When one hears talk about biopiracy," says a journalist from *O Estado de São Paulo*, "one imagines smugglers, in the still of the night, crawling in the forest to take away active substances, plant seedlings, animals, the secrets of Brazil's natural wealth that the Indians franchise for a pittance. This furtive method may actually exist, but there is another, more blatant system: many biopirates come in through FUNAI's front door in Brasília" (Sant'Anna 1998:A20).

Examples of biopiracy multiply, as attested by the frequent headlines. From one Brazilian newspaper we read: "The fight for third world genes" (*Folha de São Paulo* 1995), "United States patents virus from Indian and is accused of 'vampirism'" (1996), "Biopiracy hits Amazon forest" (1997a), "Attorney General's office in Amazonas will investigate biopiracy" (1997b), "Acre prohibits foreigners in the forest" (1997c), "Biopirates act freely in

Amazonia" (1997d), and "Biopiracy in the country is a police case, says São Paulo professor" (1998).

The effort to curb biopiracy is a rare case in which national and indigenous interests converge. Brazil, as well as other nations, has vehemently protested against such predation and has used events such as the 1992 Earth Summit in Rio de Janeiro as opportunities to elicit more equitable commitments from the Northern nations. Official reactions from Southern countries have reinforced the indigenous sense of indignation.

Through their increasing activism, indigenous peoples have become aware that their cultural patrimony is being grossly consumed for purposes totally alien to their interests. A number of international organizations have emerged to help curb abuses, although with less than ideal force and effectiveness (Posey 1991). Various meetings have served as arenas for native peoples from various parts of the world to discuss the matter, draw up plans of action, and write declarations and resolutions (IWGIA 1995).

Among these, the Coordination of Indigenous Organizations in the Amazon Basin (COICA) firmly established its position at a regional meeting held in Santa Cruz de la Sierra, Bolivia, in September 1994. The arguments emphasized the cultural embededness of indigenous knowledge, a complexity always ignored by business-oriented agents. Pointing out the "colonialist," "racist," and usurping character of commercial ventures in biodiversity, COICA members insisted on the inextricable bond of biodiversity, knowledge, and culture with the indigenous conception of territoriality, an all-encompassing notion that cannot be reduced to treating land as a mere dwelling place (Ramos 1986:13–22; Colchester 1995:6) or a potential commodity. They stressed the collective and intergenerational constitution of indigenous rights and social systems, declaring that "each generation is obliged to preserve for the next" (IWGIA 1995:24). They also repudiated the application of patents and other intellectual property rights to life forms, and maintained "the possibility of denying access to indigenous resources as well as protesting against patents or other exclusive rights over what is essentially indigenous" (24).

All this is forcing a reorientation of the premise that what is in nature is for free, but what is in the laboratory is for profit. At the 1992 Earth Summit, 170 countries signed the Convention on Biological Diversity, asserting that the products of genetic and chemical resources are no longer the "common heritage of humankind," but of the countries and communities where they are found. Nevertheless the issue of Intellectual Property Rights (IPR) is not generally accepted among the Indians. For instance, COICA has expressed its distrust of IPR because they see it as one more subterfuge for exploitation: "The system of intellectual property for indigenous peoples means the legitimation of undue appropriation of the knowledge and resources of our peoples for commercial ends" (IWGIA 1995:24). One reason for this distrust is the fact that property rights are conceptualized in terms of individual ownership, which contradicts indigenous ethics. Another reason is the very impropriety of subjecting collective knowledge and resources to the logic of Western capitalism. Far from refusing to share knowledge and resources with the rest of humanity, indigenous peoples simply object to their privatization and commodification. COICA emphasizes the capacity of indigenous peoples to manage their own traditional knowledge and declares that they are "open to offer it to humanity whenever their fundamental rights to define and control their knowledge are protected by the international community" (17).

The issue of knowledge for profit has divided anthropologists. Some defend the policy of fair compensation for the transmission of indigenous information; others feel that translating native knowledge into copyright dividends is to debase it and to subject indigenous peoples to the rapacious logic of the market: "It is not by advocating an increased commoditization of knowledge that we will alleviate the plight of tribal minorities, but rather by fighting for a world where ownership would not be the sole measure of one's ability to control one's destiny" (P. Descola, quoted in Strathern et al. 1998:119). Against the Western fixation with ownership, "we might do better to formulate our concerns in terms of accountability and responsibility rather than ownership" (P. Harvey, quoted in Strathern et al. 1998:125). Meanwhile, as we wait for such enlightenment to descend upon the Western world, native peoples of the earth are continuously despoiled.

Turning now to the cultural content of knowledge, the indivisible aspect of a cultural legacy cannot be overemphasized. What COICA members designated under the notion of territoriality encompasses a whole universe of beliefs, social relations, and practical knowledge that makes up a given society. This point is repeatedly underscored by native peoples the world over. In the words of Stella Tamang from Nepal: "In reality we are talking about the whole way in which we conceive the world, our cultures, our lands, our spirituality as indigenous peoples. All of this is joined together. We must examine this total panorama" (IWGIA 1995:14). Hence Indian knowledge cannot be mechanically transferred from one type of society to another as if by mental transfusion. What business-minded people who tout their admiration for the sustainability of indigenous ways of using natural resources fail to appreciate is that this wisdom involves much more than simply identifying, describing, and utilizing fauna and flora. Such knowledge is moored to worldviews and lifestyles so different from the Western mode as to be either undetectable or utterly baffling, and in any case, practically incompatible with the matter-of-fact, predatory vocation of industrial activities. What often passes for quaint customs—food taboos, supernatural constraints on activities like hunting and fishing, metaphysically motivated divisions of labor, elaborate rituals before economic activities—may turn out to be the cornerstone of indigenous knowledge. Pragmatic as it is by definition, business cannot be troubled with the cultural intricacies of native life, which makes one wonder if behind the frequent displays of respect for indigenous wisdom there is anything besides paying lip service to political correctness.

Genes

Through the genescapes of Amazonia the commodification of the Indian reaches its climax. One practice, in itself ethically questionable if done unilaterally, is to collect indigenous blood in the name of science, as in the case of the Human Genome and Human Genome Diversity projects. Another practice, not disconnected from the former, is the seizure of peoples' genes for commercial ends. Both of these sides of the gene rush depend on gene-rich parts of the world (Kloppenburg 1991:14), such as Amazonia. This is quite explicitly put to justify the Human Genome Diversity Project, which was created "after a plea published in *Genomics* calling attention to the vanishing of indigenous peoples and their absorption by predominant genetic clusters" (ISA 1996:22).

It is quite clear that these two fronts of genetic prospecting are not separate endeavors: "Whether collected by government, university, or corporate scientists, the genetic and cultural information extracted from the South is ultimately intended to be applied to some useful purpose" (Kloppenburg 1991:16). A 1993 RAFI (Rural Advancement Foundation International) report warned of the growing tendency to patent forms of life. It mentioned the intention of the Human Genome Diversity Project to process samples of human tissue collected from more than 700 human populations, including native peoples around the world. For instance, the U.S. government claimed universal patent rights over the cell line of a twenty-six-year-old Guaymi woman from Panama (RAFI 1995:11), but withdrew its claim after encountering mounting pressure from various organizations, including GATT (General Agreement on Trade and Tariffs); nevertheless it continued to hold the Guaymi cell line as its exclusive property. The genetic material extracted from the Guaymi woman, who had been hospitalized in Panama

City with advanced leukemia, was "immortalized" in liquid nitrogen in the labs of the American Type Culture Collection, the largest bank of patented biogenetic products in the world, with headquarters in Rockville, Maryland (Arias and López 1995:19).

The Guaymi were predictably appalled: "Never did I imagine," declared Isidro Acosta, president of the Guaymi General Congress, "that people could patent plants and animals.... To extract human DNA and patent its products violates the integrity of life itself, and our deepest sense of morality" (RAFI 1995:11). Supported by various entities, among them the World Council of Indigenous Peoples, the Guaymi General Congress launched a campaign for the repatriation of the cell line, with the result that the U.S. government withdrew its patent request (Braga 1997).

Cases such as that of the Guaymi aroused comments like those made by Alejandro Argumedo of the Indigenous Peoples Biodiversity Network, based in Canada: "The renouncement is cause for celebration on the part of indigenous peoples. At the same time, it provides us all with the opportunity to reflect upon the immorality of industrial countries that allow the commodification of human cells, genes, and other tissues" (IWGIA 1996:58).

Biogenetic technology is worthy of a "brave new world" fantasy. The cell lines are called immortal because they can be kept alive indefinitely under artificial conditions of controlled temperature, nutrition, and sterilization. Thus preserved, human cell lines are an inexhaustible source of the donor's DNA. Bioengineers "have inserted alien genes, including from humans, into the chromosomes of various animals such as hogs, sheep, goats and chickens. In the future, genetic engineering will allow scientists to combine genetic material from human beings and animals to produce animal-human hybrids" (RAFI 1995:9). The future is here, if not yet with humanoid hybrids, then with the transgenic food we now eat, even if unwillingly and most often unknowingly.

Inflamed over genetic expropriation, John Liddle, director of the Australian Aborigines Central Congress, fumed, "In the last two hundred years, non-aboriginal people have taken our land, our language, our culture, our health—even our children. Now they also want to take away from us the genetic material that makes us Aborigines" (RAFI 1995:12). The violence of the expropriation is so great that it has driven people like Liddle to fall into the genetic determinism trap set up by the West. Once native peoples succumb to the snare of biological naturalization—"genetic material that makes us Aborigines"—they indeed run the risk of meeting doom, as history attests in both old and new cases of genocide.

The issue of informed consent, complex as it is, has brought to light another set of considerations. The difficulties of explaining complicated genetic notions to laypeople, often in a language unknown to the researcher, are frequently evoked as an alibi for dispensing with informative accounts of fieldwork. In dodging the issue, researchers may simply lie about their research projects. One such case involved the Karitiana and Suruí Indians in the Brazilian state of Rondônia, who denounced researchers from Ohio State University (Braga 1997) for having taken blood samples from the villagers when they were only authorized to make a documentary film. Furthermore the researchers made the ludicrous claim to the Indians that they were interested in searching for a mythological entity, the elusive *mapinguari* sloth. The Indians took the case to a commission of the National Congress to investigate biopiracy. What happened to the blood after it was taken away is not known. The same indigenous groups from the state of Rondônia were also the object of genetic research (unauthorized by the Brazilian government) by Francis Black, from the Department of Epidemiology and Public Health, Yale University. Black and his team presented their findings from the Karitiana, the Suruí, and the Mexican Campeche. Candidly and as a matter of course, they deposited "five cell lines from unrelated individuals … in the … Human Genetic Mutant Cell Repository at the Coriell Institute for Medical Research (Camden, New Jersey) [after which the cell lines became] publicly available" (Kidd et al. 1991:778). In April 1996, Coriell Cell Repository was advertising the sale of Karitiana and Suruí DNA samples at US$500 a sample (Santos and Coimbra 1996:7; Braga 1997). Affronted by these attitudes, the Karitiana are now little disposed to cooperate with researchers. Instead they submit them to virtually impossible demands, such as requiring that they finance the paving of 100 kilometers of road surface for a graduate student whose 2003 project was to study the effects of biopiracy on Karitiana life!

The Yanomami present another case where disregard for informed consent resulted in delayed, but mighty, consequences. Stunned by the uproar triggered by the publication of *Darkness in El Dorado* (Tierney 2000), a journalist's book that offers more scandal than seriousness, the U.S. anthropological community engaged in one of the longest and ugliest confrontations in recent times. The book's author, Patrick Tierney, all but openly accused the late geneticist James Neel and his research team of having caused a measles epidemic among the Yanomami during the course of a vaccination campaign conducted in the 1960s and 1970s as part of a scientific experiment funded by the U.S. Atomic Energy Commission. A subsequent analysis of Tierney's evidence by a group of Brazilian medical doctors set the record straight by affirming that vaccines do not produce epidemics, even though they may cause lethal side effects in individuals (Lobo et al. 2001:18). There were, however, enough breaches of research ethics left in Tierney's denunciations to render the Neel episode the focus of a long and acrimonious dispute (Albert 2001). One of the manifestations of unethical behavior was the false explanation given to the Yanomami that the purpose of taking blood samples was to cure their illnesses. Bribing the Indians with piles of trade goods, Neel's team extracted blood samples that are now stored in five U.S. research institutions (CCPY 2002b:2).

In the wake of accusations and counteraccusations of unethical behavior on the part of scientists, a virtual wall was erected and fiercely disputed in what came to be known as the science versus antiscience debate. Biomedical researchers defended Neel's actions in the name of pure science, while culturally oriented anthropologists reacted critically to the notion that ends justify any means. The American Anthropological Association faced the problem by creating a fact-finding task force for the purpose of reaching a conclusion about what had happened thirty-five years earlier. As expected, the results were inconclusive and the scandal dwindled into semioblivion. But the Yanomami were startled by the news that the blood of their relatives, many of whom had already died, was circulating in foreign lands without their consent and jeopardizing the postmortem peace of its owners. The Yanomami want the repatriation of the blood samples and are considering the possibility of demanding compensation for damages—in part because they were misinformed about the real purpose of the blood-taking (CCPY 2002a, 2002b, 2003).

It is often argued that it is difficult to explain the purpose of, say, genetic research to a "monolingual" community, or worse, to people who couldn't possibly understand the complexities of Western scientific thinking, even when it is explained in their own language; this argument camouflages either the linguistic incompetence of the researchers or their indifference to what the research subjects may think. The recent experience of the Yanomami who have been successfully admitted to the official body of microscope technicians after relatively short periods of training belies such patronizing arguments. The Yanomami know enough about the Western ethology of malaria to do their technical job splendidly. There is nothing that cannot be explained satisfactorily to laypeople when researchers have an honest attitude and professional aptitude.

The controversy generated by *Darkness in El Dorado* provides a good opportunity to rekindle long overdue discussion about such issues as informed consent in the context of field research, the neutrality of "objective" science, and the social responsibility of the field-worker. There is, of course, a

significant difference between biomedical or genetic investigation and ethnographic inquiry: the former involves doing research *on* human beings, the
latter *with* human beings. Collecting samples of people's physical substances
can be more intrusive of lives and beliefs than is collecting myths or kinship
charts. The ethnographic record of the Yanomami, built on research with,
not on, them, points to the vital importance of blood and other substances
to the destiny of both their living and their dead; this has special significance
if these substances fall into enemy hands (Albert 1985). This explains why
the Yanomami were so upset when they heard that their blood was being
handled by total strangers and was completely out of their control.

The Yanomami case also points up the difference between work done in
situ and work done ex situ, where the researcher analyzes the data collected
in the field and prepares it for publication. Neel conducted genetic research
that involved collecting bodily materials, whereas his assistant, anthropologist Napoleon Chagnon, gathered data on genealogies, migrations, marriage
patterns, and such. Neel worked on research subjects, Chagnon with them;
nevertheless, each in his own way caused problems for the Yanomami. In
the case of the geneticist working in situ, the worst (but not sole) problem
occurred because of the collection and removal of the Indians' bodily substances. In the case of the anthropologist working ex situ, the worst (but
again, not sole) problem was the depiction of the Yanomami in highly de-
rogatory terms.

This example brings up yet another difference between biomedical and
ethnographic research. If, for instance, the ethnographer incurs a breach
of etiquette by demanding to know secret personal names, the Indians may
simply refuse to cooperate but will not necessarily curtail the research. But if
the geneticist asks for blood, saliva, or hair samples, the Indian's refusal may
put an end to the research project. Thus the Indians' control over genetic
research is limited to the field phase.

The Yanomami are far more concerned with the proper treatment of the
physical substances of their dead than they are with what is written about
the blood that was collected. The impact of the research upon indigenous
values is essentially the same whether the researcher deals with the blood
itself or with cell lines produced in a lab. However the issue of possible royalties generated from pharmaceutical products derived from their blood raises
questions that are not directly related to the physical or cultural integrity of
the research subjects, but concern, rather, the Indians' economic interface
with the outside world.

Conversely the end product of ethnographic data-gathering directly affects
not so much the core of indigenous values as the Indians' relationship with

the majority society. The products of ethnographic results—descriptions, analyses, representations, hypotheses, or theories—are constructed outside the context in which empirical data were collected. Once published, ethnographic writings take on a life of their own, out of the subjects' control and often out of the researcher's control as well (as in cases of misuse or misrepresentation). In sum, both biomedical and ethnographic research—for different reasons—must always be considered from an ethical perspective.

Nevertheless, the issue of informed consent is not easy to resolve and raises more questions than it answers. How informed must consent be in order to ensure that what is done in the field is not simply subtle coercion or friendly persuasion? How is informed consent constructed in the field: Is it a vacuous protocol, or is it the object of prolonged negotiations? Is it established the day the researcher, who knows nothing of the local language, arrives in a village, or months later, when researcher and hosts can communicate, albeit minimally? Can consent be given verbally or must there be a written and signed document? Is a written form of consent a sure warranty against abuses? Who regulates the process: Is it the host community, the host country, the researcher's professional association, or the researcher's government? Taken to its logical extreme, would informed consent inhibit and in due course even obliterate research? What would the absence of research mean for the people studied? At this point one can only say that if things are bad *with* informed consent, they are surely worse *without* it.

Images

Intimately tied to the issues of knowledge and genetic exploitation, but separate for analytical purposes, is the problem of the commodification of indigenous art forms and images when they are conveyed in media such as photographs, paintings, and films. Serving a different market from pharmaceutical companies and the like, these indigenous objects of desire appeal to a variety of Western interests ranging from New Age believers, to green business, to art dealers, interior decorators, recording companies, and museums. Though image commodification produces profit margins much more modest than genetic or pharmaceutical research, it can nevertheless generate a significant amount of money for the Western world. One must, however, make a distinction between material objects that the Indians themselves intend for private ownership, and other artistic manifestations, such as music, that indigenous peoples do not regard as commodities. Although due payment is appropriate in the case of crafts produced for sale, the privatization of photographs, film footage, oral literature, music, and so on, is totally alien to most indigenous traditions. Let us see first the problems involved in the commercialization of material objects and then turn to other modalities of image exploitation.

Every year, on April 19 (National Indian Day), Artíndia, the commercial outlet of FUNAI, organizes a special event known as *moitará* (a Kamayurá word for trade as it is practiced in the Upper Xingu region) (Galvão 1979:83 n20). The event includes talks, exhibitions, and the sale of indigenous crafts of superior quality. A clientele familiar with the choice pieces of the moitará gathers outside the FUNAI building in Brasília well before opening time. In less than an hour, the place is cleared of the best specimens of pots, stools, weapons, masks, featherwork, basketry, and the like. The majority of the early arrivers, usually foreigners, are sales agents for expensive shops in São Paulo, New York, Paris, or London. The price they pay at Artíndia more than compensates for the hours of waiting at FUNAI's door. For example, a Yanomami basket that sells in the field for less than US$2 is worth twice as much at Artíndia and can be seen advertised in interior decorating magazines for nearly US$100. A large Yawalapiti zoomorphic stool sells at Artíndia for US$50 at the most, whereas in posh stores in Paris it can be bought for US$700. No Yanomami I know, and very likely no Yawalapiti, has ever been reimbursed beyond the price of the original transaction.

To take another example: A São Paulo store named *Arte Nativa Aplicada* (Applied Native Art) is known for its attractive fabrics. They are painted with indigenous motifs, mostly geometric designs. Tablecloths, napkins, cushion covers, and garments usually have the name of a Brazilian indigenous group written on a corner by way of a simulacrum of a designer signature. The managers insist that their designers research indigenous styles before printing their cloths. As to paying the original artists, they argue that the publicity the Indians get is already a form of payment. In fact, the Tukano, Kadiweu, and other ethnonyms bear little relation to the designs, as indicated by the telltale Xingu "signature." In fact, Xingu is nobody's name, but it is the name of a river and a reservation. Pressed about the ethnographic accuracy of the designs, the managers confess that any design will do and that the native names are actually chosen at random. For their purposes it matters little whether the "Tukano" crisscross design is actually made in the Xingu or anywhere else. In effect, these merchants are applying the bioengineering rationale to the field of arts and crafts by behaving according to the ex situ argument: you mix styles, print them on a nonindigenous medium, and the product is no longer indigenous property. As long as the "ethnic" style remains in fashion, it is important to vendors to maintain the appearance of cultural authenticity; but it is equally desirable to keep real Indians at a distance, particularly from the profits made from their aesthetic ideas. One thus benefits from these images, transfigured through metonymy into pieces of craftsmanship, without having to deal with real people and their real demands.

In the case of cultural manifestations such as music, Western logic seems to follow the same course as in the gene market. In entering the commercial domain, indigenous music becomes commodified. "When music is owned by indigenous people, it is seen as 'public domain.' If it becomes popular in its 'mainstream' form, though, it suddenly becomes 'individual property.'" (Seeger 1991:38). The unequal treatment given to indigenous and Western musicians is well illustrated in Seeger's following remark: "I wonder at the freedom with which Brazilian television has used [Suyá's] recordings as background music—a policy that would be rigorously policed if the music were performed by, say, the Beatles, but which cannot be policed when it is 'only' performed by the best living Suyá musicians" (37).

Here, too, the ex situ rationalization operates as a shield against royalties that traditional musicians may demand. Again, indigenous peoples object not to the use of cultural expressions by non-Indians, but to their commodification and the privatization of the proceeds. Hybrid pieces of music are produced with high technology: computerized samples of traditional and nontraditional pieces are mixed in a process of recontextualization that pulverizes authorship; the result is a work that then enters the market as a potentially valuable commodity (P. Harvey, quoted in Strathern et al. 1998:123). Like cell lines, sounds and images can now be reproduced and transformed ad infinitum, thereby attaining technological eternity.

As to pictures—be they in films, paintings, or photographs—the ex situ argument, although applicable, follows a different trajectory and raises yet another set of questions about royalties. For example, indigenous pictures, taken by unidentified photographers, are used matter-of-factly in advertising: Davi Yanomami's face stamped along a row of other "united colors" on a Benetton shop window; an enlarged running Krahó youth on the wall of a Danish shoe store, illustrating the correct way to use the human foot; or healthy and handsome Kayapó men helping Anita Roddick sell her Body Shop cosmetics. Pictures of Indians are also used in other contexts, such as the Brazilian interior decorating magazine in which three close-ups of Yanomami faces, taken by photographer Claudia Andujar, are shown simply as decorative pieces on somebody's bedroom wall: neither the photographer (C. Andujar, pers. comm.) nor the Indians were ever consulted about the use of these images.

The use of authored photographs of Indians for commercial or other purposes raises a double issue: one involves the photographers' copyrights; the
other, the rights of the indigenous subjects to cede or deny the use of their
images. It is common practice to request permission—though not always to get
it for free—from photographers to use their products; very seldom are Indians
consulted about having their likenesses used for any purpose, whether commercial, artistic, academic, or philanthropic. Legislation on this matter is either nonexistent or incipient, perhaps with good reason: when are photographs
used for defense or exploitation? Is it possible to legislate about situations with
unpredictable outcomes? What would happen to the legal status of investiga-
tive journalism if a given picture, taken to record some injustice, for instance,
were to be disseminated beyond its author's control and ended up serving spu-
rious purposes? Too self-conscious precautions might preclude documentary
materials that are often the proof needed to defend indigenous rights.

Commercial film-making has consistently brought discontent to indigenous participants. Perhaps the most notorious case concerned director
Werner Herzog's tragic dealings with the Aguaruna in the Peruvian Amazon
during the shooting of *Fitzcarraldo*. The German movie-maker was accused
of fencing out all the manual laborers, including 500 non-Indians, and protecting his professional staff in an exclusive haven within the filming area.
In his quest for realism he had the Indians haul a real ship, weighing many
tons, up a steep hill, causing the death of one of the men, who was crushed
when the ship slipped. In 1979 Herzog had to interrupt the shooting because
the Aguaruna set fire to his camp to protest maltreatment, forced labor, and
ridiculously low salaries. Herzog's behavior mobilized indigenous organizations and public opinion, but there is no information to suggest that repara-
tive indemnities were made to the Indians. On the contrary, Herzog claimed
he was being persecuted, the victim of a defamation campaign (*Jornal do
Brasil* 1982:B-1).

In late 1999, the all-powerful Globo television network in Brazil began
negotiations with FUNAI to hire Indians from various parts of Amazonia to
appear in a "documentary" feature on shamanism for the weekly program
Globo Reporter. The network requested that logistics and provisions—boats,
gasoline, lodging, and so on—be taken care of by both the Indian agency
and the Indians themselves. As for paying the Indians, the producer, Camilo
Tavares (1999), wrote FUNAI that "As contribution (*gratificação*), we are
making arrangements directly with the leaders, given the limits of our budget
for a program of journalism such as *Globo Reporter*."

Three aspects of this letter are worth pondering. The first is the producer's attempt to bypass FUNAI's fair payment requirements by negotiating

directly with the Indians; apparently he expected that they would be satisfied with a bunch of trinkets, in keeping with the popular image of "Indian
want whistle" (a caricature of the Indian found in an old carnival song). The
Globo crew decided to abandon the plan to film in the Upper Rio Negro
region when the local Indians fixed their payment around US$8,000. After
moving the production elsewhere, the producer offered a mere US$450 per
shamanic performance.

The second aspect has to do with the use of the Portuguese term *gratificação*. Akin to a donation, it is never used in the context of payment for
merchandise or services. In public contexts, interviewees do not receive a
donation, contribution, or tip, but a payment, a *cachet*. The Globo producer
thus tried to hide his contempt for the Indians behind a frugal budget.

The third aspect is the same double standard, but seen from another
angle. It is the pretense that the performance of a shamanic session put on
for the benefit of a television program is not staged, but is the same as a real
shamanic session. Clearly the negotiations between producer and Indian
leaders do not involve the crew's "participant observation," for which they
have neither time nor know-how, but are a short-circuiting of real life, with
the Indians acting as shamans in staged virtuality. To then pretend it is journalism verges on fraud. Could an autobiographical movie about and starring
Frank Sinatra be classified as a journalistic documentation of his real life?
Would anyone dare suggest that Sinatra be given a donation or a tip for his
work in such a film? In short, the letter by the Globo producer says worlds
about the disdain with which indigenous images and cultural products are
treated in this ravenous New World.

Agents, Not Commodities

We have seen some reactions from indigenous peoples to all this tamper-
ing with cultures and resources. The Kuna Indians of Panama provide a
good New World example: they have one of the most effective systems of
control over the entry and activities of foreign scientists in indigenous territory. The Project for the Study of the Management of Wildlife Areas of Kuna
Yala (PEMASKY), created in 1983 to set up a forestry program, gradually
devised rules and regulations about non-Kuna scientific work in the area.
The Kuna require not only that researchers be briefed but also that they
provide their hosts with copies of all reports, photographs, and specimens
of fauna and flora. Five years after the project was created, these procedures
became formalized for researchers in the manual "Research Program: Sci-

entific Monitoring and Cooperation," which explicitly limits the geographic areas open and closed to research, encourages the transfer of results to the Kuna themselves, and urges scientists to take Kuna assistants on their expeditions. Thus, while keeping control over scientists in their territory, the Kuna do not deter scientific research (Chapin 1991:17).

It is an unfortunate consequence of genetic and resource plunder that indigenous peoples are being put on the defensive and are therefore holding researchers at arm's length (as in the Karitiana case mentioned earlier). In the Amazon, indigenous leaders are reacting more bluntly toward scientists and other professionals than are the Kuna. The Ecuadorian Shuar reject research projects that bring no clear benefit to them. Restrictions to non-Indian research activities in the Brazilian Amazon are beginning to appear. To put all research in jeopardy is to throw the honest baby out with the unscrupulous bath water. In the end, "we will all lose out" (Cunningham 1991:5). Seeger (1991) also expressed his concern "about the perception that 'someone is getting rich on our music' and the effect it has had on music research" (37).

On the brighter side, we saw research being politically redeemed when various indigenous groups became interested in participating in projects to evaluate the pressures they have felt from the world market economy. Indian researchers from Bolivia, Ecuador, Peru, Colombia, and Brazil jointly participated in research teams with national and foreign colleagues in a series of projects sponsored by COICA and Oxfam America. The projects were intended to take stock of the impact of development schemes on traditional gift economies after those economies had been bombarded with development projects imposed upon them by both national and international interests. Caught between the logic of the gift and the logic of the commodity, "the members of modern indigenous communities often get confused as to which responsibility they should attend to, and how to take care of their interests and personal security in the long run: to either accumulate wealth or strengthen the social bonds of mutual duties and support" (Smith 1996a:7). Weighing successes and failures, the researchers raised crucial questions that speak directly to the issue of productivity of indigenous lands. The pressure to produce or else lose land has been so great that many native peoples have been driven into destructive economic activities, such as cattle raising in the rain forest. The COICA research teams ask,

> Why have most projects failed?... Why has there been no capital accumulation in indigenous communities? Why has rubber tapping brought about a stable economy while coffee has not? How have the

projects affected subsistence activities?... What impact have private property regimes had in resource distribution and management?... Is generosity still praised in local economies? Does it have any value in a monetary and market context? What happens to the traditional open access to resources when boundaries are set around properties and certain resources obtain market value?... What are the cultural, social, financial, and physical-biological factors that must be taken into account in future economic strategies of a developing community?

(Smith 1996b:40–41)

The newly acquired knowledge that comes from the work of these research teams stimulates the political vigor and sophistication with which regional, national, and international indigenous organizations fight to transform the Indians from commodities for the benefit of non-Indians into the agents of their own interests. To do our part as researchers in this transformation it is not enough for us simply to improve our field ethics: we must be open to having the accepted routines of our "normal science" challenged by indigenous peoples, who, as they participate in the research process and even become researchers in their own right, will no longer be simply the objects of our inquiry.

References

Albert, B. 1985. Temps du sang, temps des cendres: Représentation de la maladie, système rituel et espace politique chez les Yanomami du Sud-est (Amazonie brésilienne). Ph.D. dissertation, University of Paris X.

Albert, B., ed. 2001. Research and Ethics: The Yanomami Case. Brazilian Contributions to the Darkness in El Dorado Controversy. Brasília: Comissão Pró-Yanomami, Document No. 2.

Arias, M. and A. López. 1995. La propiedad inmemorial y la propiedad intellectual de los pueblos indigenas. Asuntos Indígenas (Grupo Internacional de Trabajo sobre Asuntos Indígenas) 4:19–20.

Ávila, T. A. M. de. 2001. Biopirataria e os Wapichana: Análise antropológica do patenteamento de conhecimentos indígenas. Undergraduate thesis, Departamento de Antropologia, Universidade de Brasília.

Basso, K. 1979. Portraits of "the Whiteman": Linguistic Play and Cultural Symbols Among the Western Apache. Cambridge: Cambridge University Press.

Braga, P. H. 1997. A patente que veio do índio. Folha de São Paulo. June 1, section 5:15.

Buarque de Holanda, S. 1986. O Extremo Oeste. São Paulo: Brasiliense.

CCPY (Comissão Pro-Yanomami). 2002a. Boletim 25. Available at http://www. proyanomami.org.br/boletimMail/yanoBoletim/htm/Bulletin_25.htm (accessed December 27, 2004).

CCPY [Comissão Pro-Yanomami]. 2002b. Boletim 26. Available at http://www. proyanomami.org.br/boletimMail/yanoBoletim/htm/Bulletin_26.htm (accessed December 27, 2004).

CCPY [Comissão Pro-Yanomami]. 2003. *Boletim* 27. Available at http://www.proyanomami.org.br/boletimMail/yanoBoletim/htm/Bulletin_32.htm (accessed December 27, 2004).

Chapin, M. 1991. How the Kuna keep scientists in line. Special issue, "Intellectual Property Rights: The Politics of Ownership." *Cultural Survival Quarterly* 15(3):17.

Colchester, M. 1995. Algunos dilemas referentes a la reivindicación de los "derechos de propiedad de los pueblos indígenas." *Asuntos Indígenas* (Grupo Internacional de Trabajo sobre Asuntos Indígenas) 4:5–7.

Cunningham, A. B. 1991. Indigenous knowledge and biodiversity. Special issue, "Intellectual Property Rights: The Politics of Ownership." *Cultural Survival Quarterly* 15(3):4–8.

Davis, S. 1977. *Victims of the Miracle: Development and the Indians in Brazil.* Cambridge: Cambridge University Press.

Farage, N. 1991. *As Muralhas dos Sertões: Os povos indígenas no rio Branco e a colonização.* Rio de Janeiro: Paz e Terra.

Folha de São Paulo. 1995. A luta pelos genes do 3° Mundo. January 29, section 6:15.

Folha de São Paulo. 1996. EUA patenteiam vírus de índio e são acusados de 'vampirismo.' June 16, section 5:13.

Folha de São Paulo. 1997a. Biopirataria atinge floresta amazônica. June 1, section 1:1.

Folha de São Paulo. 1997b. Procuradoria no Amazonas vai investigar biopirataria. June 6, section 1:11.

Folha de São Paulo. 1997c. Acre proíbe estrangeiros na floresta. July 4, section 1:12.

Folha de São Paulo. 1997d. Biopiratas agem livremente na Amazônia. July 13, section 1:18.

Folha de São Paulo. 1998. Biopirataria no país é caso de polícia, diz professor da PUC-SP. July 15, section 3:4.

Galvão, E. 1979. *Encontro de Sociedades: Índios e brancos no Brasil.* Rio de Janeiro: Paz e Terra.

ISA. 1996. *Povos Indígenas no Brasil, 1991/1995.* São Paulo: Instituto Socioambiental.

IWGIA (International Workgroup for Indigenous Affairs). 1995. *Asuntos Indígenas,* no. 4.

IWGIA (International Workgroup for Indigenous Affairs). 1996. *Asuntos Indígenas,* no. 5.

Kidd, J. R, F. L. Black, K. M. Weiss, I. Balazs, and K. K. Kidd. 1991. Studies of three Amerindian populations using nuclear DNA polymorphisms. *Human Biology* 63(6):775–794.

Kloppenburg Jr, J. 1991. No Hunting! Biodiversity, indigenous rights, and scientific poaching. Special issue, "Intellectual Property Rights: The Politics of Ownership." *Cultural Survival Quarterly* 15(3):14–18.

Latour, B. 1993. *We Have Never Been Modern.* Cambridge, Massachusetts: Harvard University Press.

Lobo, M. S., K. M. P. Rodrigues, D. M. de Carvalho, and F. S. V. Martins. 2001. Report of the medical team of the Federal University of Rio de Janeiro on accusations contained in P. Tierney´s *Darkness in El Dorado*. In B. Albert, ed., *Research and Ethics: The Yanomami Case (Brazilian Contributions to the* Darkness in El Dorado *Controversy)*, pp. 15–42. Brasília: Comissão Pró-Yanomami, Document No. 2.

McGou'an, J. 1991. Who is the inventor? Special issue, "Intellectual Property Rights: The Politics of Ownership." *Cultural Survival Quarterly* 15(3):20.

Pinto, L. F. 1980. *Amazônia: No rastro do saque.* São Paulo: Hucitec.

Posey, D. 1991. Effecting international change. Special issue, "Intellectual Property Rights: The Politics of Ownership." *Cultural Survival Quarterly* 15(3):29–35.

Pratt, M. L. 1992. *Imperial Eyes: Travel Writing and Transculturation.* London: Routledge.

RAFI (Rural Advancement Foundation International). 1995. La patente de material genético humano. *Asuntos Indígenas* (Grupo Internacional de Trabajo sobre Asuntos Indígenas) 4:8–13.

Ramos, A. R. 1984. Frontier expansion and Indian peoples in the Brazilian Amazon. In M. Schmink and C. H. Wood, eds., *Frontier Expansion in Amazonia*, pp. 83–104. Gainesville: University of Florida Press.

Ramos, A. R. 1986. *Sociedades Indígenas.* São Paulo: Editora Ática.

Ramos, A. R. 1991. Amazônia: A estratégia do desperdício. *Dados* 34(3):443–461.

Ramos, A. R. 1994. The hyperreal Indian. *Critique of Anthropology* 14(2):153–171.

Rivière, P. 1972. *The Forgotten Frontier: Ranchers of Northern Brazil.* New York: Holt, Rinehart & Winston.

Rivière, P. 1995. *Absent-Minded Imperialism: Britain and the Expansion of Empire in Nineteenth-Century Brazil.* London: Tauris Academic Studies.

Sant'Anna, L. 1998. Brasil perde com biopirataria na selva. *O Estado de São Paulo* (August 9): A20.

Santilli, J. 1996. A proteção aos direitos de propriedade intelectual das comunidades indígenas. In C. A. Ricardo, ed., *Povos Indígenas no Brasil, 1991/1995*, pp. 17–21. São Paulo: Instituto Socioambiental.

Santos, R. V. and C. Coimbra Jr. 1996. Sangue, bioética e populações indígenas. *Parabólicas* (Instituto Socioambiental) 20(3):7.

Seeger, A. 1991. Singing other peoples' songs. Special issue, "Intellectual Property Rights: The Politics of Ownership." *Cultural Survival Quarterly* 15(3):36–39.

Smith, R. C. 1996a. Prefacio. In R. C. Smith, ed., *Amazonía: Economía indígena y mercado: Los desafíos del desarrollo autónomo*, pp. 7–8. Quito, Ecuador: COICA/Oxfam America.

Smith, R. C. 1996b. Introducción a los estudios de caso. In R. C. Smith, ed., *Amazonía: Economía indígena y mercado. Los desafíos del desarrollo autónomo*, pp. 39–45. Quito, Ecuador: COICA/Oxfam America.

Strathern, M., M. C. da Cunha, P. Descola, C. A. Afonso, and P. Harvey. 1998. Exploitable knowledge belongs to the creators of it: A debate. *Social Anthropology* 6(1):109–126.

Taussig, M. 1987. *Shamanism, Colonialism and the Wild Man: A Study in Terror and Healing*. Chicago: University of Chicago Press.

Tavares, C. 1999. Letter to FUNAI, December 5.

Tierney, P. 2000. *Darkness in El Dorado: How Scientists and Journalists Devastated the Amazon*. New York: Norton.

Viveiros de Castro, E. and L. M. M. de Andrade. 1988. Hidrelétricas do Xingu: Oestado contra as sociedades indígenas. In L. A. O. Santos and L. M. M. de Andrade, eds., *As Hidrelétricas do Xingu e os Povos Indígenas*, pp. 7–23. São Paulo: Comissão Pró-Índio de São Paulo.

Wapichana, C. A. 1999. Seminar panelist. Biodiversidade, Justiça e Ética. *Revista do Centro de Estudos Jurídicos* (Brasília) 8:41–44.

World Bank. 1981. *Economic Development and Tribal Peoples: Human Ecologic Considerations*. Washington, DC: World Bank, Office of Environmental Affairs (OEA/PAS).

17 Euphemism in the Forest

Ahistoricism and the Valorization of Indigenous Knowledge

Stephen Nugent

Historically the Brazilian Amazon's natural wealth has been measured in terms of its minerals (gold, iron ore, bauxite), plants (mahogany, palm fruits), and animals (shrimp, fish). Since the reevaluation of the region's resource potential, beginning with its abrupt "integration" into the national development plan around 1970, there has been a significant addition to that repertoire: indigenous knowledge. The significance of indigenous knowledge (or native, or folk, science) has been revealed in a number of detailed studies, but overall characterization has proved evasive. In this discussion, rather than start from a particular Amazonian example and then amplify—say from a discussion of the way in which forest management by *caboclos* (riverine inhabitants of mixed European and Amerindian descent) illustrates a dynamic manipulation of biological systems in the name of desirable social ends (a story that is, in this volume, presumably noncontroversial)—it may be more useful to approach the issues obliquely and rather more generally.

There are two reasons to take this approach. The main one is that indigenous knowledge—Amazonian or not—has become one element among many that have been drawn into a complex and confusing discussion of intellectual property rights, a discussion that is a key feature in the hotly negotiated recalibration of globalization in the twenty-first century.

A second reason for beginning this way is that the discussion of indigenous knowledge has a history poorly represented in many current debates. It is significant not only that this provenance is nearly absent from discussion, but also that its exclusion may actually subvert the goals of ensuring the livelihoods and existence of Amazonian peoples.

Some Missing Histories

The first strand of these provenances is represented in a literature that includes the rationality debate, modes of thought, Lévi-Straussian mentalism, cognitive anthropology (in both its ethnoscience and evolutionary-psychology phases), and more recently, the epistemologically pluralist arguments of Atran (1990), Berlin (1992), and Sperber (1996). With the possible exception of Lévi-Strauss's work—and albeit in bowdlerized form—this literature is outside the mainstream of contemporary anthropological concerns. Nevertheless it does represent an important contribution to an appreciation of the very notion of science as conducted by peoples outside canonical or so-called Western science. At the very least, it represents in anthropology a shift away from the relativized mainstream in which hermeneutic fixations disavow any relevance of the idea that scientific preoccupations of anthropology significantly overlap the scientific predilections of anthropological subjects. Instead of acknowledging the similarities between culturally heterodox sciences, the convergences that command the most attention are represented in the confused rhetorics of sustainability, globalization, and neoliberal virtue.

It is curious that recent attempts to valorize indigenous knowledge as a significant contribution to sustainable development should ignore, for example, the provocative sketch long ago provided by Lévi-Strauss for whom the "science of the concrete" was a gloss on what later came to be known as ethnoscience. Additionally, Lévi-Strauss clearly made connections between indigenous knowledge and the politics of indigenism and anthropology:

> It is the outcome of a historical process which has made the larger of part of mankind subservient to the other. During this process millions of innocent human beings have had their resources plundered and their institutions and beliefs destroyed, whilst they themselves were ruthlessly killed, thrown into bondage, and contaminated by diseases they were unable to resist. Anthropology is daughter to this era of violence.
>
> (Lévi-Strauss 1978:54–55)

The Valorization of Science

The designation "indigenous knowledge" does have historical depth, and not the least of the issues raised in earlier discussions (Tyler 1969) is that the exploitation of such knowledge entails political and moral considerations in addition to the commercial ones that now command so much attention.

This second provenance or history may appear far removed from the direct concerns of anthropologists and their indigenous collaborators, but is no less relevant. Its focus is the socialization of intellectual production within the academic communities of the West, and a few broad points on the subject might clarify the larger sociohistorical context in which an interest in indigenous knowledge—previously confined to fairly narrow scholarly pursuits—now garners greater popular and commercial attention.

A number of commentators (Bowles and Gintis 1976; Chomsky et al. 1997) have observed that one of the underlying factors in the economic stability of the United States (to take a key example of intellectual hegemony) since World War II has been the successful management of intellectual labor, accomplished in no small measure through a university system that achieves economies of scale that are beyond the resources of private capital. For example, the general climate in the United States, according to Lewontin (1997), is one in which "economic wars are replacing armed struggle as a major impetus for state intervention in the economy" (32), and with the increasing tendency toward the privatization of knowledge—even knowledge previously regarded as nonsense or exotic native belief—the upsurge of interest in intellectual property rights and indigenous knowledge is a not a terribly surprising corollary. Nor is it terribly surprising that the dramatic increase in Amazonian research commencing with the Plan for National Integration and the construction of the Transamazon Highway (c. 1970), has been largely driven by interest in resources and systems defined as strategic from the perspective not only of Amazonians, but of big consumers elsewhere. Such pressing local concerns as land reform and environmental conservation pale by comparison with the bright packaging of global, sustainable development plans. It is hardly news that support for particular kinds of research is not shaped by purely scholarly interests: the current enthusiasm for indigenous knowledge achieves a degree of prominence that significantly obscures crass commercial motives (in part because the very definition of indigenous knowledge is so insubstantial). While cancer-killing plant preparations garner headlines, the actual products of native, forest wisdom are more likely represented in cosmetics, fad foods (guaraná, açaí) or knickknacks (stuffed-piranha ashtrays).

These two histories lie outside most current discussions of indigenous knowledge[1] (although see Overal and Posey 1990), but could usefully in-

---

1. For the purposes of this paper, the parameters are (1) development-studies-led search for new resources and (2) historical ecology as outlined in Balée 1998.

form several aspects of the current debates in Amazonia. They could provide a far more reputable and historical context in which to appreciate why it is that general public concern with indigenous knowledge and intellectual property rights achieves such prominence now. They may also help explain why the apparently progressive thrust of indigenous knowledge promotion may prove to be a poisoned chalice if it turns out that the current fascination with indigenous knowledge is more a testimony to the commercial value of research than it is an effective defense of the increasingly disenfranchised people whose knowledge is sought.[2] In short, the emergence of indigenous knowledge as a focal issue in Amazonia carries with it more than a trace of what Galtung (1967) referred to as "scientific colonialism."

While the Cold War prerogatives may have faded, that hardly means that a global strategy has receded. It is difficult, for example, to ignore the role that ecology and environmentalism have played in readjusting and redefining spheres of influence initiatives (witness the ardent embrace of the SIVAM satellite monitoring program in Amazonia). In light of the current emphasis on intellectual property rights, the co-occurence of neoliberal terms of reference and "ecologism" is so ubiquitous as to have become almost unnoteworthy—but it should be noted. That indigenous peoples may be able to carve out an adequate economic space through systematic exploitation of indigenous knowledge has plausible features, but in the absence of rigorous documentation and analysis of that knowledge—and absent factors such as territorial security, legal defense of human rights, and medical and educational provision, as well—there is cause for skepticism. And the historical record is not encouraging. A program for the defense of indigenous rights based on the promotion of intellectual property rights attracts support from diverse quarters and has many appealing aspects, but it may suffer from a misreading of the distance between conceptual elegance and grassroots reality. Examples from Amazonia and one from more remote realms may prove instructive.

In her account of the revival of the Waiapí of the northeast Brazilian Amazon, Dominique Gallois (1998) makes two vital points. Her first point is that Waiapí notions of autonomy (and hence the capacity to pursue strategies beneficial to the aims of the Waiapí themselves) do not correspond with official notions of autonomy (178). That Waiapí desires may converge with official ones is a moot point, but it seems quite clear that what non-Waiapí

---

2. In one study (Morsello 2002) of the effects of high-profile market integration of indigenous peoples (Body Shop International and the Kayapó), the payment of research fees to those studying the effects of market integration appears to have been more divisive than was the accelerated commercialization of Brazil nut production.

regard as a significant advance (e.g., official recognition of Waiapí territory) does not correspond with the social needs of the Waiapí. Her second point is that the economic benefits achieved by selective commercialization are modest, if at the moment adequate.

The second Amazonian example is provided by the agreement between fractions of the Kayapó and Body Shop International (BSI). In this widely documented case, a native product (Brazil nut oil) has served as a marker of two forms of cultural validation. From the Kayapó point of view, it is claimed, participation in the commodification of previously noncommercial products provides the Kayapó with income that permits their extended social reproduction. At the same time, it allows the Kayapó's commercial agents—BSI—to exploit their association as part of a distinctive marketing strategy: we (BSI) bridge the gap between well-remunerated native producers and socially aware metropolitan consumers. (See Rabben 1998 for an overview and Morsello 2002 for a detailed analysis.)

Slightly further afield, Ramos (1991) has argued that, in Brazil, the *image* of the Indian is as potent as the *reality*. The Indian is summoned forth to be addressed by national political discourse while being repudiated by that same discourse, for the existence of the Indian both permits deliberation over the distinctive character of the modern Brazilian state and simultaneously challenges the legitimacy of the state. The *subject* of indigenous knowledge replicates the dilemma at a higher level of abstraction. The hypervalorization of indigenous knowledge—a cure for cancer may lie in the hands of savages, for instance—is not mysterious. As socially appropriate behavior comes to be defined in terms of access to commercial knowledge, it is not surprising that those few peoples who have evaded normalization should now be perceived and represented as possessors of valued esoteric knowledge.

Same Knowledge, New System

Between about 1970 and the present, the status of Amazonian native, or indigenous, and peasant science has undergone a remarkable transformation: indigenous practices once regarded as near-desperate coping with untoward circumstances (e.g., Blaut's critical analysis of "The Doctrine of Tropical Nastiness," 1993:69–80) have become the armory of "wise forest management." That the emphasis is now placed on "knowledge systems" should not blind one to the fact that the appeal of these systems lies in the material possibilities they afford to those capable of exploiting them. The massive germplasm transfers that followed the expansion of European agrarian systems into the New World (Crosby 1986; Juma 1989; and the well-documented role of imperial botanical gardens in Brockway 1979) were more narrowly commercial in their dispositions, but were no less about systems management than is the U.S. folktale about "red" Indians showing Pilgrims how to fertilize maize plantings with fish.

Yet when one reads in the *Indigenous Knowledge and Development Monitor* (IKDM) that "the value of local systems in facilitating development is gradually being recognized by national and international development agencies" (Quiroz 1996), the need for a reality check becomes compelling.

Although most contributions to IKDM draw attention to the technical resources embedded in indigenous knowledge, Agrawal (1995) slightly disturbs the normative, static conceptions of indigenous knowledge "systems." In the introduction to his article, Agrawal gives an account of the stages of the "rhetoric of development": economic growth, growth with equity, basic needs, participatory development, and sustainable development. That sequence adequately represents a folk view of the progression of development discourse, but what it crucially doesn't represent is an accurate connection with historical reality. In fact, each stage of the process has been overshadowed by events and attitudes that are incompatible with the stated objective (table 17.1).

Agrawal (1995) writes: "The focus on indigenous knowledge clearly heralds a long overdue move. It represents a shift away from the preoccupation with the centralized technically oriented solutions of past decades, which failed to improve the prospects of most of the world's peasants and small farmers."

TABLE 17.1 Stages of Development Versus Cold Reality

Stage of Development[1]	Overshadowed by (examples)
Economic growth as the declared priority	Cold War spheres of influence (emergence of World Bank group, Council on Foreign Relations)
Growth with equity	Defusing of local resistance (Project Camelot; Vicos; overthrow of Allende)
Basic needs	Famine (Sahel; Ethiopia/Eritrea)
Participatory development	"Accept full administration or suffer" (structural adjustment)
Sustainable development	"We're all in this together" (Brundtland Report; Earth Summit)

[1]*Source:* Agrawal 1995.

However, an account of a stepwise involvement with indigenous knowledge does not acknowledge higher-level institutional redefinitions of landscape and resource management. Orlove and Brush (1996) note that a systematic approach to conservation and protection—following long on such precedents as royal and imperial game reserves in Europe and the colonies (330)—is represented by policies functioning at three levels: targeting individual species (population ecology), focusing on protection of the habitats of threatened species (ecosystem ecology), and managing assemblies of ecosystems (landscape management) (331)—all policies that have emerged in response to the threat of species extinction (conventionally marked by publication of Carson's *Silent Spring* in 1962).

It has been suggested that "the indigenous knowledge perspective"—local (esoteric) management of landscapes (ecosystem assemblies)—is a fourth stage:

Micro-scale landscape assessment can be compared with the peasant's perception of his/her space and resources. This integration may indicate a way forward in the search for sustainability, by matching scientific and peasant knowledge and practices.

(Bocco and Toledo 1997)

Despite the claims on behalf of knowledge systems, what is centrally at issue is the set of ends to which such knowledge is put. The homepage of the *Indigenous Knowledge and Development Monitor* notes:

In other words, advocates and propagators of indigenous knowledge are being challenged by the development enterprise to put indigenous knowledge to good use. The time has come to show that there is no rhetoric in John Madeley's statement that "indigenous knowledge is the largest single resource not yet mobilized in the development enterprise."

(IKDM 1998)

Repeatedly in the article attention is drawn to the priority of finding the most cost-effective way that indigenous knowledge can be employed—as though this criterion is transparent and cross-culturally relevant. And Umans (1997, box 1), in his appraisal of the health system of the Guarani Indians in Bolivia, reaches a nadir in this discussion, proposing mission statements—the apotheosis of management science—for native peoples in the form of four action points for what amount to community line-managers:

1. For the *payes*: "maintain our culture"
2. For the *elders*: "to look after the well-being of the community and see to it that the culture is not lost"
3. For the *leaders*: "to look after the community and promote the work of all health-care workers"
4. For the *midwives*: "to reduce the mortality rate, and to prevent illness through the use of herbs"

Indigenous knowledge is not a new concept; yet current usage of this term—in the name of many laudable and progressive goals—tends less to promote indigenous conceptions of indigenous knowledge than to promote assumptions about how indigenous knowledge can be exploited by others (with the complicity, of course, of imputed beneficiaries). It is widely and uncritically assumed that the exploitation of indigenous knowledge is a good thing for all concerned. That is an argument that bears further examination.

Sleeping with the Past

A heated exchange in the early 1990s between Stephen Corry of Survival International and Gordon Roddick of Body Shop International dealt directly with the question of ownership and control of indigenous knowledge (Corry 1993; Roddick 1993). That debate took place in the aftermath of what had been—in the late 1980s and early 1990s—an unprecedented assault on natural and social landscapes in Amazonia, and was informed by a wide array of forces: a multilateral program to manage environmental depredation in the face of rising demand for scarce resources; an environmentalist movement split along diverse lines (e.g., Gaia, social justice, native knowledge, cancer cures, fourth-worldism); increased commercial pressure on resources previously disregarded in favor of more accessible alternatives; and academic cadres for whom basic research and advocacy have created, at times, convenient partnerships.

In the disputes focused on the future of Amazonian peoples and Amazonia, what has rarely been addressed is the question of the provenance of what are casually referred to as knowledge systems. At one, easily dismissed level, it is a foolish question, the answer to which is: what Indians and other Amazonians themselves systematically engage in during the course of securing conditions of existence they regard as standard, average, normal, and desirable. But at another level it may not appear so foolish a question; its working answer might be: a construction placed on Indians in order to valorize their knowledge in commercial terms.

Valorizing Others' Knowledge

For the period between the early phases of conquest and the present time it is possible to construct a crude account of the development of interest in Amazonian indigenous knowledge systems. Following the failure to find El Dorado, ersatz tropical preciosities served to maintain imperial interest in the exploitation of the region. The systematic explorations by the Victorian naturalists Bates, Wallace, and Spruce provided inventories of flora and fauna, but there was little recognition that these scientific activities might be replicating knowledge that was already available locally.

The emergence of the rubber industry did produce—inadvertently—an acknowledgment of local botanical and geophysical expertise that could enhance the continued growth of extractive enterprise (e.g., Wright 1912). The commercial collapse of the region was followed some decades later by ethnographic exploration that produced, among other things, knowledge-focused (as opposed to the more narrowly resource-focused) accounts of Amazonian social systems. A cardinal feature of modern Amazonian ethnography has been (following functionalist, structure-functionalist, structuralist, and human ecology protocols) the documentation of coherence among those historically dismissed as forest scrabblers.

The designation "indigenous knowledge" is frequently little more than a neologism for tropical preciosity and reflects not so much the discovery of something new as a change in the global division of labor. Anthropology—a late modern science—can now stake some kind of authoritative claim to official discovery procedures that in earlier times were the responsibility of others, like Herbert Wright, editor of *The India Rubber Journal* and author of the standard reference work on *Hevea brasiliensis* (1912).

Wright, it could be fairly said, is not really talking about indigenous knowledge systems, but is only concerned with the commercial value of certain items when moved outside their place of origin. While it is certainly the case that his agenda is explicitly commercial, or in the service of empire, it is not so clear that the instrumentalism of his analysis varies all that much from what is proposed by anthropological and other defenders of indigenous knowledge: the efficient exploitation of tropical preciosities. Although Wright and his commercial colleagues may have had little concern for the maintenance of those social systems whose members' knowledge facilitated the exploitation of such products as rubber, neither is it clear that the attempt to link the ethical exploitation of indigenous knowledge to environmental and social justice is plausible. Current holders of (potentially) esteemed esoteric knowledge are conventionally portrayed as isolated relics of prehistorical societies, which, despite the best efforts of colonial overseers, have persisted into the early twenty-first century, and the prevailing indigenous knowledge perspective stakes much on the claim that indigenous scientific knowledge is homologous with culture, that the two are inseparable—yet this is precisely what is not acknowledged in the pursuit of esoteric indigenous knowledge. Instead, the promise is that freeing up indigenous knowledge will provide salvation to indigenous peoples in the form of market share, just as under capitalism there is a radical disjunction between control of knowledge and rights as a citizen.

The example of Amazonian rubber (the extraction of which, over a period of about one hundred years, represents something quite different from the "boom" gloss with which it is typically characterized) draws attention to some important factors relevant to contemporary interest in the knowledge and resource systems of Amazonia. First, the current phase of indigenous exploration and exploitation is a reprise, not a novelty: the extractive economies that have characterized Amazonian integration into the global economy (Bunker 1985) present many examples of the incorporation of indigenous knowledge into systems of commodity production, at great cost to indigenous and peasant peoples of Amazonia. Second, the relationship between Amazonia and the global economy during the period of intense rubber extraction was a highly asymmetrical one from the perspective of overall economic and social development: once the rubber trade collapsed, there was little in the way of economic infrastructure that remained in Amazonia. Third, as in the case of the rubber phase, current exploitation of Amazonian resources and knowledge is a pretty brutal affair. Certainly there are many researchers carrying out environmentally, as well as socially and culturally, attuned work, but they are doing so against a backdrop of clear-cutting, open-pit mining, large-scale commercial fishing, and unregulated placer mining. The gentle evocations of "indigenous knowledge" are romanticized glimmers in the midst of base predation. Indigenous knowledge may be the source of a revitalized indigenism, but can never be more than a token expression if the decisions about how to exploit it are divorced from the social and cultural priorities of the possessors of that knowledge—yet those priorities receive scant and derisory attention from official quarters.

Conclusions

The idea of indigenous knowledge as a "resource" is flawed in several ways, yet persists as an appealing rallying point. That it should be proposed as the main platform for advancing indigenous rights prompts skepticism. The emergence of extractive reserves (as advanced by the Rubber Tappers Union, for example), the struggle to establish territorial rights by indigenous peoples, and the organizing of river peoples against the intensive despoliation of the *várzeas* (floodplains) may be enhanced by appeals to the knowledge resources that would be lost were the frontier-conquest mentality to prevail, but there is little doubt that the strength of such movements derives in the main from political action, not from the promotion of folk science. The link between indigenous knowledge and indigenous rights is paramount, but the maintenance of this link is likely to be sustained only if the fetishized resources are recognized not as free-floating values waiting to be seized and efficiently exploited, but as cardinal features of societies whose last bargaining chips are in danger of being frog-marched to auction.

References

Agrawal, A. 1995. Indigenous and scientific knowledge: Some critical comments. *Indigenous Knowledge and Development Monitor* 3(3). See www.nuffic.nl/ciran/ikdm/3-3/contents.html (accessed January 5, 2005).

Atran, S. 1990. *The Cognitive Foundations of Natural History*. Cambridge. Cambridge University Press.

Balée, W., ed. 1998. *Advances in Historical Ecology*. New York: Columbia University Press.

Berlin, B. 1992. *Ethnobiological Classification: Principles of Categorization of Plants and Animals in Traditional Societies*. Princeton, New Jersey: Princeton University Press.

Blaut, J. M. 1993. *The Colonizer's Model of the World*. New York: Guilford Press.

Bocco, G. and V. Toledo. 1997. Integrating peasant knowledge and geographic information systems: A spatial approach to sustainable agriculture. *Indigenous Knowledge and Development Monitor* 5(2). See www.nuffic.nl/ciran/ikdm/5-2/contents.html (accessed January 5, 2005).

Bowles, S. and H. Gintis. 1976. *Schooling in Capitalist America: Educational Reform and the Contradictions of Economic Life*. New York: Routledge and Kegan Paul.

Brockway, L. 1979. *Science and Expansion: The Role of the British Royal Botanic Gardens*. New York: Academic Press.

Bunker, S. 1985. *Underdeveloping the Amazon: Extraction, Unequal Exchange, and the Failure of the Modern State*. Chicago: University of Chicago Press.

Carson, R. 1962. *Silent Spring*. London: Penguin Books.

Chomsky, N., I. Katznelson, R. C. Lewontin, L. Nader, D. Montgomery, R. Ohmann, I. Wallerstein, R. Siever, and H. Zinn. 1997. *The Cold War and the University: Toward an Intellectual History of the Postwar Years*. New York: New Press.

Corry, S. 1993. The rainforest harvest: Who reaps the harvest? *The Ecologist* 23(4):148–153.

Crosby, A. 1986. *Ecological Imperialism: The Biological Expansion of Europe, 900–1900*. Cambridge: Cambridge University Press.

Gallois, D. 1998. Brazil: The case of the Waiapí. In A. Gray, A. Parellada, and H. Newing, eds., *Indigenous Peoples and Biodiversity Conservation in Latin America*, pp. 168–184. Copenhagen: AIDESEP (Inter-Ethnic Association for the Development of the Peruvian Jungle), The Forest Peoples Programme, and IWGIA (International Working Group for Indigenous Affairs).

Galtung, J. 1967. After Camelot. In I. Horowitz, ed., *The Rise and Fall of Project Camelot: Studies in the Relationship Between Social Science and Practical Politics*, pp. 283–296. Cambridge, Massachusetts: MIT Press.

IKDM (International Indigenous Knowledge Network). 1998. Editorial. *Reactions* 6(3). *See* www.nuffic.nl/ciran/ikdm/6-3/index.html (accessed January 5, 2005).

Juma, C. 1989. *The Gene Hunters: Biotechnology and the Scramble for Seeds*. London: Zed Press.

Lévi-Strauss, C. 1978. *Structural Anthropology*. Vol. 2. Harmondsworth, England: Penguin.

Lewontin, R. C. 1997. The Cold War and the transformation of the academy. In Chomsky et al., *The Cold War and the University: Toward an Intellectual History of the Postwar Years*, pp. 1–34. New York: New Press.

Morsello, C. 2002. Socio-economic sustainability of forest management by indigenous groups in Amazonia: A case study of the Kayapó. Ph.D. dissertation, University of East Anglia, UK.

Orlove, B. and S. Brush. 1996. Anthropology and the conservation of biodiversity. *Annual Review of Anthropology* 25:329–352.

Overal, W. and D. Posey, eds. 1990. *Ethnobiology: Implications and Applications*. Proceedings of the First International Congress of Ethnobiology. Belém, Pará: Museu Paraense Emílio Goeldi.

Quiroz, C. 1996. Local knowledge systems. *Indigenous Knowledge and Development Monitor* 4(1). *See* www.nuffic.nl/ciran/ikdm/4-1/contents.html (accessed January 5, 2005).

Rabben, L. 1998. *Unnatural Selection: The Yanomami, the Kayapó and the Onslaught of Civilisation*. London: Pluto Press.

Ramos, A. 1991. A hall of mirrors: The rhetoric of indigenism in Brazil. *Critique of Anthropology* 11(2):155–169.

Roddick, G. 1993. Reply to Corry. *The Ecologist* 23(4):198–200.

Sperber, D. 1996. *Explaining Culture: A Naturalistic Approach.* Cambridge, England: Blackwell.

Tyler, S., ed. 1969. *Cognitive Anthropology.* New York: Holt, Rinehart & Winston.

Umans, J. 1997. The rapid appraisal of a knowledge system: The health system of the Guarani Indians in Bolivia. *Indigenous Knowledge and Development Monitor* 5(3). *See* www.nuffic.nl/ciran/ikdm/5-3/contents.html (accessed January 5, 2005).

Wallerstein, I. 1997. The unintended consequences of Cold War area studies. In Chomsky et al., *The Cold War and the University: Toward an Intellectual History of the Postwar Years*, pp. 195–231. New York: New Press.

Wright, H. 1912. *Hevea Brasiliensis or Para Rubber: Its Botany, Cultivation, Chemistry and Diseases.* London: Maclaren.

18 What's the Difference Between a Peace Corps Worker and an Anthropologist?

A Millennium Rethink of Anthropological Fieldwork

Joanna Overing

> Asserting the superiority of western, or global, knowledge re-
> quires ignoring much of what people actually do and say, declaring them ignorant
> and incapable of commenting on their own actions. This seems rather silly, not to
> say narrow-minded.
>
> Mark Hobart, "As I Lay Sleeping"

This essay highlights the transformation in anthropological fieldwork conditions in Amazonia which have been developing over the past few years, and draws principally upon the recent experience of researchers at the Centre for Indigenous American Studies and Exchange at the University of St. Andrews, Fife, Scotland. First, however, we need to understand that anthropology today is not what it was a decade ago, and as a result the ethnographic "eye" has shifted its focus. To some extent we anthropologists are in a period of confusion, indeed breast beating, for as each new postcolonialist treatise is published, we find that the anthropologist is replacing the missionary as the "bad guy" of the Western world (e.g., Asad 1975; Said 1978; Fabian 1983; McGrane 1989; Inden 1990; Thomas 1994; Fardon 1995). Although the breeze of postcolonialism is for many of us a refreshing one (a point to

A shorter version of this essay was previously published under the same title in K. N. Geza, ed. 1999. *Menyeruwa*. Budapest: Szimbiozis 8, Yearbook of the Department of Cultural Anthropology, Eotvos Lorand University.

which I shall return), there is no simple answer to the accusation that anthropology has served as handmaiden to colonial conquest and government (Rapport and Overing 2000a).

Although most anthropologists working among the colonized have viewed their programs as a means to alleviate the weight of colonialism by reducing the effects of its mistakes (e.g., James 1975; Goody 1995), anthropology—to the extent it has been considered a social science and not an art—has nevertheless inadvertently served colonialism's aims. It has also served the hegemony of modernist programs of development and ways of thinking. To be sure, we have always had the benefit of voices warning against the dangers of scientism with its wrongheaded assumptions about sociocultural progress (e.g., Boas [1886] 1973; Evans-Pritchard 1964). It was very early in anthropology's history that Boas firmly placed all cultures (as ever-changing rationalizing systems) on a par, and scoffed at notions that wed technological might with social or cultural superiority (also see Rapport and Overing 2000a). However, since Boas there has also been the potent influence of the likes of Durkheim (1938), Radcliffe-Brown (1965), and Lévi-Strauss (1981). Their message was the modernist one: it stressed the superiority of the scientific investigator and the progress of science, modern science being the yardstick by which all anthropological (and indigenous) accomplishment should be assessed. A primary concern was with the "objective" abstract structures created by the anthropological expert, rather than with living, experiencing human beings who follow particular lifeways (Brady 1991; Overing and Passes 2000a). We find that the more usual theoretical approaches within modernist anthropology (structural functionalism, evolutionism, structuralism, structural Marxism) have been dismissive of much of what the rest of the world had to say. In this act of silencing the Other, the ethnographic Other becomes object, a resource for the self, as Anne Salmond (1995:23) describes it. She goes on to suggest:

> The fundamental error in anthropological thinking about the "other" may not be, as Johannes Fabian (1983) has suggested, the creation of artificial distance between self and other in space/time, but rather the transformation of "other" into anthropological object. For objects have these negative properties in western thought—they cannot speak, they cannot think, and they cannot know. "Objectivity" creates an immediate epistemological privilege for the "observer"—only he/she can truly know.
>
> (Salmond 1995:41)

As a result, anthropologists, in their describing and measuring, created Others as "exotic curiosities for European consumption" (Salmond 1995:23; also see Hiller 1991; Mason 1998; Rapport and Overing 2000b)[1] and in the process gave further force to the determination of science to define nonscientific ways of knowing as local, illegitimate, insufficient, wrong. Needless to say this a serious point, if even for the dismal selfish reason that we are on the verge of losing these knowledges.

Some Examples of "Engaged" Research

The present-day researcher does not have the option of assuming the stance of the objective specialist, a "value-free" research strategy that might free him or her from the contamination of political desire and engagement. In our postcolonial world the move has been to discard the notion of "objectivist description." More important, the Amazonian peoples themselves reject being treated as objects for study. They are deeply suspicious of the integrity of and reasons for such value-free investigations. They have a distaste for being treated as specimens. Today's anthropologists must be engaged, and persuasive in this engagement, in order to be allowed as guests within an indigenous community. They also must be politically streetwise in a way not previously expected of anthropologists, and they must use managerial and diplomatic skills that previously would not have been of much use. They must know how to deal with the myriad organizations in which the indigenous community is, or wishes to become, involved, ranging from the grassroots level to that of the nation-state and the multinational. Nowadays a community hosts anthropologists often in anticipation of access to their managerial talents, which can then be put to good use in the community's dealings with the multitude of agents of change who wish it well or ill.

Amazonian peoples are well aware of the dangers of rapid economic and social development (Rapport and Overing 2000b). They have suffered conditions of extreme change over the past forty years. They have seen strangers entering their territories to take their land and destroy it. They have seen the building on their lands of hydroelectric dams and roads, the mining of gold and the extracting of oil, the burning of forests, and the creation of large

---

1. Salmond (1995:41–42) suggests that "it is possible that anthropology has been the specific mode in which western epistemology assimilated the knowledges of others, silencing them and converting them into curiosities that need not be taken seriously, even on the matter of their own experience in the world."

cattle ranches and monoculture plantations. They have been displaced from their up-river small villages to down-river highly populated communities. Their young people have had to enter an educational system that uses a foreign language and teaches an alien knowledge of the world. The indigenous peoples have good insight into their problems; they know the enemy and its effect upon them. They know about the social costs of rapid change, and about the economic and personal costs of the market economy. They increasingly do not have the land to sustain the indigenous practices and the type of community life that they value. These experiences have effect upon the anthropological endeavor: nowadays, anthropologists who wish to be granted entry into indigenous communities (and thus tolerated as strangers from the outside, hostile world) must demonstrate that they both desire and are able to work with the community toward *its* goals, which are often political.

As a result, anthropologists cannot go to Amazonia today without considerable and constant reflection upon their own social and political positions regarding the people with whom they wish to live and work. In fact, it is a real step forward when anthropologists do not take it for granted that they will automatically be accepted by their hosts. In the examples that follow, I will introduce you to some young anthropologists whose plans for working with indigenous peoples included making themselves acceptable to the communities they wished to study by showing themselves to be people who wanted to engage not only in worthy academic research but also in contributing to the social and economic health of the subjects of their research.

• *Example 1.* Paul Oldham was a young researcher when he worked within the Piaroa communities of the Venezuelan Orinoco Basin during the early 1990s. Piaroa people have experienced accelerated change since the early 1970s, when they became incorporated as citizens into the nation-state. This is an interesting case, for the researcher found the rhetoric used by indigenous peoples about anthropologists to be strikingly harsh; many Piaroa indicted anthropologists not only for lacking in intelligence but also for being completely "useless," especially to the indigenous cause (Oldham 1996). They also reproached anthropologists' tendency toward arrogance: telling indigenous peoples what to do without first listening to what the people said they desired or needed. It is pertinent that, throughout the previous decade, many of the agents of change for the Venezuelan state who worked among Piaroa people introduced themselves as anthropologists, even though they had not been formally trained in the field (Paul Oldham, pers. comm.). Such rhetoric from the Piaroa may have been a strong form of moral persuasion, for in this case it served them particularly well by promoting constructive conversation with the researcher. Each community with which Oldham wished to live demanded formal procedures prior to his acceptance, including a trial period of residence, during which he could prove his worth by earning people's trust and demonstrating the ways in which he could improve their political relations with the nation-state. His roles were first to *listen*, and only then to enter into *dialogues* with both leaders and young people, and in so doing contribute his knowledge of Western political theory. Piaroa were particularly fascinated by his distinction between direct and indirect democracy, and they decided, through their extensive debates on the merits of each, that they, in their dealings with the nation-state, could accept direct, but not indirect, representation. Direct democracy, they decided, would provide the procedures that would allow them to engage actively and on their own terms with agents of the state (Oldham 1996). An important concern for both Piaroa leaders and their followers, given their own egalitarian and atomistic political process (Overing Kaplan 1975), was how to organize themselves effectively, but in a nonhierarchical way. In the end, Oldham helped the Piaroa set up the first national meeting in Venezuela of indigenous organizations, another process that involved much discussion within the community (Oldham 1996). In his doctoral dissertation, Paul Oldham (1996) focuses on the practical organization of Piaroa communities for (1) political action vis-à-vis the nation-state and (2) their reflections and debates on responses to their ongoing problems with the nation-state and its agents.

• *Example 2.* As a research student, Stephen Kidd worked among Enxet people, hunter-gatherers of the Paraguayan Chaco. He initially became involved with Enxet in the 1980s while working as an Anglican missionary on a development project. During that time he quickly became converted to the indigenous point of view. He speaks Enxet fluently and has long worked toward regaining land for various peoples of the Chaco. In the 1990s, as an anthropologist, he ran a German-financed nongovernmental organization (NGO) for three years; with the help of Paraguayan lawyers he fought for the land-rights claims of 14 indigenous communities, all either Enxet, Sanapaná, or Angaité peoples of the Paraguayan Chaco, and won cases for six of them. He always cooperated with and worked through indigenous leadership, advising and directing the leaders in their approaches to governmental bodies. In the late 1990s he wrote about the necessity of his own contribution to the success of organizing the communities and their leaderships in the land-rights project (Kidd 1997). He also noted, however, that there had been a general tendency for many anthropologists, who wished to preserve a public image of the "authenticity" or "purity" of present-day indigenous organizations, to mask in their writing the central role of white people within indigenous organizations. This point, to which I will return, speaks to the problems of those highly egalitarian peoples who are now dealing with the hierarchically structured institutions of the nation-state. Stephen Kidd completed his doctoral dissertation on the aesthetics of emotions and community life among Enxet peoples in 1999.

• *Example 3*. Research student David Menell participated in a project in Peru that included a number of NGOs and multinationals who were active in, or impinging upon, indigenous affairs. In this case, Menell added to his anthropological expertise a knowledge of geographic information systems (GIS), and created programs that wed the anthropological concern for the particular and the local to GIS technology. He was looking toward the possibility of giving greater strength to legal battles and land-rights claims for indigenous communities. His research concern was with the effect of NGO activities and rationalist-situated knowledge (with the emphasis on a particular type of high-tech sustainability) on indigenous ways of knowing and on the communally situated practices that are attached to such knowledge. Menell completed his doctoral dissertation on that topic in 2003.

• *Example 4*. The last example is the research of Carlos Londoño Sulkin, who worked with two communities among the Muinane of lowland Colombia in their attempts to create an indigenous-friendly education system that would include both Muinane and Western knowledges in its program of study. Originally involved with Muinane education through an NGO, Londoño Sulkin later incorporated the topics of NGO activities and attitudes into his doctoral research on Muinane peoples. The following point is critical. He found that he himself had to go through a "conversion" process to begin to understand both the reasons for the Muinane's attachment to their own ways of knowing, and the vital relationship between their own knowledge and their values with regard to the proper ways of living a human sort of life. Unlike Western knowledge, Muinane knowledge pertains to the capabilities and skills required to create and maintain a particular type of community, an existence they refer to as "cool," which in their view is healthy and moral. To their concern, the Western knowledge taught to their schoolchildren undermined their own knowledge about healthy living within a community of relationships. Muinane people therefore demanded a school curriculum that included both systems of knowledge. The question was, of course, How could the two types of knowledges be viably integrated? As is usual nowadays, Londoño Sulkin entered the field as an anthropologist (as opposed to his earlier role as an NGO worker) only through formal and diplomatic negotiations with the communities with which he lived. His doctoral dissertation (2000) centered on Muinane knowledge: its acquisition and its situated practice.

With these case studies as examples, we can see that the key to present-day anthropological research no longer comprises skills for objective observation, which require the researcher to maintain a critical and skeptical distance from the "object" of study (also see Wagner 1991). The vision of proper research has shifted toward an emphasis upon the anthropologist's ability to establish *dialogical, equitable,* and *useful* relationships with indigenous hosts. The practical is wedded to the academic, and the principles for interaction are premised upon *diplomacy* and *civility,* where coevality must be achieved (also see Fabian 1983 and Velho n.d.). Although this insight into sociality may be new to anthropology, it is not for indigenous peoples, who fully understand and also put into practice the principle that coevality can only be achieved through the process of establishing convivial and dialogical relationships (Overing and Passes 2000a). Piaroa, in their rhetoric against the "useless" anthropologist, were demanding that Paul Oldham follow this new set of rules, and they were teaching him how to do so. Their own understanding of knowledge is that to be worthwhile it must be situated; that is, it must be intimately connected to particular practices and to a specific type of social engagement through which coeval relations are assumed or can be created. The index of sociality for Piaroa people demands the dialogical and the equitable (also see Hobart 1995 and Salmond 1995). It is therefore wrongheaded of the anthropologist to think that the gathering of a vast corpus of origin myths or the statistics of garden production would in themselves lead to an understanding of indigenous knowledge, which, to the contrary, is always fixed to social practice. It is a relational matter. To learn indigenous knowledge is to learn how to engage socially and skillfully in particular sorts of ways.[2] This type of engagement on the part of the anthropologist does not allow for the creation of an "objectified other," which at any rate would now be impossible because of the refusal of indigenous peoples to be so objectified. Thus a new type of fieldwork is being called for, one where the anthropologist is working *with* the so-called objects of study, who instead of being transformed into dissected (ignorant) objects must remain knowing, intentional, and social subjects.

---

2. The linking of knowledge to both dialogue and practice is not unique to Amazonia. See Hobart (1995:58–59, 60) on the Balinese, who also situate knowledge in practice and dialogue.

What Is the Difference Between the Anthropologist and the Peace Corps Worker? Or, the Anthropologist as Hybrid

Having discussed the need for researchers to engage with indigenous communities in terms more meaningful to the subject, my next questions are: How do we go about evaluating the aims of this "new" anthropologist? How is a present-day young field researcher any different from the Peace Corps worker? Perhaps both, in an idealistic pursuit of the betterment of the Other, are equally clothed in all the colonialist, hegemonic trappings of modernism. Indeed, this is certainly how some postcolonialist writers would view them (e.g., Asad 1975; Said 1978). However it is my understanding that the main job of the Peace Corps worker, at least from the point of view of the government he or she is serving, is to take Western scientific knowledge, as well as good will, to the underprivileged peoples of the world. This role of the Peace Corpsman (to use a term from the prefeminist political climate) more efficiently fulfilled the scientist's and politician's dream of what in former times was the duty of the missionary, that is, to tame "the native" for future political and capitalistic endeavors. The rhetoric declares that people with an "inferior" lifeway should "assimilate" the knowledge of a "superior" people in order to have a "better life." As Zygmunt Bauman has noted (1990:158), the premise of such a relationship is one that has made inequality—political, social, epistemological—the axiomatic starting point of all argument. Such assumed inequality thereby becomes secured against challenge and scrutiny. Hobart (1993) notes that most development projects, particularly those premised upon "dependency theory," are based on assumptions of an assimilation paradigm of social change, and therefore also assume the absolute ignorance of so-called Third World peoples and the absolute superiority of their own Western knowledge. Inequality is the bottom line of the "developer's" relationship with those people who are to be "developed."[3]

The remainder of this essay will be a discussion of the consequences for anthropology—with respect to present-day debates about knowledge diversity and certain issues surrounding the notions of "the global" and "the local"—of a very different type of overt engagement on the part of the anthropologist. Here I argue that to be successful the new anthropologist must assume the

---

3. In *An Anthropological Critique of Development*, Hobart (1993) concludes that development projects that underestimate the value of local knowledges have been from the start doomed to failure.

status of "the reflexive hybrid" (Overing 1998; Rapport and Overing 2000b). In so doing, he or she sheds, to the extent possible, the status of representative (and certainly spokesperson) of a powerful nation-state and academy. Representative status should be resumed only if strategically necessary to the process of dealing with governmental, multinational, and funding institutions.

It is gratifying to hear research students and those who have recently achieved their doctorates proclaim, "We today have different aims from our mentors." "We want to empower," they say, "and not to prove our own power." Yet we must even ask ourselves: Is it our job to empower? Indeed, would we not be approaching our job with the condescension of missionary or Peace Corps assumptions and zeal? Who is our audience: the West, or the Rest? Do we preach to the system (trying to convert all those academic modernist readers), as has been our tendency up until now, or do we speak for "the unhomely," to use a Bhabha (1994) turn of phrase that summarizes all the migrants, refugees, colonized, women, gays—the hybrids—of the world (see also Overing 1998)? Is it our primary duty as anthropologists to create the voices that might adequately express the lives and the experiences of all those people who now dwell at the interstices and margins of nation-states? Our primary goal in anthropology is "translation." But of what? And for whom? I shall make some suggestions.

One of our strengths as anthropologists lies in our understanding of the "people" element; the extent to which we are successful in this endeavor enables us to combat all those grand narratives of our times that tend to be dismissive of "the people." We focus upon the *particular* (careful anthropologists do not usually go around making global, universalistic, sweeping statements). We contextualize specific ways of knowing by keeping intact the systems of values and practices to which each is tied (e.g., the values of community, of personal autonomy, and of work; and the specific understandings of what work is). Indigenous sharing practices in matters of production and consumption, and the value attached to them, cannot be untangled from indigenous technological "know how," which is why indigenous peoples often find Western development propositions offensive (MacLullich 2003). Another of our strengths is that we appreciate multiple perspectives: there is a multiplicity of colonial and postcolonial discourses, each having a number of potential and active subversions. Our recognition of plurality is surely anthropology's real strength (Ardener 1985), and we have as examples the research students working with an assortment of NGOs and learning the many and various indigenous responses to the effects of such organizations.

Within this universe of pluralities, there are also plenty of hybrids and states of hybridity. What is the aim of anthropological translation in expressing such hybridity? Homi Bhabha (1994) offers some good advice. He calls for a literature "of recognition," suggesting that by attempting to destabilize "traditional relations of cultural domination," translation can be a revisionary force. The point of translation then is to get people to think differently. It is not a value-free task (Overing 1987). The very act of hybrids demonstrating in their writing another territory—their "unhomeliness," that is, the fact that they have no home within "the system"—helps them to recognize, to signify, their own historic desire (Homi Bhabha 1994). Homi Bhabha is talking, for instance, about the hybrids' empowerment and the discovery of their own voices. He is talking about the "migrant writers"—black Americans of migrant background, Latin American creoles and mestizos of migrant stock—who express (translate) the oppositional Other (also see Mason 1990, and Corbey and Leerssen 1991) and their extreme disgust with (their "nausea" with) linear, progressivist claims. It is clear that new histories must be written, but where does anthropology fit in? There is also the question of who has the power to label the unhomely? And for whom? For the unhomely of the world, the answer to these matters (i.e., who is writing for whom, and for what reason) makes all the difference to the game of empowerment, for the flip side is disempowerment, a strategy of fixing identities (e.g., types of ethnicities, types of races or classes) that are in the interests of centralized control and domination long used by the nation-state (Overing 1998; Rapport and Overing 2000a).

Homi Bhabha (1994) writes as if the hybrid, the unhomely, is a product of Western civilization. Although most of us would not disagree with his examples of hybridity, there is one anthropological view that sees *all* cultural activity as belonging to hybridity (Overing 1998; Rapport and Overing 2000c). In this way of thinking, there is no such thing as a pure type or pure system. Just because hybridity typifies much of personal and social interaction within our globalized universe does not mean that the condition of multiple identifications is unique or relevant only for today. We might say that there is no such thing as "a culture," and that "culture" is less a noun than a verb: it does not exist as a thing but as an activity (Ingold 1994; Rapport and Overing 2000c). Westerners who place "culture" into museums—centering and reifying it—turn it into the matter of evolutionists' dreams. For most peoples around the world, the activity of culturing is endless, ongoing, and overtly visible or expressed, which ill fits the social scientist's (or museum curator's) fixed categories of belonging, or cultural identifications. In contrast, and from an Amazonian perspective, culture refers to those skills that a people have, mold, and use to live a life particular to their time and place (Overing 2003).

The question of whether we anthropologists can become one of the voices for the unhomely, for indigenous peoples, is an interesting one: I think we can, but only to the extent that we too consider ourselves one of the unhomely. Perhaps, as I have already proposed, the perspective of a good anthropologist is one that also recognizes his or her own hybridity and unhomeliness (Overing 1998; Rapport and Overing 2000b). Does that mean we can also take on the point of view of the unhomely—all those blurred categories, the migrants' wanderings to and fro, the unease with Western capitalism, and the disgust with the structures of domination created through modernist thought and action. To translate those blurred categories of human existence: now that is a project worthwhile.

An Anthropology of the "Everyday." Or, Making the Unhomely "Homely"

One way, acceptable to both ourselves and the unhomely of the world, that we can undertake this project is to systematically undermine the exotica of Western obsessions: the use and display of things and peoples as exotic. To displace Western use and concepts of exotica is a laudable anthropological ambition. This is not to underplay the extent that people can differ, for perspicacious difference, where the relation between self and other becomes centered upon the mutual conversation, is the object of this message. Exotica and difference do not need to be conjoined (Rapport and Overing 2000b). We can diminish the exoticism rampant in anthropology (all that magic, all that ritual, taken out of context) by accepting the "everyday" of indigenous life, and then translating it in such a way that it becomes familiar to us (Overing 2003; Overing and Passes 2000a). In other words we can become at home with the unhomely. What about situated practices, the everyday ways of knowing and doing things, of acting and responding? What about the sentiments (e.g., of love, humor, or honor) and body styles (e.g., scarification or physical stance) that are considered attractive, and those that offend? What about the personal relationships of everyday life? What about the "homeliness" of the unhomely? I do know that we would understand little about Piaroa social and political ways of thinking if we did not concentrate on the minutiae of their daily life—and their discussions about them—for it is the everyday that these people celebrate rather than some grander design of existence (Overing 2003).[4] As an aspect of their egalitarianism, it is the everyday they praise, not their gods and spirits of the jungle (even their gods cannot command them). It is wise to remember that they see what they do as ordinary; for them, what *we* do is exotic. With good reason the everyday is increasingly becoming the focus for some of us in anthropology. As Ivan Brady (1991) has described it, the everyday is that "polyphonic, symbolically layered environment" that has elusiveness as a permanent attribute, and where there is no uniform institutional or performative closure. The everyday is dynamic, continuously reopening itself, an existence that has no definite, singular meaning. Working on seeing the everyday is just one more way we anthropologists may be able to overcome the obscuring weight of grand narratives, and linked to this emphasis on the everyday is the question of how to let other voices speak for themselves, as for instance, through direct quotes.

Attached to everyday behavior are ways of knowing and an ontology very different from our own as anthropologists. To understand this everyday practice, to make it familiar, requires a type of approach that I would summarize as "engaging in conversational translation." The first rule in creating a conversational translation is to consider one's partner in the dialogue as a philosophical equal. For example, I have been impressed with the intellectuality of much of Piaroa social philosophy, and with the depth of their psychological insight into the human condition. But here we must be careful. It would be condescending of me to comment on the sophistication of their understanding, as if to imply, Hey, they are as smart as we are! Or even smarter! These are patronizing pronouncements that we might use to praise schoolchildren who perform well. A better way to establish a conversation between equals is to focus upon concerns that are common to both indigenous peoples and ourselves. For example, we might ask, What does it mean to be a human alive and well in this earthly existence of ours?—a good anthropological concern if there ever was one. Or we can ask, What is the relationship of the individual to the community, and its link to his or her ways of doing things? In fact, many of the questions about the nature of a proper moral or political life that are meaningful to us in the West are meaningful to Amazonian peoples as well; it is the answers that are different. This is why, while indigenous peoples might well want to engage in development projects for the good things they could bring to the community, they might at

---

4. Throughout 1968 and for six months in 1977, I lived and conducted research with the Piaroa of the Venezuelan Amazon Territory.

the same time be much offended by the rationalizations of Western developers, who emphasize efficiency while ignoring the knowledge practices of the
peoples with whom they are working.

A Collaboration Between Local Knowledges

To the extent that the discourses between anthropologists and indigenous
peoples overlap, they can affect each other, for instance in their respective
understandings of the notion of personal autonomy, or egalitarian practices.
We can share knowledges, take seriously the dialogical aspect of knowledge,
and thereby be aware of what we can learn through the hybridity of our inter-
change. Our learning can then have mutual effect. For the anthropologist,
the aim should also be the achievement of a mutual, if partly incommen-
surable, critical and multiculturally situated anthropology, through which
to create an anthropology for the unhomely. We on our side could then
transmit the message that *we too are local*, and from the perspectives of other
peoples, often bizarre in our solutions. Amazonian peoples, who tend to be
more relativistic in outlook than we are in many matters, certainly do not
need to be taught that knowledge is local, whether the knowledge is their
own or ours. It should be unnecessary at this point to observe that projects,
economic or otherwise, that are initiated with indigenous communities can
have a successful outcome only insofar as agents of change pay heed to this
message: all local knowledges, including our own, can in many respects appear outlandish and inappropriate to others. On the other hand, indigenous
peoples of the Amazon are usually very comfortable with the idea of shared
knowledges, but for obvious reasons are not so keen on the idea of assimilation. Not having our hegemonic ideas about knowledge, and being more
tolerant about difference, they are usually much more open than we, epistemologically speaking (see Salmond 1985, on the Maori).

Contrary to popular ideas about the matter, indigenous peoples are often
quite open to change. If I say that they tend to be epistemologically open, it
follows that they are not opposed to accepting new practices, since knowledge and practice are not separate for them. It is not unreasonable also to
say that they themselves prefer having a voice in the matter of deciding what
is good for and what is damaging to their communities and ways of living.
The Piaroa insisted upon this right in their discussions with the researcher
Paul Oldham (Oldham 1996). However there are many among academics,
bureaucrats, and politicians who believe strongly in the purity, and the authenticity, of indigenous lifeways, and therefore view any sign of openness to

change with disdain, especially if self directed and self motivated.[5] The idea of self-directed change is too much of a challenge to the axiomatic premise of the assimilation paradigm, in which all change, to be considered appropriate, must be decided upon by a representative of the superior-knowledge system, who by definition "knows best." In other words the idea of authenticity is a political stance, which can be used against the unhomely of the world as a means of keeping them under control.

A word to the wise for all of us working with indigenous peoples comes from the Caribbean author and Nobel Prize winner Derek Walcott (1996:271). He warns us about the dangers of the "patronizing gaze" that insists upon the purity of culture. Walcott reminds us that a lot of defensive, aggressive academics—as well as aggressive, defensive politicians—have seized upon the definition of "folk." He notes that there is something dangerous about "the property of reaching people and preserving what belongs to the people, all that stuff"—there is a "curious" kind of patronage in it. He says that he himself does not write "folk":

> When you talk about *folk* as a writer, then the danger there is you tend to say: "Well, we've got to preserve what we have, you've got to be rootsy, X or Y, you've got to talk that way." You know, that kind of thing; it's all very dangerous and ephemeral, that kind of aggressiveness. It turns into anthropology; and you can't patronize genuine people by making them anthropological specimens, like saying: "Oh, you are a great representative of the folk. Now you keep doing that." Right? While in the meantime you've been watching a good soap opera, or singing country songs.
>
> (Walcott 1996:271)

Walcott goes on to insist that each person has the right to go to the cinema, "instead of being a damned representative of the *folk* for the rest of his or her life. So anyway, there's that, that we have to look out for" (1996:271).

Stephen Hugh-Jones (1992) makes a similarly powerful plea for the right of indigenous peoples of the Northwest Amazon to make their own

---

5. See Oldham (1996) on the topic of politicians blocking indigenous leaders in Venezuela from accomplishing projects for their communities, and Salmond (1995) on a similar story about the Maori. See Hobart (1993:12) on the resistance of developers to self-motivated change. Also note the land-rights rhetoric in Brazil, where the government wants only "authentic" Indians to have a legal right to community land. Of this we may well ask: What makes an Indian "authentic"?

decisions about the acquisition of consumer goods. His argument goes as follows: there is a certain hypocrisy in the assumption that the integrity of Amazonian culture is in jeopardy as it comes into contact with the allure of greedy market forces, which are now entering contemporary Amazonia from myriad directions. The idea behind this assumption is that Amazonian people must not partake in, they must in fact be protected from, the capitalist vision of humankind's limitless needs, and its program for propelling humanity "to its own benefit" into an unbounded spiral of progress. Unlike people of the Western world, who benefit from and seem not to be beguiled by the necessary ruthlessness of market forces, indigenous peoples are seen as passive victims of the market economy (Hobart 1993). Yet in the name of progress, they have been drawn into it willy-nilly at the hands of missionaries, merchants, and government agents. As Hugh-Jones (1992) observes, the possibility is rarely considered that these are people who are also capable of reflecting upon their relationships with the market economy, and that indeed they often fully understand its risks, its dangers, and its allure. There is a certain danger in holding to the notion of these people as passive victims, an idea that assumes that they do not know the damage they are doing to themselves when they buy a shirt, a pair of trousers, a gun, or a radio. In fact they are active and creative participants in a two-sided process of exchange—think of the long-distance networks of trade that crisscrossed the preconquest Guianas and of the consumer goods that for centuries have had a place in the Guianese system of bartering and gift exchange. Of course people can lose their heads and be exploited and thereby suffer drastically; they can also understand risk, and handle it. Furthermore, as Hugh-Jones (1992) notes, it is unwise to assume that consumer goods have the same social and economic significance to Amazonian peoples as they have to us (also see Renshaw 1988). Although Yanomami of Venezuela and Tukano of the Northwest Amazon may enthusiastically wear white people's clothes, it does not mean that they gain status within the community for doing so, or that they become egoistic and selfish entrepreneurs the moment they put shirts on and shed loincloths for skirts and trousers. Rather, most consumer goods usually "undergo redefinition and may be put to novel uses" (Hugh-Jones 1992) as they are moved from one "regime of value" (Appadurai 1986) to another; that is, are transferred from a market economy sphere of value to an indigenous one. For instance, in Piaroaland many of the large or expensive items that a leader took home from his long-distance trading expeditions became community property ("owned by the community") (Overing Kaplan 1975); thus machetes and guns took their place alongside the community canoe and cassava graters. In incorporating expensive consumer goods—

particularly those for productive use—into their economy, Piaroa peoples' sense of ownership did not fit the set of values relating to personal property that is typical within the market economy.[6]

To wave the flag of authenticity is also dangerous for highly political reasons. Imagine this scenario. Indigenous peoples have their ordinary everyday life and practices, but they are increasingly being denied them as their status becomes transformed by the nation-state into the category of troublesome ethnic minorities—hybrids, outsiders who are damaging to the goals of national progress. They are an irritating force, resistant to the building of the nation's hydroelectric dams, the agricultural development of its lands, and the extraction of its precious minerals. Thus the indigenous peoples lose their lands, a process that makes them increasingly dependent upon consumer goods. They no longer have the land on which to follow their own knowledge practices. To survive and feed their families they must engage in wage labor, where they can neither look nor act in accordance with their ordinary (authentic?) ways of doing things. They have many skills, but not those for dealing with big government or big multinationals, skills that would be necessary to regain their land and thus their freedom for leading, in one way or other, their ordinary, everyday life. To then question the authenticity of indigenous peoples who accept help from white people who are skilled in relating strategically with governmental bodies would border on the vacuous. But in Amazonia, the pomposity of accusations of inauthenticity from bureaucrats and agents of development goes unheeded.

Stephen Kidd (1997) writes that it was for this reason, the possibility of the indigenous organizations of the Paraguayan Chaco being branded inauthentic, that he was initially reticent about writing of his crucial role in their land-rights battles. Kidd also worried that his participation might cast a questionable light upon his integrity as an anthropologist, for anthropologists (it is often mooted) are people who should protect the authenticity of indigenous action. The most glaring irony here is that the charge of instigating inauthenticity would be leveled against an anthropologist who assists indigenous peoples in organizing themselves to regain the economic means for leading a life in which they would be free to make better use of their own knowledge practices. Certain government officials did use this argument against Kidd.

Kidd (1997) goes on to note that by playing his part in the organizing of indigenous leaders he was in fact acting on behalf of the indigenous peoples

---

6. This was true in 1968. I do not know whether or to what extent the emphasis on community-held property still exists, or if it exists whether it applies to the same items.

in accordance with their own set of values. The Paraguayan Enxet, who have a more tolerant view of their relation to the Other than do white people and thus are not so choosy about "authentic" color of skin, were perfectly willing for Kidd to use his skills (which they did not have) on their behalf vis-à-vis the government and legal system of the nation-state. They therefore not only allowed but actively encouraged him to assume the role of an indigenous leader—as long as he also obeyed their own rules for leadership (Kidd 1997). And this he did, as did Paul Oldham (see example 1) when he helped the Piaroa in their organization for dealing with the nation-state. Throughout Amazonia the role of the indigenous leader is not the Western one that endows the leader with power of a coercive sort. Amazonian leaders cannot give orders to followers who staunchly value their own personal autonomy, nor are they supposed to make decisions on their own but only as a representative of their people after long discussions with them (Oldham 1996). Leaders are considered acceptable only insofar as they display appropriate skills, which always include a social astuteness and a strong capability for dialogical relations. To be considered a leader, a person, male or female, must first and foremost be judged as using his or her skills well in ways that are beneficial to the community. In sum, what is valued in a leader is the use of knowledge as dialogue, an idea for the most part alien to Western understandings of knowledge. This is the value that Piaroa leaders insisted that Paul Oldham learn before they would accept him into their communities. Why should indigenous peoples not use the skills of the white person—so long as he or she learns how to act on behalf of the community in the manner of an indigenous leader (Overing 1996).

Conclusions

"Hybridity" can take many forms. The message here is that of the novelist Walcott (1996), who argues that everyone has a right to his or her hybridity. My own interpretation of hybridity would include the anthropologist and not just the migrants and the indigenous peoples of the world. Anthropologists are increasingly deciding that they need to involve themselves in interdisciplinary research, an academic hybridity necessary for the success of present-day research goals. The conference upon which this volume is based began with the goal of advancing greater understanding of and support for Amazonia and its peoples, while its conclusion has been to stress the need to generate and be encouraging of greater interdisciplinary research. With regard to the latter aim, we should learn from indigenous peoples the advantages of dialogical and shared knowledge practices. In other words, each of those different knowledges—the academic disciplines; Western technological advances; indigenous knowledge practices, both technical and social—should be considered local and as such shared when possible. Projects within the Amazon will be successful insofar as there is a pointed collaboration between these knowledges, the ideal situation being to follow the procedures of "participatory research," where the aim is to facilitate the possibilities for peoples to conduct their own research. The main point is that the ideal of interdisciplinary cooperation takes on a new light if by definition it includes the local knowledges of indigenous peoples. Hybridity in research and program should be celebrated, not denied.

References

Appadurai, A. 1986. Introduction: Commodities and the politics of value. In A. Appadurai, ed., *The Social Life of Things*, pp. 3–63. Cambridge: Cambridge University Press.

Ardener, E. 1985. Social anthropology and the decline of Modernism. In J. Overing, ed., *Reason and Morality*, pp. 47–70. London: Tavistock.

Asad, T., ed. 1975. *Anthropology and the Colonial Encounter*. London and New York: Ithaca Press and Humanities Press.

Bauman, Z. 1990. Modernity and ambivalence. *Theory, Culture & Society* 7:143–169.

Bhabha, H. 1994. *The Location of Culture*. London and New York: Routledge.

Boas, F. [1886] 1973. The limitations of the comparative method of anthropology. Reprinted in P. Bohannan and M. Glazer, eds., *High Points in Anthropology*, pp. 84–91. New York: Knopf.

Brady, I. 1991. Harmony and argument: Bringing forth the artful science. In I. Brady, ed., *Anthropological Poetics*, pp. 3–30. Maryland: Rowman-Littlefield.

Corbey, R. and J. Th. Leerssen, eds. 1991. *Alterity, Identity, Image: Selves and Others in Society and Scholarship*. Amsterdam and Atlanta, Georgia: Rodopi.

Durkheim, E. 1938. *The Rules of Sociological Method*. New York: Free Press.

Evans-Pritchard, E. E. 1964. *Social Anthropology and Other Essays*. New York: Free Press.

Fabian, J. 1983. *Time and the Other: How Anthropology Makes Its Object*. New York: Columbia University Press.

Fardon, R 1995. *Counterworks: Managing the Diversity of Knowledge*. London and New York: Routledge.

Goody, J. 1995. *The Expansive Moment: Anthropology in Britain and Africa, 1918–1970*. Cambridge: University of Cambridge Press.

Hiller, S., ed. 1991. *The Myth of Primitivism*. London and New York: Routledge.

Hobart, M. 1993. Introduction: The growth of ignorance? In M. Hobart, ed., *An Anthropological Critique of Development*, pp. 1–30. London and New York: Routledge.

Hobart, M. 1995. As I lay laughing: Encountering global knowledge in Bali. In R. Fardon, ed., *Counterworks: Managing the Diversity of Knowledge*, pp. 49–72. London and New York: Routledge.

Hugh-Jones, S. 1992. Yesterday's luxuries, tomorrow's necessities: Business and barter in northwest Amazonia. In C. Humphrey and S. Hugh-Jones, eds., *Barter, Exchange and Value: An Anthropological Approach*, pp. 42–74. Cambridge: University of Cambridge Press.

Inden, R. 1990. *Imagining India*. Oxford: Basil Blackwell.

Ingold, T. 1994. Introduction to culture. In T. Ingold, ed., *Companion Encyclopedia of Anthropology: Humanity, Culture and Social Life*, pp. 329–349. London and New York: Routledge.

James, W. 1975. The anthropologist as reluctant imperialist. In T. Asad, ed., *Anthropology and the Colonial Encounter*. London and New York: Ithaca Press and Humanities Press.

Kidd, S. 1997. Invisible whitemen and indigenous organisations. Paper presented at the International Congress of Americanists, Quito, Ecuador, July 6–12.

Kidd, S. 1999. Love and hate among the people without things: The social and economic relations of the Enxet people of Paraguay. Ph.D. dissertation, School of Philosophical and Anthropological Studies, University of St. Andrews.

Lévi-Strauss, C. 1981. "Finale." In *The Naked Man*, pp. 625–695 . New York: Harper & Row.

Londoño Sulkin, C. D. 2000. The making of real people: An interpretation of a morality-centred theory of sociality, livelihood and selfhood among the Muinane (Colombian Amazon). Ph.D. dissertation, School of Philosophical and Anthropological Studies, University of St. Andrews.

MacLullich, C. 2003. The moral (im)possibilities of being an applied anthropologist in development: An exploration of the moral and ethical issues that arise in theory and practice. Ph.D. dissertation, School of Philosophical and Anthropological Studies, University of St. Andrews.

Mason, P. 1990. *Deconstructing America: Representations of the Other*. London and New York: Routledge.

Mason, P. 1998. *Infelicities: Representations of the Exotic*. Baltimore and London: Johns Hopkins University Press.

McGrane, B. 1989. *Beyond Anthropology: Society and the Other*. New York: Columbia University Press.

Menell, D. J. 2003. The application of geometric technologies in an indigenous context: Amazonian Indians and indigenous land rights. Ph.D. dissertation, School of Philosophical and Anthropological Studies, University of St. Andrews.

Oldham, P. 1996. The impacts of development and indigenous responses among the Piaroa of the Venezuelan Amazon. Ph.D. dissertation, University of London.

Overing, J. 1987. Translation as a creative process: The power of the name. In L. Holy, ed., *Comparative Anthropology*, pp. 70–87. Oxford and New York: Basil Blackwell.

Overing, J. 1996. Who is the mightiest of them all? Jaguar and conquistador in Piaroa images of alterity and identity. In A. J. Arnold, ed., *Monsters, Tricksters, and Sacred Cows: Animal Tales and American Identities*, pp. 50–79. Charlottesville and London: University Press of Virginia.

Overing, J. 1998. Is an anthropological translation of the "unhomely" possible, or desirable? In *Brief: Intellectual Traditions in Movement* (ASCA Yearbook), pp.101–116. Amsterdam: Amsterdam School for Cultural Analysis, Theory and Interpretation.

Overing, J. 2003. In praise of the everyday: Trust and the art of social living in an Amazonian community. *Ethnos* 68(3):293–316. (Orig. pub. 1999 as Elogio do cotidiano: A confiança e a arte de vida social em uma comunidade Amazônica. *Mana* 5(1):81–108.)

Overing, J. and A. Passes. 2000a. Introduction: Conviviality and the opening up of Amazonian anthropology. In J. Overing and A. Passes, eds., *The Anthropology of Love and Anger: The Aesthetics of Conviviality in Native Amazonia*, pp. 1–30. London and New York: Routledge

Overing, J. and A. Passes, eds. 2000b.*The Anthropology of Love and Anger: The Aesthetics of Conviviality in Native Amazonia*. London and New York: Routledge.

Overing Kaplan, J. 1975. *The Piaroa, a People of the Orinoco Basin*. Oxford: Clarendon Press.

Radcliffe-Brown, A. R. 1965. *Structure and Function in Primitive Society*. New York: Free Press. (Orig. pub. 1952.)

Rapport, N. and J. Overing. 2000a. Alterity. In *Social and Cultural Anthropology: The Key Concepts*, pp. 9–18. London and New York: Routledge.

Rapport, N. and J. Overing. 2000b. The unhomely. In *Social and Cultural Anthropology: The Key Concepts*, pp. 363–374. London and New York: Routledge.

Rapport, N. and J. Overing. 2000c. Culture. In *Social and Cultural Anthropology: The Key Concepts*, pp. 92–102. London and New York: Routledge.

Renshaw, J. 1988. Property, resources and equality among the Indians of the Paraguayan Chaco. *Man* 23(2):334–352.

Said, E. 1978. *Orientalism*. New York: Viking Press.

Salmond, A. 1985. Maori epistemologies. In J. Overing, ed., *Reason and Morality*, ASA Monograph 24. London and New York: Tavistock Publications.

Salmond, A. 1995. Self and other in contemporary anthropology. In R. Fardon, ed., *Counterworks: Managing the Diversity of Knowledge*, pp. 23–48. London and New York: Routledge.

Thomas, N. 1994. *Colonialism's Culture: Anthropology, Travel and Government*. Cambridge, England: Polity Press.

Velho, O. n.d. Globalization: Object—perspective—horizon. Unpublished manuscript.

Wagner, R. 1991. Poetics and the recentering of anthropology. In I. Brady, ed., *Anthropological Poetics*, pp. 37–50. Maryland: Rowman-Littlefield.

Walcott, D. 1996. Animals, elemental tales, and the theater. In A. J. Arnold, ed., *Monsters, Tricksters, and Sacred Cows: Animal Tales and American Identities*, pp. 269–280. Charlottesville and London: University Press of Virginia.

19 Traditional Resource Use and Ethnoeconomics

Sustainable Characteristics of the Amerindian Lifestyles

Clóvis Cavalcanti

Although growth should not be confused with development, the fact is that economic development has usually meant the persistent increase in per capita income of a country or economy. This, at least, is how it is defined in a place like Brazil (and more broadly in Amazonia, as well as in all of Latin America).[1] It is in such a context that ecological economics has come about as a discipline shaped by the need to reconcile material progress with the sound management of the environment. This requires that some sort of consistency be attained between two conflicting tendencies: one tendency is the increasing demand for resources caused by the expansion of the economy and by population growth, and the other is the unavoidable constancy of the ecosystem (meaning ultimately the invariance of the amount of matter and energy at the disposal of humans). This consistency is the essence of the idea of sustainability, whether it is defined in theoretical or operational terms. The problem with this idea is that it requires us to maintain indefinitely the potential productivity of the system that supports not only economies but life itself. It is evident now that the model of economic

---

An earlier, somewhat different, version of this essay appeared under the title "Patterns of Sustainability in the Americas: The U.S. and Amerindian Lifestyles" in F. Smith, ed., 1997, *Environmental Sustainability: Practical Global Implications*, pp. 27–45. Boca Raton, Florida: St. Lucie Press.

1. "Western industrial societies are often called 'consumer societies,' presumably because it is perceived that in these societies consumption is the most important contributor to human welfare. Certainly the principal objective of public policy in these societies is the growth of the gross national product (GNP)" (Ekins 1995:5).

development—or growth, for that matter—as it is practiced in the modern, Westernized world does not lead to the required compatibility of the ecological base with the goals set for the economic system. On the other hand, an improvement in the living conditions of the poor of the world, which is a moral duty of society, is normally associated with the augmentation of aggregate product. It is said that growth is necessary so that we can eliminate or reduce extreme poverty. In this context nature is simply asked to provide the resource layer needed to sustain the expansion of the economy. Almost no one discusses the extent to which nature can fulfill this function. Some people (e.g., Simon 1987:19) even contend that material progress must not be stopped because our entire modern life-support system is composed of artifacts made by man.

But that perception is not the only one that has existed. Even today, people living in distant, far-removed areas—for example, traditional and native populations like some Amazonian *índios* (Indians)[2]—have known for a long time that natural resources are meant to be used without jeopardizing the ability of future generations to employ them for their own benefit. Certainly this is a very different understanding of the problem raised by the modern money economy. It is the possibility of the existence of a "traditional economy" (Binswanger 2000:20) based on recycling that creates the need for a discipline that we may call "ethnoeconomics." Lionel Robbins (1932) famously defines economics as "the science which studies human behaviour between ends and scarce means which have alternative uses (15)." His approach, however, deals with "different ratios of valuation" (15) in money terms (market prices). A different situation arises in a natural, self-sustaining, self-regenerating economy based on exchange (Binswanger 2000:20–23). In contrast with what happens in a market economy, different ratios of valuation must be considered as a very distinct concept in a natural (traditional) economy that utilizes, for instance, the knowledge of shamans for decision-making concerning resource use (Reichel-Dolmatoff 1976). This is one more justification for the field of ethnoeconomics, whose objective should be to understand how primitive societies have learned to exploit nature sustainably, that is, by allocating resources (their means) efficiently.

If we accept Sachs's observation that "a global monoculture spreads like

---

2. The natives of the Americas have mistakenly been called "Indians" ever since the "discovery" of the continent. This ambiguity of the English language does not exist in Latin America, where the Portuguese and Spanish words (*índios* and *indios*, respectively) are used. The word for people from India is *indiano* in both languages.

an oil slick over the entire planet" (1992:102), we should admit the inevitability of economic development as preached since the start of the Cold War. To develop, therefore, one would have to follow the guidelines established by the experience of the industrialized countries, allowing oneself to be sucked into the homogenizing pool of cultural traits (market, state, science, technology) peculiar to the West. That is what is expected from a world that embraces (and is embraced by) multifarious cultural elements and traditions, some of which are simply incompatible with the idea of growth. Furthermore, the modern understanding of truth—which is not the only possible understanding of truth (e.g., Faber, Manstetten, and Proops 1994:7)—constitutes the sole framework of ideas that is adopted to rule decisions related to technological progress, economic performance, and social change. We are led to think that the options for a decent survival of man on earth are reduced to the paradigm offered by the first world, the Organization for Economic Cooperation and Development (OECD) countries, and some exceptional cases of high economic achievement. However, although incipient, the findings of research in the fields of anthropology and, especially, ethnobiology reveal indigenous perceptions about ecology and the utilization of natural resources that show that "there are options for the survival of Man in the Biosphere" (Posey 1990:57, 1992a:17). These options, which can be found in the lifestyles of native peoples, serve to caution against the tendency to promote economic development at such a pace that it cannot be halted in time to prevent the sometimes irreversible destruction it can cause (1992a:17).

It is my contention that the approach to economic issues which supposes the existence of real ecological boundaries (e.g., the planet as a nongrowing entity, the constancy of matter and energy), is something that can be conducted with the support of traditional knowledge and the practices of indigenous peoples, like some of those we still find in Amazonia. In other words, the accepted ecological treatment of economic problems grounded in modern Western thinking tends to elide important perspectives and to see nature from a Cartesian, less holistic perspective in which nature is dominated by man. Traditional knowledge—it is being more and more widely accepted now—offers sound alternatives for resource use and management, as shown by the experiences with and the close monitoring of native practices that have occurred over very long periods. Traditional knowledge can supplement modern science and open new horizons of understanding. It can contribute a lot to the solution of ecological-economic problems, chiefly because it is the only source of alternative models of development which has been shown to be ecologically and socially sound (Museu Goeldi 1987:33).

It is in traditional knowledge, apparently, where we can find a place for ethnoeconomics.

The purpose of the present essay is to delve into the question of sustainability by employing evidence from anthropology and ethnoscience that helps us comprehend how a society can live within the limits of the possible and still have a joyous life. The support for this task is provided by an anthropological literature that is not specifically designed to study the sustainable features of given social groups. However that is a problem because the anthropological evidence is sparse, and there is no systematic way of showing how sustainability is achieved. Moreover, I, as an economist, am not well trained to deal with either the issues or the methods of anthropologists and ethnoscientists. I am aware, however, of the traps that exist when we enter fields of inquiry different from our own. As a tentative practitioner of ecological economics, and as an apprentice in ethnoeconomics, I think that we cannot avoid doing inter- and transdisciplinary work within these new areas of study. And I think that we should follow Georgescu-Roegen's advice that "venturing into territories other than one's own" is an endeavor "definitely worth undertaking" (1971:4). When he set out to analyze the relationship between the entropy law and the economic process (which he did successfully, by the way),[3] what he did was "to build on the writings of the consecrated authorities" in the field of physics, while remembering to add that "even so, one runs some substantial risks" (4). I face the same risks and challenges when I invade the fields of anthropology and ethnoscience. But I find it an effort worth making, not the least because I judge that economists in general have much to learn from ethnology.

I will use here research undertaken mainly with Amerindian tribes still living in the Amazon, whose lifestyles are worth examining in the context of this essay. It is not my intention to convey a picture of the Brazilian Indians as inhabitants of a countercultural utopia facing the straitjacket of progress, but to call attention to a form of knowledge that can be extremely helpful in devising a sustainable future for humanity. The Indians exhibit a harmonious way of life in terms of man-nature relationships and thermodynamic thrift. For this reason, they deserve much more attention than has been given them until now. It is unfortunate that some people (supposedly outside the academic world) still view the Indians as being an inferior culture—as was claimed by a Brazilian authority speaking about the Yanomami Indians in the late 1980s, and even earlier by the Catholic Church, which held that

---

3. See Cleveland and Ruth (1997).

the Amerindians were not human beings—until Pope Paul III (1534–1549) removed that characterization in his papal bull *Sublimis Deus* (Gronemeyer 1992:56).

Paradigms of Sustainability

The Economist Intelligence Unit regularly publishes a chart of risk ratings, by country, for 26 "emerging markets" (e.g., Economist Intelligence Unit 2004) which shows a summary of national credit-risk indexes based on strictly economic and political factors. What if it charted those countries in terms of global environmental soundness or ecological sustainability; or what if it listed countries according to ecological sustainability and ascribed a rating to different lifestyles? For some time I have been trying to compare two very different life paradigms in terms of ecological sustainability (Cavalcanti 1992). One extreme paradigm (see figure 19.1) is found in the United States, with its high rates of per capita resource consumption. At the other end of the spectrum is the paradigm of frugality: it is the lifestyle of the Brazilian Indians, who in the wild state still inhabit portions of the Amazon, and whose consumption needs are satisfied by much more moderate consumption, which is regulated by austere and unchanging community standards. Ecological sustainability is naturally much higher in the case of the Indians, who live within what we can call "the limits of the possible," without causing social or ecological stress. This seems to fulfill the condition of the steady-state economy (Daly 1980), which slows down the energy flow or the throughput of matter and energy.

If ecological economics is the science and management of sustainability, then the Indian paradigm cannot be ignored in terms of what it teaches. It is precisely this point that is underscored by the late Colombian-born anthropologist Gerardo Reichel-Dolmatoff when he says that "the Indians' way of life reveals to us the possibility of an *option*, of a separate strategy of cultural development," which in his view offers "*alternatives* on an intellectual level,

More Sustainable		Less Sustainable
Amazon Indians	————————————————————	U.S.A.

FIGURE 19.1 Paradigms of Ecological Sustainability Measured in Terms of a Hypothetical Scale of Degrees of Sustainable Lifestyles

on a philosophical level," or alternative cognitive models that "we should keep in mind" (1990:14). The assumption is that the Amazonian Indians copy the patterns of nature by assimilating the principles they observe in natural ecosystems. Their lifestyle thus reflects a basic systemic wisdom (Bateson 1972), one that is inherent in nature (Branco 1989). The Indians' worldview is based upon their knowledge, whereas the Americans' relies on modern science (the scientific *logos*). The natives of Amazonia observe sustainability insofar as they plan according to the needs of future generations and maintain the living conditions of other species, thus assuring the preservation of biodiversity. The Indians, with their strong sense of community, do not pursue the interests of the individual unrestrictedly; this contrasts with the American paradigm in which man-nature relationships are defined according to Western thought, that is, from an anthropocentric standpoint. Reichel-Dolmatoff, referring to the Tukano people's worldview, says that their cosmological myths "do not describe Man's Place in Nature in terms of dominion, of mastery over a subordinate environment" (1976:318). He also remarks that the primitive tribes of the Amazon Basin, which to some people are "fossil societies"(308) that would not have anything to teach us, are not incomplete in the sense that they have not evolved, but rather have developed highly adaptive behavioral rules for survival "framed within effective institutional bodies" (308). The ecological principles elaborated by the Indians are combined with a system of social and economic rules leading to "a viable equilibrium between the resources of the environment and the demands of society" (308). It is worth noting here that Reichel-Dolmatoff, whose study of the Tukano spanned more than fifty years, found that there is little concern among them for maximizing short-term gains or for obtaining more food or raw materials than are actually needed. In the Indians' view, "man must bring himself into conformity with nature if he wants to exist as part of nature's unity, and must fit his demands to nature's availabilities" (Reichel-Dolmatoff 1976: 311; also cf. B. Commoner, quoted in Tiezzi 1988:9). This is simply the opposite of Aristotle's *pleonexia* (to have and to always wish to have more), which is "the driving force of modern productive work" (Faber, Manstetten, and Proops 1994:88). Other anthropologists have arrived at similar conclusions, as for instance when Viveiros de Castro (1992:168) alluded to the Araweté in Pará state, Brazil, whose contacts with the white man occurred only in the late seventies, and whose culture he found to be wholesome, strong, cheerful, original, and imaginative. The focus of the Indians' interests is conservation of their territory. This was clearly expressed by a Yanomami tribesman in a letter to Brazil's president, José Sarney, dated September 1, 1989: "Our thought is our land. Our interest is to preserve the land, not to create diseases for the people of Brazil, and not [to preserve the land] only for the [sake of the] Indians"[4] (CCPY/CEDI/CIMI/NDI 1990:43).

Contrasting with modern perceptions, and the American paradigm as well, the Indians' understanding of concepts and fundamental aspects of Western civilization—like money, ownership, the nation-state, sexual taboos, division of labor, misery, domination, and so on—is extremely precarious (Viveiros de Castro 1992:166). In the letter to President Sarney referred to above, the Yanomami tribesman said, "We do not know anything about money, shoes, clothes.... The government does not know our customs, our thoughts" (CCPY/CEDI/CIMI/NDI 1990:43). When these elements of modern civilization are introduced into the Indian society they provoke serious disturbances, as is indicated by Betty Mindlin, who has studied tribes in the Amazon state of Rondônia. The results of her findings show that "the use of money modifies food habits, reduces the rhythm of agricultural work, causes undernourishment, not because of scarcity properly ... but because of a new utilization of time, new behaviors ... [and] money is not distributed with the same fairness, according to the village's laws of reciprocity. It prevails over kinship, over the previous rules for a good living: and our society understands it well" (Mindlin 1994:248). Similarly, disturbances following the contact with the white man tend to increase inequality between man and woman (246).

Ecologically sound land-use planning is a common feature of Indian societies in Amazonia (although, on occasion, the natives may contribute to the degradation of their lands, as well). Some Indians, for instance, apply adaptive rules to birth rates and harvest rates (in other words, to the exploitation of the physical environment) to ensure individual and collective survival and well-being (Reichel-Dolmatoff 1976:312). The task of determining and applying these rules is conducted by the shamans, who are the managers of resource use. Some measures traditionally undertaken by Amazonian indigenous peoples, like the Wanana's longtime practice of protecting forests along riverbanks as a resource for fruit- and leaf-eating fish (Chernela 1989:75), only recently have been considered scientifically sound. These measures are implemented with a sense of profound respect for nature, from which the Indians copy their methods of environmental management. Viveiros de Castro (1992:157), speaking of the Araweté, comments on the Indians' simple technology and high capacity for improvising. It is not surprising then to

---

4. Translations from the Portuguese have been rendered by the author of this essay.

discover that local communities and tribal groups are "the most cost-effective managers of the resources" available to them (Panayotou 1991:357). Their knowledge of how to live within the limits of the possible extends to the control of socially disruptive behavior (such as aggression in interpersonal relations), which among the Tukano, for example, is submitted to rules that counterbalance it (Reichel-Dolmatoff 1976:312).

The legacy of centuries of balanced environmental management by the native societies was appreciated by the first Portuguese to arrive in Brazil in 1500 (Cortesão 1943). What they found was a magnificent, beautiful country (Ribeiro 1987:11) with abundant vegetation, pristine water, and plentiful game and fruits (Gandavo 1924:43–48, 82), the same environment that can be encountered today in parts of the Amazon. The primitive inhabitants of Brazil in 1500 were good looking, healthy, and strong (Cortesão 1943); these are the same attributes noticed by anthropologists who have done research among Indians in this century. In talking about the Araweté, Viveiros de Castro points out that in 1981 they were "visibly well nourished" (1992:155). Seeger (1980) arrives at the same conclusions about the Suyá, attributing their good health to their adequate diet. Baldus (1970:440) comments that the Tapirapé (men and women, adults and children) were used to enduring long journeys of 40 to 50 kilometers through the forest and savanna without becoming exhausted; he took this to be proof of the Indians' vitality, in spite of their average short life span. Writing in the sixteenth century, Gandavo (1924:124) makes analogous remarks. The Tupinambá, according to Sousa (1971:313–314), also a sixteenth-century writer, were excellent divers, swimmers, runners, and rowers, and showed great ability to climb trees and to jump.

Besides being strong and healthy, Indians today seem to be very happy with their lifestyle. This is stressed by Viveiros de Castro, who has studied the Araweté for more than fourteen years. In his words: "To live with the Araweté is a fascinating experience. Few human groups, I imagine, are so easy to deal with, so joyful in their daily life ... absolute in giving and asking, unrestrained lovers of the pleasures of life" (1992:154). An equivalent state of affairs could be found among the Tapirapé by Baldus, who in 1935 came upon a constant atmosphere of joy in the village where he lived. "All the environment is tenderness. No one yells at anyone and even the dogs that bark at me on my way are taught discreetly to respect me. Everywhere I find gladness and laughter" (1970:449). He adds in the same passage that "Courteousness ... manifested itself in various degrees as a general pattern of behavior" (449), concluding that the Tapirapé "were the most joyful people" he found in his life (464). The Yanomami Indian Davi Kopenawa gave an interview (after the white man had invaded his peoples' territory), in which he said:

> Now you tell the other white men ... how we were, with good health.... How we did not die easily, we did not have malaria. Tell how we were really happy. How we hunted, how we gave parties.... You saw that.... Today the Yanomami do not build their big houses anymore ... they live only in small shanties in the woods, under plastic sheets. They do not even grow crops, they do not go hunting anymore, because they become ill all the time.
>
> (Davi Kopenawa, as told to Bruce Albert, in
> CCPY/CEDI/CIMI/NDI 1990:14)[5]

This is in stark contrast to the situation of the still isolated Araweté, about whom Viveiros de Castro concludes, "This is not a desperate, culturally demoralized people, composed of sick, alcoholic, hungry and fearful persons—up to now" (1992:168).

Other characteristics of the Indians who inhabit portions of Amazonia today—and who inhabited Brazil in 1500—suggest not only that they are well adapted to the environment and enjoying good health and a joyous life, but that they are peaceful and courageous (Baldus 1970:440), that they do not accumulate material things (Gandavo 1924:130), that they do not worry about locking their belongings and are not familiar with stealing (Cardim 1939:112), that they are hospitable (Sousa 1971:316), and that they have a strong sense of community, generosity, and sharing (313). Sousa sums up his observations by saying that the Indians he described, the Tupinambá, were like Franciscan friars in their propensity to give away their possessions (314). The dissimilarity between the Indians' lifestyle and one centered on the craving for all kinds of gadgets is striking. We face here two very different perceptions of life, with serious implications for environmental health and social equilibrium. It is no surprise then that Faber, Jöst, and Manstetten (1994) would raise the following question: "Are not the increasing problems of social disorder, violence, drugs, etc. consequences of the level of our present standard of living?" (16). I do not intend to resuscitate the Rousseauesque myth of the *bon sauvage*. But indigenous experiences with sustainable development deserve to be seriously considered by researchers, for the Indians are "a diligent, intelligent and practical people who have adapted successfully

---

5. See also Chagnon (1992). The title of his book *Yanomamö: The Last Days of Eden* speaks for itself.

for thousands of years in the Amazon" (Posey 1992b:43), making their livelihood in many different ways according to local constraints. In a word, the
Amerindians' lifestyle is clearly a sustainable one: the evidence of their living
patterns, accumulated over a long stretch of time—as opposed to the insufficient historical evidence offered by modern, industrialized countries—is
enough to demonstrate the effectiveness of their methods to protect ecological sustainability. The evidence also seems to demonstrate a clear command
by the Amazonian natives of the principles of ethnoeconomics.

The Ethnoeconomy of the Amazon Indians' System

Using evidence provided by anthropology and ethnoscience, I have prepared a list of some of what I regard as the chief characteristics of Indian
societies still living in Amazonia. The list that follows is a summary of what
one finds sparsely in the literature, where the subject of sustainability springs
up mostly in an implicit, unsystematic way, and is mixed with such topics
as kinship, material culture, rituals, descriptions of daily life, customs, traditions, native knowledge, and myths:

- Very clear man-nature relationships
- No energy problem; no use of fossil fuels; basic source of energy—the
  sun
- Ignorance of money and ownership; no wealth accumulation
- Complete observation of the laws of nature; nature not used, but re-
  vered
- Simple, soft technology; no use of inorganic chemical products of any
  sort
- Satisfaction of basic needs
- Daily consumption of materials per person remains constant over
  time
- Life supported by the biological product of photosynthesis, water, for-
  ests, and clearings
- Populations held within given limits
- Inexistence of income inequalities (idea of poverty ignored); inter-
  generational equity
- Respect for biodiversity; maintenance of environmental quality
- No economic development in the modern sense (no growth, of
  course)
- Itinerant agriculture; nomadism; dispersion

- Simple material culture; extreme thermodynamic thrift
- Small villages; small production units
- Sustainable and efficient use of natural resources; preserving productive ecosystems
- Scale of activities within the carrying capacity of their territory
- Absence of technological improvements
- Long-run perspective
- Holistic, integrated view of life, reality, and the environment
- Apparent enjoyment of life

Although the picture offered in the foregoing list seems to contradict the evidence put forth by Lewis (1992), which shows that premodern peoples do not live in harmony with their surroundings, it conforms with the remarks of the late anthropologist Berta Ribeiro (1987:9), who found that the Indians treated their surroundings with respect, love, and care to ensure the permanence of nature as a source of resources for food, human welfare, and the cure of their illnesses. It also reflects what enthobiologist Darrell Posey and other researchers have discovered in their important work at Belém's Goeldi Museum, namely, that the basic aspect of the natives' management of natural resources is "a long-term perspective, with emphasis on preservation, and not on the destruction of [the] native resources of Amazonia" (Museu Goeldi 1987:65). It reflects, moreover, recent findings from prehistoric archaeology in the Amazon: the Amerindians' health conditions got much worse after the conquest (Roosevelt 1991:127). In fact, according to anthropological sources, the Amerindians' beliefs about and attitudes toward life—combined with the hundreds of little things they do, think, or avoid; their perception of the universe; and so forth—"form a highly structured order" (Reichel-Dolmatoff 1990:13). The knowledge of the Kayapó (who call themselves Mebêngôkre), for example, constitutes an integrated system of beliefs and practices, so that "each and every *Mebêngôkre* believes that he or she has the ability to survive alone in the forest for an indefinite time" (Museu Goeldi 1987:15).

One aspect of the Amazonian Indians' view of nature, as exemplified by the Tukano, is its "remarkable semblance to modern systems analysis," according to Reichel-Dolmatoff (1976:310), who points out that the Tukano's ecological theory "conceives of the world as a system in which the amount of energy output is directly related to the amount of input the system receives" (310). Energy in such a scheme should never be used without being restored as soon as possible. The restitution to nature of the energy potential utilized involves complex rules, practices, and rituals "whose totality corresponds to a way of life, to an integrated system" (1990:12–13). This way of life represents a sharp contrast to a lifestyle dependent on the ever-increasing consumption of goods and nonrenewable energy sources.

It is worth noting that the native peoples of the Amazon showed a very different geographical distribution before conquest, when human occupations of large proportions ("paramount chiefdoms") were established. With the arrival of the first colonizers, the natives were dislodged to the interriver forests of the Amazon Basin (Roosevelt 1991), which had some of the world's most nutrient-poor soil (Posey 1992a:15). But the preconquest Indians adapted their techniques for living in harmony with nature, obtaining favorable results without degrading or exhausting the environment (17), a pattern of behavior that is still witnessed among present-day remnant groups. The paramount chiefdoms of Amazonia developed intensive food production, urban-scale settlements, and monumental earth constructions, "including the earliest pottery-age cultures in the hemisphere" (Roosevelt 1991:134). Dispersion and the formation of smaller communities occurred after the sixteenth century. This recent archaeological finding reaffirms the enormous ability of early Amazonian natives to relate in appropriate ways to their natural surroundings, applying rules of conduct that sustained life without disturbing nature. The Indians have demonstrated an ability to take advantage of the possibilities at their disposal by remaining healthy despite the assumption that they would be susceptible to illness because of inadequate protein consumption. Instead one finds that the Indians are an example of vigorous physical strength (Ribeiro 1987:35).

Serious soil deficiencies have been overcome by elaborate systems of agriculture and intensive soil management. It has been demonstrated by Hecht and Posey (1990) that the Kayapó agricultural system, for example, is superior to modern Amazonian agricultural methods, which are characterized by pasture and short-cycle crops "which are notorious for their lack of sustainability and low rates of return" (79). The Kayapó system does not need purchased inputs and is naturally much richer. The comparison Hecht and Posey (1990) make of Kayapó, colonist, and livestock production patterns in eastern Amazonia reveals that Kayapó yields per hectare over five years are 183 percent higher than the yields of the colonist system, and 176 times that of livestock (81)! In terms of protein yields from vegetable sources over five years, Kayapó figures are roughly double those of colonist agriculture and more than 10 times the protein production from livestock (81). "In ten years ... 1 ha of pasture has produced less than a ton of meat, and slightly more than 100 kilos of protein or roughly 5% of the protein generated by the Kayapó system" (81). The conclusion is clear: without damaging the resource base—which modern systems noticeably do—the Kayapó produce many more calories and proteins per hectare than do either of the alternatives in the region. The irony of the situation is exposed by Hecht and Posey: "Hundreds of millions of dollars have been funneled into surveys and experiments which have not made [the] colonist's agriculture more stable, or livestock more productive" (84). It seems obvious, therefore, that land uses by Amazonian indigenous peoples must reflect an assessment of soil capabilities and practical measures involving all aspects of land use, in an approach we might call ethnoagronomics. "Researchers should also recognize that there is a complex intellectual system that underlies the native management of soil resources, the ensemble of which is 'ethnopedology'" (76).

Thus it is not surprising to discover through ethnoscience that the kind of itinerant agriculture undertaken by the Indians does not constitute a primitive and incipient method; that it is, on the contrary, a specialized technique conceived as a response to the specific conditions of climate and soil encountered in the rain forest (Meggers 1977). Crop diversification, as found in Kayapó territory, also represents a rational form of land use. So does the creation of forest "islands" (*apêtês*), which the Mebêngôkre have developed in tropical savannas to modify the ecosystem and increase biodiversity (Museu Goeldi 1987:18). These forest islands are a notable feat of ecological engineering, accompanied by precise knowledge of insect behavior. For example, the Kayapó restrict leaf-cutting ants from gardens and fruit trees by introducing nests of ants of a genus that repels the leaf-cutters by physical and chemical means (Overal and Posey 1990).

Indigenous natural-resource classifications are not aimless. To the contrary, they are not only systematic and based on theoretical knowledge, but they are also comparable, from a formal point of view, to those that zoology and botany use (Lévi-Strauss, quoted in Ribeiro 1987:66). The Kayapó classify their natural resources within various ecosystems:

> Each ecosystem is perceived by the Indians to exist with a specific association of plants and animals. Having a profound knowledge of animal behavior, the Kaiapó know which plants attract each animal. On the other hand, they associate several species of plants with varieties of soils. Consequently, each ecosystem is a harmonious union of interactions between plants, animals, types of soil and the Kaiapó themselves.
>
> (Posey 1992b:23).

Improving soil fertility and productivity is one of the beneficial consequences of classifying by ecosystem. When one remembers that modern agricultural practices in Amazonia have exhausted soil fertility and caused serious ecological problems (Uhl 1992), the superior ability of the Amazon Indians to deal with their environment must be acknowledged. Anthropologist Carmen Junqueira (1984:1285) has found that all the productive activities of the Cinta-Larga tribe obey complex cultural rules that range from the organization of work teams to the different modes of distributing the produce. That complex system of rules and institutions is a counterpoint to technological simplicity and constitutes the pillars of the Indian communal organization (1285). Such an elaborate cultural structure is also what explains the natives' ability to limit population size (the abhorrent practice of infanticide observed in some groups notwithstanding). Plants like *Curarea tecunarum* (Ribeiro 1987:57) are used by the Deni as a contraceptive, and abstention from sexual activity for a long period after delivery occurs in tribes like the Xamakôko and the Taulipáng as a means to reduce birthrate (Baldus 1970:277).

In the Indians' understanding of nature, man is only one part of a complex network of interactions that include both society and the entire universe. This is illustrated by Reichel-Dolmatoff's analysis of the meaning of animal behavior to the Indians. He indicates that animal behavior represents a model for what is possible, for what can be done to successfully adapt to the environment. "Animals then are metaphors for survival. By analysing animal behaviour the Indians try to discover an order in the physical world—[an] order to which *human* activities can then be adjusted" (1976:311). The importance of animals to the natives of Amazonia has deep foundations. Game, fish, and wild fruits are viewed as food resources that can satisfy protein needs. That nutritional evaluation is arrived at with the help of shamans. The Indians equate environmental degradation not with soil exhaustion, but with the eventual depletion of game and the increased walking time needed to obtain food (314). From a shamanistic point of view, it is the upsetting of ecological balance, such as occurs with overhunting, that explains disease (315). For a Tukano, illness results from a person's interfering with a certain aspect of the ecological order. Incidentally, Reichel-Dolmatoff remarks that the Tukano, as well as several other Amazonian tribes, "believe that the entire universe is steadily deteriorating" (317), a clear indication of the Indians' sense of entropy. This tendency can be counterbalanced, according to the Tukano, by a continuous cycle of ritual creation and reestablishment of order and purpose. This is done during ceremonies in which the universe and its components are renewed, and links with past and future generations are reaffirmed.

Ribeiro (1987:39) informs us that, despite more than three hundred years of contact between the Desâna and the national society—and the corresponding loss of cultural goods, symbols, and values brought about by this contact—the Desâna continue to foster subsistence by adapting wisely to an ecosystem they profoundly comprehend. As part of the interaction they perceive between the material sphere and the spiritual world, the Desâna and other Indian societies practice austerity in the satisfaction of their consumption needs. The close relationship the Indians have with the principles, cycles, and limits of nature teaches them how environmental stress can be avoided so that nature is not disturbed, and so that a continuous flow of resources sufficient for the individual's well-being is guaranteed. Such a complex system of ecological engineering corresponds to planning a life within the limits of the possible, and involves both present and future generations in the process (this refers to intergenerational equity or sustainable development in the sense of the Brundtland Report, WCED 1987:43). That attitude, in turn, amounts to a negation of the nonsatiation principle, postulated by economists as a normal trait of human character (Faber, Jöst, and Manstetten 1994:87–91). It amounts likewise to a holistic way of understanding the world, in acute contrast to the perception of modern man and science. It is interesting to observe that austerity for the Indians does not lead to penury (even speaking of poverty is out of the question because it is not applicable as a sociological category in Indian societies). Just the opposite, for an abundance of staples is usually found in the Indian villages. Baldus reports how lavishly he was received at the Tapirapé village when he arrived there for the first time:

> Just to give an idea of the food variety of the Tapirapé in a determined period of the year, I want to list the dishes they offered me when, in June 1935, I arrived for the first time at Tampiitaua [23 different dishes are then listed]. Unwilling to offend anyone, I ate in the same afternoon, in all [the] village's houses, great quantities of each of these delicious dishes.
>
> (Baldus 1970:232, 448)

That abundance had disappeared by 1947, when Baldus returned to the village, in large measure because the cultural shock brought about by the contacts with the white man following his first visit modified the formerly unlimited Tapirapé hospitality (Baldus 1970:233).

Conclusions

Following the advice of Seeger (1980:15) that we should avoid both the evolutionist's ethnocentrism and the romantic's view of the noble savage, I conclude that the Amerindians' economy seems to be a concrete demonstration of how to live sustainably. Certainly it is an extreme example of compliance with the rules for a sustainable society, and a very difficult one for modern man to adopt. However the other extreme—epitomized in the statement that "our primary objective is growth. In my view, there is no longer any ambiguity about this. It is toward growth that [the International Monetary Fund's] programs and their conditionality are aimed" (M. Camdessus, statement before the United Nations Economic and Social Council in Geneva, July 11, 1990, quoted in Przeworski and Vreeland 2000:385–386)—cannot be considered as a serious goal. Georgescu-Roegen (1971) has already proposed that no elaborate argument is needed for one "to see that the maximum of life quantity requires the minimum rate of natural resources depletion" (21). Thus to grow forever and simultaneously—not to mention healthily—cannot be the global objective. The question, then, is how to imagine development within the context of the Indian paradigm of developing within the limits of the possible. As already discussed, the Indians study animal behavior precisely as a model for what is possible. Possibilities mean physical constraints. But they also mean the acknowledgment of the second law of thermodynamics, which is an actual limitation even beyond unlimited supplies of resources. The prevailing notion of development associates the pace of natural resources utilization with the rate of progress: the faster the pace, the quicker progress takes place (Tiezzi 1988:32). But our way of life, of consumption, also determines the speed of the entropic process, the velocity with which available energy is dissipated. Indian behavior clearly softens the tendency of dissipation. The natives of Amazonia apply naturally, instinctively, the principles of ecology. These same principles could be at the root of the design "of an economic system that can essentially last forever" (Brown 1991:354)—last, *not grow*, forever.

How do the Indians define the categories of their social life? Not by permitting an idea like growth to occupy the center of their preoccupations. Development is a purely Western concept (Esteva 1992:9) that robs peoples of different cultural frameworks of the opportunity to design their own societal objectives. Sustainable development, on the other hand, can only occur if "productive capacity" can be preserved "for the infinite future" (Solow 1992:4), that is, if future generations are assured a standard of living not inferior to the present one. Does the Indians' ethnoeconomy preserve productive capacity? Of course, it does: the natives of the Amazon have done that for *millenia*, not centuries, as the discovery of the paramount chiefdoms in Amazonia has unearthed (Roosevelt 1991). But the economic performance of the Indians has nothing to do with Western concepts. This conclusion exposes the absurdity of the methods of rainforest exploitation that have leveled the model adopted by the Indians. With the exception of the methods contained in the idea of "extractivist reserves," all Western methods have been shown to be unsustainable. For example, "for each cubic meter of wood taken from the forest" with so-called modern methods of production, "almost two cubic meters are destroyed" (Uhl 1992:57–58). This can be explained by the inconsistent configuration of markets and policies, which leaves fundamental resources of life outside the market place, "unowned, unpriced, and unaccounted for—and more often than not it subsidizes their excessive use and destruction despite growing scarcities and rising social costs" (Panayotou 1991:357). In Gérald Berthoud's words: "With money as a supreme value, life counts less" (1992:81). Or as Gustavo Esteva says: "Establishing economic values requires the disvaluing of all other forms of social existence" (1992:18). Study of the Indians' lifestyles shows how different the whole picture becomes when life is the supreme value. In this landscape, the emphasis that mainstream economics puts on economic growth before everything else, including distribution, cannot be maintained. One may look with scorn at a primitive way of life like the Indians' and consider that it is simply unrealistic or a utopian fantasy in the modern world. Nevertheless, nothing in nature or society demonstrates that a law of transformation guarantees that any given society is in a process of evolution toward "ever more perfect forms" (Esteva 1992: 22–23). Or in Georgescu-Roegen's view: "No social scientist can possibly predict through what kinds of social organization mankind will pass in its future" (1971:15).

This brings us to the discussion about the need for a paradigm shift, away from the dominant model of natural-resources use (including use of matter and energy) and environmental management and toward a system of resources use within the earth's carrying capacity and in compliance with the principles of ecology. No doubt the Amazonian Indians' paradigm offers an alternative—a proven one. This is convincingly illustrated by the example of the Kayapó, about whom ethnobiological research has been conducted since 1977 by the Belém's Goeldi Museum; this research has revealed that their "traditional knowledge offers some of the most viable and promising options for sustained resource use in the tropics" (Posey 1992b:19). The commitment of the Indian model to the well-being of future generations in accordance with the accepted notion of sustainability (cf. Taylor 1989:11) is another point to be emphasized. It is also relevant to remember that the Indian paradigm contains an appreciation of the practical wisdom (or *phronesis* in Aristotle's terminology) (Faber, Manstetten, and Proops 1994:18) that is so meaningful for the solution of environmental problems and the promotion of conservation. It is well known that the market is not reliable in terms of conservation of natural systems. Nothing in its structure induces real sustainability. But not only is sustainability a requirement of the new concept of development, it is also a general prerequisite of life. Goodland (1990) posits that a voluntary return to sustainability is unavoidable "before global selection does it for us at a much lesser steady state value" (xiv).

Particularly to those who live in the Americas, it is extremely important to work with the Amerindian paradigm in mind and to try to grasp the workings of its system of ethnoeconomics—not necessarily to adopt it literally, but to look at it, scrutinize it, and understand it. It is the opinion of botanists and zoologists doing research with the Indians that the complex ways in which primitive cultures relate to their surroundings will assume a growing significance to the process of devising policies for the preservation of threatened ecosystems like the Amazon (Ribeiro 1987:65). The emerging body of ethnobiological information in Amazonia shows that ecological sustainability can be attained with the help of indigenous knowledge. It also serves, in my view, to demonstrate the necessity for the study of ethnoeconomics. Systems of resource management conceived on the basis of such knowledge can promote sustainability "and may generate levels of income that exceed the regional average" (Hecht and Posey 1990:77). So there is ample reason to pay the greatest attention to the details and intricacies of the Amazon Indians' lifestyle and economy: they serve as admirable examples of coexistence with nature, or more precisely, of knowing how to live within the limits challengingly fixed by nature.

References

Baldus, H. 1970. *Tapirapé: Tribo Tupí no Brasil Central*. São Paulo: Cia. Editora Nacional/Editora da Universidade de São Paulo.

Bateson, G. 1972. *Steps to an Ecology of Mind*. New York: Ballantine.

Berthoud, G. 1992. Market. In W. Sachs, ed., *The Dictionary of Development: A Guide to Knowledge as Power*, pp. 70–87. London: Zed Books.

Binswanger, H. C. 2000. Towards Sustainable Development. In C. Cavalcanti, ed., *The Environment, Sustainable Development and Public Policies: Building Sustainability in Brazil*, pp. 19–29. Cheltenham, England: Edward Elgar.

Branco, S. M. 1989. *Ecossistêmica: Uma Abordagem Integrada dos Problemas do Meio Ambiente*. São Paulo: Edgar Blücher.

Brown, L. 1991. Is economic growth sustainable? Roundtable discussion at Proceedings of the World Bank Annual Conference on Development Economics, 1991. In supplement to *The World Bank Economic Review and The World Bank Research Observer*, pp. 353–355.

Cardim, F. 1939. *Tratados da Terra e da Gente do Brasil*. São Paulo: Cia. Editora Nacional. (Orig. pub. 16th century.)

Cavalcanti, C. 1992. The path to sustainability: Austerity of life and renunciation of development. Paper presented at the Second Meeting of the International Society for Ecological Economics (ISEE), Stockholm University, August 3–6.

CCPY/CEDI/CIMI/NDI. 1990. *Yanomami: A Todos os Povos da Terra*. São Paulo: Ação pela Cidadania/OAB, July.

Chagnon, N. A. 1992. *Yanomamö: The Last Days of Eden*. New York: Harcourt Brace Jovanovich.

Chernela, J. M. 1989. Os Cultivares da Mandioca na Área de Uaupés (Tukano). In B. G. Ribeiro, coord., *Etnobiologia*. Vol. 1 of *Suma Etnológica Brasileira*, pp. 151–158. Rio de Janeiro: Vozes/FINEP.

Cleveland, C. and M. Ruth. 1997. Capital Humano, Capital Natural e Limites Biofísicos no Processo Econômico. In C. Cavalcanti, ed., *Meio Ambiente, Desenvolvimento Sustentável e Políticas Públicas*, pp. 131–164. São Paulo: Cortez.

Cortesão, J. 1943. *A Carta de Pero Vaz de Caminha*. Rio de Janeiro: Edições Livros de Portugal.

Daly, H. 1980. Introduction to the steady-state economy. In H. Daly, ed., *Economics, Ecology, Ethics: Essays Toward a Steady-State Economy*, pp. 35–37. San Francisco: Freeman.

Economist Intelligence Unit. 2004. Emerging-market indicators. *The Economist* (Feb. 14) 370(8362):98.

Ekins, P. 1995. *Programme for a Sustainable Economy*. Photocopy. Cheltenham, England: Forum for the Future/The Sustainable Economic Unit, May.

Esteva, G.1992. Development. In W. Sachs, ed., *The Dictionary of Development: A Guide to Knowledge as Power*, pp. 6–25. London: Zed Books.

Faber, M., F. Jöst, and R. Manstetten.1994. *Limits and Perspectives of the Concept of a Sustainable Development*. Discussion Papers, no. 204 (January). Heidelberg: Universität Heidelberg.

Faber, M., R. Manstetten, and J. L. R. Proops. 1994. *Knowledge, Will and the Environment*. Discussion Papers, no. 205 (February). Heidelberg: Universität Heidelberg.

Gandavo, P. de Magalhães. 1924. *Tratato da Terra do Brasil*. Rio de Janeiro: Edição do Annuario do Brasil. (Orig. pub. 1570 or earlier.)

Georgescu-Roegen, N. 1971. *The Entropy Law and the Economic Process*. Cambridge, Massachusetts: Harvard University Press.

Goodland, R. 1990. *Race to Save the Tropics: Ecology and Economics for a Sustainable Future*. Washington, DC: Island Press.

Gronemeyer, M. 1992. Helping. In W. Sachs, ed., *The Dictionary of Development: A Guide to Knowledge as Power*, pp. 53–69. London: Zed Books.

Hecht, S. and D. A. Posey. 1990. Indigenous soil management in the Latin American tropics: Some implications for the Amazon Basin. In D. A. Posey and W. L. Overal, eds., *Ethnobiology: Implications and Applications*, vol. 2, pp. 73–86. Proceedings of the First International Congress of Ethnobiology, 1988. Belém, Pará: Museu Paraense Emílio Goeldi.

Junqueira, C. 1984. Sociedade e cultura: Os Cinta-Larga e o exercício de poder do estado. *Ciência & Cultura* (August) 36(8):1284–1292.

Lewis, M. W. 1992. *Green Delusions: An Environmentalist Critique of Radical Environmentalism*. Durham, North Carolina: Duke University Press.

Meggers, B. J. 1977. *Amazônia, a Ilusão de um Paraíso*. Rio de Janeiro: Civilização Brasileira.

Mindlin, B. 1994. O aprendiz de origens e novidades. *Estudos Avançados* (Universidade de São Paulo) 20 (January/April) 8(20):233–253.

Museu Goeldi. 1987. *A Ciência dos Mebengôkre, Alternativas Contra a Destruição*. Belém, Pará: Museu Paraense Emílio Goeldi.

Overal, W. L. and D. A. Posey. 1990. Uso de formigas *Azteca* spp. para controle biológico de pragas agrícolas entre os Índios Kayapó. In D. A. Posey and W. L. Overal, eds., *Ethnobiology: Implications and Applications*, vol. 1, pp. 219–225. Proceedings of the First International Congress of Ethnobiology, 1988. Belém, Pará: Museu Paraense Emílio Goeldi.

Panayotou, T. 1991. Is economic growth sustainable? Roundtable discussion at Proceedings of the World Bank Annual Conference on Development Economics, 1991. In supplement to *The World Bank Economic Review and The World Bank Research Observer*, pp. 355–358.

Posey, D. A. 1990. The application of ethnobiology in the conservation of dwindling natural resources: Lost knowledge or options for the survival of the planet. In D. A. Posey and W. L. Overal, eds., *Ethnobiology: Implications and Applications*, vol. 1, pp. 47–59. Proceedings of the First International Congress of Ethnobiology, 1988. Belém, Pará: Museu Paraense Emílio Goeldi.

Posey, D. A. 1992a. Introduction to the relevance of indigenous knowledge. In A. E. de Oliveira and D. Hamú, eds., *Kayapó Science: Alternatives to Destruction*, pp. 15–18. Belém, Pará: Museu Paraense Emílio Goeldi.

Posey, D. A. 1992b. Kayapó Science: Alternatives to destruction. In A. E. de Oliveira and D. Hamú, eds., *Kayapó Science: Alternatives to Destruction*, pp. 19–43. Belém, Pará: Museu Paraense Emílio Goeldi.

Przeworski, A. and J. Vreeland. 2000. The effect of IMF programs on economic growth. *Journal of Development Economics* 62:385–421.

Reichel-Dolmatoff, G. 1976. Cosmology as ecological analysis: A view from the rain forest. *Man* (n.s.) 2(3):307–318.

Reichel-Dolmatoff, G. 1990. A view from the headwaters: A Colombian anthropologist looks at the Amazon and beyond. In D. A. Posey and W. L. Overal, eds.,

Ethnobiology: Implications and Applications, vol. 1, pp. 9–18. Proceedings of the First International Congress of Ethnobiology, 1988. Belém, Pará: Museu Paraense Emílio Goeldi.

Ribeiro, B. G. 1987. *O Índio na Cultura Brasileira*. Rio de Janeiro: UNIBRADE/ UNESCO.

Robbins, L. 1932. *An Essay on the Nature and Significance of Economic Science*. London: Macmillan.

Roosevelt, A. C. 1991. Determinismo ecológico na interpretação do desenvolvimento social indígena da Amazônia. In W. A. Neves, ed., *Origens, Adaptações e Diversidade Biológica do Homem Nativo da Amazônia*, pp. 103–142. Belém, Pará: Museu Paraense Emílio Goeldi.

Sachs, W. 1992. One world. In W. Sachs, ed., *The Dictionary of Development: A Guide to Knowledge as Power*, pp. 102–115. London: Zed Books.

Seeger, A. 1980. *Os Índios e Nós: Estudos sobre Sociedades Tribais Brasileiras*. Rio de Janeiro: Editora Campus.

Simon, J. 1987. Now (I think) I understand the ecologists better. *The Futurist* (September–October):18–19.

Solow, R. 1992. An almost practical step toward sustainability. An invited lecture on the occasion of the Fortieth Anniversary of Resources for the Future, Oct. 8. Washington, DC: Resources for the Future.

Sousa, G. S. de. 1971. *Tratado Descritivo do Brasil em 1587*. São Paulo: Companhia Editora Nacional/Editora da USP. (Orig. pub. 1587.)

Taylor, P. W. 1989. *Respect for Nature: A Theory of Environmental Ethics*. Princeton, New Jersey: Princeton University Press.

Tiezzi, E. 1988. *Tempos Históricos, Tempos Biológicos: A Terra ou a Morte: Problemas da "Nova Ecologia."* Trans. F. Ferreira and L. E. Brandão. São Paulo: Nobel. (Orig. pub. 1984 in Italian.)

Uhl, C. 1992. O desafio da exploração sustentada. *Ciência Hoje* (May/June) 14(81):52–59.

Viveiros de Castro, E. 1986. *Araweté: Os Deuses Canibais*. Rio de Janeiro: Zahar.

Viveiros de Castro, E. 1992. *Araweté: O Povo do Ipixuna*. São Paulo: Centro de Documentação e Informação (CEDI).

WCED (World Commission on Environment and Development). 1987. *Our Common Future* (Brundtland Report). New York and Oxford: Oxford University Press.

20   Enhancing Social Capital

Productive Conservation and Traditional Knowledge in the Brazilian Rain Forest

Anthony Hall

      Traditional ecological knowledge (TEK), as a component of social capital for promoting economic progress and supplying environmental services, has until recently been almost totally neglected by official planners and policymakers. Indeed for decades the presence of human populations has been perceived as incompatible with natural-resource conservation, and nowhere more so than in the Amazon rain forest. Yet during the 1990s there was a move away from total reliance on the preservationist, centralized, command-and-control strategy that largely excludes local populations, and toward a decentralized approach that seeks to utilize governance potential to create more sustainable models (Hall 1997b, 2000). That change has already found expression in official regional development and environmental policy (Brazil 1995, 1997, 2000).

    Inevitably any such discussion must come to terms with the notion of "sustainable development." Although the Brundtland Report (WCED 1987) set broad parameters for development that meets the needs of both present and future generations, our purpose requires a more specific definition. In the present context, sustainable development may be defined as "development which allows for the productive use of natural resources for economic growth and livelihood strengthening, while simultaneously conserving the biodiversity and sociodiversity which form an integral part of this process" (Hall 1995:4). This variant of sustainable development could be called "productive conservation" (Hall 1997a). Its attainment depends on meeting a number of complementary and interdependent goals. Economically viable and ecologically sound productive conservation requires a strong measure of local organization that taps into and builds upon the expertise and knowledge of resource users for project and program design and implementation.

Traditional knowledge is central to the future of Amazonia from two standpoints. First, a substantial proportion of the population is still dependent to some degree on traditional knowledge for meeting its livelihood needs. Those needs are currently being met in whole or in part by the extraction of nontimber forest products (NTFPs) such as rubber, nuts and fruits, traditional fishing, agroforestry, and small-scale farming. All of those productive activities are based on adapted forms of local knowledge. Without continued access and improvements to the application of such knowledge in devising development solutions, the livelihoods of many forest dwellers would be undermined. Second, because traditional knowledge is based (generally if not always) on nonpredatory forms of resource use, the stock of natural capital is far more likely to remain intact. In turn, wider environmental services (such as biodiversity conservation, climate regulation, and watershed management) that are provided by the rain forest for local, national, and even global benefit will be retained. Although still perceived by many as contradictory, these two objectives—production for livelihood strengthening and resource conservation—are viewed here as being potentially compatible rather than conflictive.

Legitimate Knowledge and the Development of Amazonia

Until the nineteenth century, Amazonia—perceived as a marginal and exotic hinterland—remained isolated from the mainstream of Brazilian national life. During the late 1990s the rubber industry brought an influx of labor to the region and economic prosperity for a few, but the region became an abandoned frontier once more during the first half of the twentieth century. Amazonia came seriously into the development spotlight only with the advent of military rule in 1964 and the birth of a strategy of modernization that sought to integrate the region into the national economy. Multiple development aims were pursued:

- Utilization of the region's natural-resource base to produce an economic surplus through heavily subsidized cattle-ranching enterprises, logging, and large-scale mining.
- Defusion of social tensions arising from land conflicts in northeast and southern Brazil by transferring small-scale farmers to the

Amazon frontier along the Transamazon Highway, and especially to
Rondônia.
- Guarantee of territorial integrity by occupying the region's vast
"empty" spaces, supposedly to reduce the risks of foreign invasion.

In that scenario, what counted as legitimate knowledge was outsiders'
knowledge. Amazonia was perceived as a time warp, the anachronistic remnant of a preindustrial past which had to be dragged as quickly as possible—
screaming if necessary—into the modern age. Its transformation would be
induced by the introduction of capital investment packages and technical
knowledge concentrated in "development poles," which would, according to
the theories of the day, spread progress and prosperity. Central planners under the influence of the Sorbonne Group of Brazil's Higher War College (*Escola Superior de Guerra*) deemed Amazonia a free resource pool that could
be exploited without economic, social, or ecological cost in order to serve the
needs of regional and national development. The production methods of *caboclos* (mixed-race riverine inhabitants) and indigenous groups would have to
be rapidly replaced by modern farming and production techniques. Through
the intervention of a highly centralized state machine it was, in the words of
General Golbery do Couto e Silva, necessary to "inundate the Amazon forest
with civilization" (Schmink and Wood 1992:59). Traditional knowledge was,
in that view, synonymous with backwardness and underdevelopment.

During the 1970s and 1980s the notion that patterns of development
should be shaped to minimize environmental damage was in its infancy.
Environmental policy in Brazil and elsewhere consisted essentially of ex-
ercises in limiting damage, either by detecting and punishing ecological
abusers or by preserving hitherto untouched ecosystems, such as protected
areas where human action was prohibited or severely restricted. Environmental monitoring and control through agencies such as the Institute of
Forestry Development (IBDF) and the Superintendency of Fisheries Development (SUDEPE) were of extremely limited effectiveness. Conservation
efforts since then have been more significant, with almost 9 percent of Legal
Amazonia (45 million hectares) currently set aside in some 120 federal- and
state-administered "conservation units" (Capobianco 1996). More than 70
percent of that area comprises fully protected units that receive indirect use
(*uso indireto*), such as national parks and biological reserves in which anthropogenic disturbance is prohibited. Underlying this strategy has been the
premise that local populations constitute a threat to natural resources and
are incompatible with conservation aims, and must therefore be physically

excluded from conservation units. That notion was reinforced by a number of theories and views during the 1980s and 1990s.

Controversial notions such as the Pleistocene refugia theory were used to back the idea that biodiversity preservation could be maximized by protecting key "islands" of rain forest (Foresta 1991). Government ideologues backing the agribusiness and land-owning lobby placed the responsibility for ecological damage exclusively on the shoulders of smallholder settlers. By blaming the victims, they created the stereotype of small-scale producers who had nothing positive to offer and thus merited no official support, thereby enabling resources to be redirected toward corporate interests. In a similar vein, Hardin (1968) suggested in his seminal "The Tragedy of the Commons" that common-pool resources such as forests and lakes were all open access and doomed to destruction as individual users sought to maximize personal gain with no regard for the collective good. Neo-Malthusian "limits to growth" theories reinforced the idea that demographic pressures were the prime cause of environmental destruction (Ehrlich 1972). Finally, both modernization and neo-Marxist theories tended to portray the Amazonian peasantry and indigenous groups as temporary or residual social phenomena doomed to disappear as economic development advanced.

Thus, until the 1990s, official development and environmental policy for Amazonia either ignored (at best) or despised (at worst) TEK, viewing it as an irrelevance to the modern era. Yet at the turn of the decade, a number of key events prompted a reexamination of this blinkered vision. A mounting wave of domestic and international concern with the fate of Amazonia's forests during the "decade of destruction" in the 1980s culminated in December 1988 with the murder in Xapurí (Acre) of the rubber tappers' leader Francisco "Chico" Mendes. In response, President Sarney introduced the country's first comprehensive environmental program (*Nossa Natureza*) and set up a single federal environmental protection agency, IBAMA, amalgamating forests, fisheries, and rubber; the first extractive reserves were decreed in 1990. The National Environment Program (PNMA) was introduced by the government, and some environmental monitoring and control responsibilities were decentralized to the state level. In the run-up to the Earth Summit of 1992, President Collor de Mello grasped the environmental nettle, so to speak, by setting up a special secretariat for the environment (SEMA), attached directly to the president's office, and by appointing radical ecologist José Lutzenberger as its head. A national environmental council (CONAMA) was established, formally bringing together for the first time representatives of government and civil society.

The setting up of a Ministry of the Environment in 1993 with a special secretariat for Amazonia signaled a new concern for addressing regional issues more systematically. Commencement of the US$350 million G7 Pilot Program to Conserve the Brazilian Rain Forest (PPG7) in 1993 and publication of the *Integrated National Policy for the Legal Amazon* two years later also indicated that a much more broadly based approach to environmental policy, which would recognize the interests of all stakeholders in the region, was beginning to take hold. Concepts that a few years before would have been ridiculed in official circles now found clear expression in policy documents. Government action was now to show "respect for regional diversity," while "education and training should be directed towards the generation of low-cost technologies capable of solving problems [and] conserving the environment" and, significantly, growth should bring together "the combined knowledge of all social agents" (Brazil 1995).

As the above brief overview indicates, official thinking on what constitutes appropriate environmental policy for addressing Amazonia's problems has progressed significantly over the past decade. Yet there is undeniably a continuing need for conservation which is recognized and reflected in recent moves. In 1997, for example, the World Bank, World Wildlife Federation, and Brazilian government declared a goal of setting aside 10 percent of Amazonian land in fully protected units by the year 2000. Conservation "corridors" are to be set up through the G7 Pilot Program, and monitoring will be reinforced through the Information and Surveillance Service for Amazonia (SIVAM) project, while the classification and management of protected areas has been revised under the new National System of Conservation Units (SNUC). Within this broader and more flexible conservationist strategy, local populations are viewed not simply as a source of environmental degradation but also as providers of solutions (Hall 1997b). Extractivists such as rubber tappers, small-scale farmers, fisher people, and indigenous groups perform a number of valuable roles in the design and execution of productive conservation strategies as a form of sustainable development.

Social Capital and Productive Conservation

The term "social capital" refers to the broad idea that social relationships and values constitute an asset (as do physical or human capital), which may assist in the attainment of development goals. Discussions of social capital have originated in different academic disciplines and reflect complementary emphases. Political sociologists and anthropologists approach the concept through the analysis of social norms, networks, and organizations. Robert Putnam (1993), in his study of democracy in Italy for example, focused on the key role of nonformal, horizontal relationships linking individuals and civic associations in promoting cooperation within society. James Coleman (1990), originator of the concept, refers to social networks and relationships, both horizontal and hierarchical, that allow actors to meet their objectives. Economists have developed a more institutional focus that, in addition to acknowledging the importance of local and informal arrangements, includes the role of formal institutional relationships and structures. North (1990), for example, explores the relationships between formal institutional structures and the rate and pattern of economic development.

Whatever the definition adopted, social capital has a significant impact on the development process at all levels. Certain common themes may be identified within these diverse approaches:

• *Human relationships.* All social capital theories link the economic, social, and political dimensions of human activity. They stress the importance of interpersonal relationships as a key variable in determining the effectiveness and efficiency of both individual and collective action in achieving a given development objective.

• *Public benefits.* Social capital is, generally speaking, considered beneficial to the public good and thus merits public support; for example, in the management of common-pool resources.

• *Positive impacts.* Research tends to focus on social relationships that have a positive impact on the development process; for example, in mobilizing communities for action, reducing selfish behavior (e.g., "free-riding"), and improving the flow of information for planning purposes (e.g., incorporating local knowledge). However, the potentially negative impact of some forms of social capital is also acknowledged.

Outsiders' stereotypical perceptions of traditional populations' environmental sensitivities are polarized. At one extreme, local populations may be seen as "green guerrillas," fighting unselfishly to save the rain forest. At the other end of the spectrum, they are seen as potential destroyers of flora and fauna, only too willing to overexploit natural resources if it serves their personal interests. As usual, the truth usually lies somewhere along this continuum. The theme of this essay is that community-level social capital can be harnessed to promote more sustainable forms of resource use; this theme is based on the premise that traditional groups have the interest and the power, when engaging jointly with other social-institutional actors, to influence the course of events in their favor (Hall 1997a). The rationale behind such collective action is sustained by two main factors: (1) livelihoods that are underpinned by access to and productive use of natural resources and (2) some notion of group identity and mutuality of interests.

There is clearly a strong instrumentalist motivation underlying collective action by populations trying to protect their environment. The very survival of those groups may be dependent upon the continued extraction and protection of forest products (timber and nontimber forest products, fish and subsistence crops). The livelihood issue is especially critical in the early stages of confrontations, when the threat posed by external aggressors can bind geographically dispersed and socially fragmented local populations. Yet once external threats have been dissipated, it can be difficult if not impossible to sustain collective action for longer-term resource planning. In the long run, although economic-utilitarian factors are still important, group action requires additional bases of solidarity. This is especially so when it becomes necessary to develop a systematic environmental management strategy that must reconcile *individual needs* (which encourage the tendency to free ride or maximize income) with *the collective and wider interests of the community* (which require maintaining the resource base as a source of natural capital for the group and as a supplier of environmental services). It therefore becomes necessary to develop and strengthen group identity and commitment to the *common* cause, whether or not gains accrue to individual members.

Boosting social capital clearly has a major role to play in supporting collective action: it sustains the momentum of movements by enhancing the human relationships within them. Local populations now play a pivotal role in the design and implementation of environmental management strategies in Amazonia. Without such involvement, it is highly doubtful that there can be any effective environmental policy in the region, certainly at the local level and perhaps more broadly. It is therefore important to examine the links between social capital and the achievement of more sustainable development in the region. Key questions are: (1) What kinds of social capital already exist among traditional populations in Amazonia, and what is the role of traditional knowledge in sustaining such capital? (2) How can development policymakers and practitioners help to build upon existing social capital at the community level for the purpose of promoting more sustainable forms of development?

Existing Social Capital and Traditional Knowledge

Many traditional communities of caboclos and small-scale settlers possess characteristics that constitute a partial foundation upon which the edifice of an environmental management strategy may eventually be constructed. (Indigenous communities merit a somewhat separate discussion, which will not be pursued in this essay.) Concentrated along Amazonia's river margins and highways, the sheer *physical presence* of those populations constitutes a major asset in terms of project and program implementation. Those communities may be organized into small settlements or villages, but more often than not they comprise dispersed individuals or small clusters of households. Fishing groups, small farms, and rubber tappers' landholdings (*colocações*) are examples of such a pattern. Households may be loosely united by extended kinship ties and brought together by labor exchange agreements for mutual benefit (*mutirão*). Religious activities, especially through the Catholic Church and the practice of Liberation Theology, may also encourage a certain community spirit and lay the foundations for future action.

Yet it is easy to exaggerate the degree of solidarity conferred by those kinds of social organization. As already mentioned, traditional populations tend to be distributed in small households or nuclear settlements, often over huge geographical areas. Demographic instability is reinforced by the seasonal, rural-urban migration needed to maintain the diversified livelihood strategies that compensate for the frequently limited returns from extractivism and other traditional activities. Lacking adequate means of transport, contact with the world outside the immediate vicinity is often limited to occasions such as feast days. Physical isolation is one of the main barriers faced by such groups and presents all kinds of problems in terms of access to basic health and education services as well as to markets for their produce. Furthermore, there is an almost total lack of any tradition of political mobilization. Establishing horizontal organizational links is frequently undermined by strong traditions of patron-clientage and debt-bondage in dealings with intermediaries, landowners, and other sources of authority.

When galvanized into action by external threats, however, these obstacles may be overcome if circumstances permit, and social capital may be harnessed for the purpose of self-defense. A reactive stance against threats posed by outside invaders has been a recurrent theme as the thrust of mainstream development has encroached upon the livelihoods of traditional populations. Rubber tappers mounted hundreds of standoffs (*empates*) in southern Acre against loggers and ranchers in the 1970s and 1980s, reportedly saving over one million hectares of forest (Gross 1989; Hall 1997a). During the same period, as the fishing frontier moved westward from Belém and Manaus, fishing communities along the middle and upper reaches of the Amazon River resorted to staging aquatic empates against commercial fishing vessels that were invading traditional reserve areas (Goulding, Smith, and Mahar 1996; Hall 1997a).

In addition to social organization, another form of capital is the TEK acquired by communities over generations, which has enabled them to live in a state of relative equilibrium with the environment. Although the Amazon rain forest at first appeared to early settlers and even government planners to be a virgin, untouched landscape, it has for centuries been sustainably managed by indigenous and other long-settled groups. Research has shed light on the subtle complexities of a vast range of techniques, developed in the fields of natural-forest management and agroforestry, which have supported quite large populations (Anderson 1990; Emperaire 1996). In the case of indigenous groups, these informal management systems were and still are, in many instances, intricately bound up with belief systems that placed nature and the need to preserve it at the center of the universe. However short-term economic interests almost invariably prevail over conservation per se. Even in the case of Amazonia's indigenous populations, the desire for profit (through the sale of logging and mineral-exploration rights, for example) has in some well-known and many less famous cases prevailed over other considerations. In the case of caboclos and smallholder settlers, livelihood and survival issues are paramount. As will be discussed below, social capital in the form of ecological knowledge is being adapted for the purpose of providing alternative strategies to deforestation associated with pasture formation and swidden agriculture.

Enhancing Social Capital

Existing social capital based on community organization and traditional knowledge is a necessary but not a sufficient prerequisite for implementing environmental management plans that attempt to reconcile economic use with resource conservation. The challenge lies in building upon that foundation to create a more effective tool for implementing a productive conservation policy. Community assets in the form of organization and knowledge contribute decisively toward the process. Although such arrangements are locale specific, conditioned by the nature of the ecosystem and social situation, those roles would, typically, include one or more of the following: (1) vigilance and monitoring of the resource base, (2) articulating local needs, (3) providing ecosystem knowledge inputs into research and development, and (4) cost sharing.

• *Vigilance and monitoring.* As already noted, Amazonian populations have enjoyed a degree of success in spontaneously defending their territorial integrity against external enemies, due in large measure to the effectiveness of stands taken against predatory commercial interests by rubber tappers, fisher people and small-scale farmers. Yet involvement in long-term resource governance requires more than guerrilla tactics. In order to systematically protect a given area, an infrastructure for permanent vigilance and monitoring must be established and maintained so that the appropriate agency (IBAMA) can be notified of infractions and legal sanctions imposed. Given the limited capacity of government to effectively police the environment, even with the assistance of sophisticated satellite technology, the most effective guardians of the forests and waterways are without doubt the local people themselves.

On federal extractive reserves, for example, locals are recruited as environmental monitors (*colaboradores fiscais*). It is their task to watch out for illegal activities such as logging and to notify IBAMA so that punitive action can be taken. On extractive reserves local vigilance has led to a significant reduction in illegal timber extraction. At Mamirauá, on the upper reaches of the Amazon, floating guard posts (*flutuantes*) placed at strategic entrance points to lake systems have been highly successful in keeping out commercial fishing boats (Hall 1997a). Providing basic communications equipment to the guard posts (such as solar-powered shortwave radios that link them with project headquarters and IBAMA offices) greatly enhances their effectiveness. Although not without their own problems, such vigilance structures build significantly upon the existing presence and ecosystem knowledge of local populations.

• *Needs articulation.* A second key role performed by local resource users is the articulation of their own interests, based upon their ecosystem knowledge and the need to insert these demands into the planning process. Conventionally, official planning in Brazil—and elsewhere for that matter—has not involved the direct consultation of impacted groups except in cases where grassroots political pressure has been considerable. Since Amazonian populations have not on the whole tended to be politically organized, nongovernmental organizations (NGOs) have frequently been instrumental in facilitating the process of articulation, acting as go-betweens for community

associations and external organizations such as government agencies and donors. This is evident in a number of diverse fields (Hall 1997a).

Amazonia's four federal extractive reserves (Chico Mendes and Alto Juruá
in Acre, Rio Ouro Preto in Rondônia, and Rio Cajarí in Amapá) together
cover more than two million hectares and are currently implementing
management plans with financial and technical support from the G7 Pilot
Program. Building upon existing social and organizational networks among
rubber tappers, decentralized community associations with locally elected
officers who often have prior rural union links have been set up on each reserve. Notwithstanding the innumerable obstacles to undertaking collective
action, those associations are beginning to facilitate the participation of the
population in designing and implementing initiatives in areas such as vigilance and the introduction of new production systems for rubber and other
forest products. A central association on the Mamirauá reserve in Amazonas,
which brings together representatives from nine sectors, performs a similar
function.

Needs articulation as a form of harnessing local ecosystem knowledge
need not be confined to "traditional" populations of caboclos and indigenous groups. Newer small-farm colonists from northeastern and southern
Brazil, whether arriving spontaneously or with the support of the settlement
agency INCRA (National Institute of Colonization and Agrarian Reform),
have often been painted as environmental criminals who adopt slash-and-
burn techniques with no regard for resource conservation. There is no reason to suppose that migrant farmers are inherently destructive, but they may
become so when their options are limited by factors such as land conflicts
or lack of access to official support (e.g., credit, marketing channels, seeds,
technical advice). The situation for migrant farmers is changing, however,
and there is now a growing number of new initiatives in forest management
and agroforestry offering alternative production technologies to small-scale
farmers which do not entail entering a vicious circle of deforestation. Along
the Transamazon Highway at Marabá and Altamira, small-scale settlers participate in programs that seek to help communities articulate their needs
through their own organizations. Yet there is an ever-present tension when
trying to reconcile the short-term demands of farmers wanting a quick economic return with the longer-term perspective needed for resource conservation (Hall 1997a; Smith et al. 1998).

• *Research and information.* In addition to vigilance and needs articulation, another stage in establishing an appropriate institutional framework for
resource management is the process of information gathering and research
that is based to some extent on local or traditional knowledge. TEK has, of

course, constituted the basis for informal management practices by Amazonian populations over many centuries. It is a sine qua non of the productive conservation approach to environmental management that TEK be as effectively harnessed as possible to provide technologies that are nondestructive of resources (RAFI 1994). An important planning prerequisite is an understanding of fragile ecosystems: how they function, the limits to extracting resources from them, their carrying capacity, and the viable resource-management technologies available to govern their resources. Understanding those factors represents a utilization and adaptation of TEK within a new context. For example, the ancient Amazonian practice of traditional agroforestry for subsistence purposes—which is based on forest enrichment, managed fallows, and home gardens—has provided valuable technological knowledge for the development of modern, commercial agroforestry (Smith et al. 1998; Smith 2000).

In the cases already cited, local knowledge has fed directly into research and project design. On extractive reserves, rubber tappers have informed the process of undertaking technical improvements to develop higher quality products. Communities at Mamirauá have collaborated with research scientists in mapping out resource distribution and ecosystem functions, terrestrial and aquatic. They have not only provided valuable insights into migratory movements of fish and aquatic mammals, for example, but have also been directly involved in drawing up (and enforcing) an internal zoning plan that demarcates areas for breeding as well as for subsistence and commercial uses (Hall 1997a; McGrath 2000). Similarly, small-scale settlers on the Transamazon Highway have provided agricultural researchers with information in an effort to develop more appropriate sedentary farming systems that reduce rates of deforestation (Hall 1997a; Smith et al. 1998).

• *Cost sharing.* Cost sharing by project beneficiaries is often dismissed as manipulative "development on the cheap." A counterargument to this is that cost sharing helps to avoid the dependence created by the paternalistic practices often adopted in the process of external funding. Notwithstanding that particular debate, it should be remembered that local communities are often obliged to bear heavy nonmonetary costs along the frequently conflictive path toward sustainable development. Community contributions of time and labor to the processes of policing and resource management, for example, are essential if such systems are to function properly. Land struggles have often been necessary to gain territorial stability for small-scale producers, thus effectively underpinning subsequent production initiatives. During the process, there has been a constant history of struggle accompanied by physical and psychological trauma, all representing a severe social cost. In a

broad sense, that cost is a direct consequence of the recognition that securing territorial integrity is a fundamental prerequisite for protecting people's
livelihoods and for promoting more sustainable patterns of development in
Amazonia.

Conclusions

This essay has attempted to show how local knowledge constitutes a form
of social capital that is playing a decisive role in the design and execution of
a range of new strategies for productive conservation in Amazonia. There are
many problems associated with this approach in terms of its organizational
and economic feasibility which have not been detailed here but which must
be borne in mind (Hall 1997a, n.d.). These include local populations' limited management capacity, their social fragmentation, and their sometimes
overdependence on external organizations such as NGOs. Other difficulties
concern the small and often volatile commercial markets for forest products
and the high transaction costs of setting up alternative production systems.
More time, effort, and money must be invested in strengthening organizational capacity, exploring and expanding new markets, and investing in
appropriate production and transport infrastructure to support productive
conservation.

As far as the harnessing of traditional knowledge for development purposes
is concerned, there are additional potential pitfalls. The use of such knowledge to serve a development agenda imposed largely by outsiders is clearly
an ever-present risk. Crude knowledge extraction by planners and policymakers to benefit other interests (regional, national, or global) at the expense
of local populations is a danger that should be fought against (Thompson
1996). A particularly pertinent case involves the intellectual property rights
of indigenous Amazonian and other traditional populations over biodiversity and the extraction of natural substances for use in the pharmaceutical
industry (Posey 2000). There is also the danger of exaggerating the inherent propensity of traditional groups to automatically adopt environmentally
friendly practices. Care must be taken not to romanticize the issue or project
an environmental concern onto populations whose major preoccupation is
essentially that of survival. The broader issue then becomes one of how to
provide appropriate organizational, management, and incentive structures
to minimize those risks and to encourage conservation alongside production
to meet people's livelihood needs.

Notwithstanding such problems, traditional knowledge clearly does have a future in the context of productive conservation as it increasingly complements more conventional forms of environmental management in Amazonia. Although the majority of grassroots project experiences are relatively embryonic, and their financial if not ecological sustainability has still to be proven, there can be no doubt that vital lessons are being learned. Furthermore, by helping to establish new and promising initiatives, such experiments pave the way for wider policy reforms. Grassroots movements such as those of the rubber tappers and inland fisher groups have been instrumental in prompting the reformulation of Brazil's protected-area policy under its new law on conservation units (SNUC). That law recognizes the integrated role of local populations in protected-area management and will help cater to the needs of many groups that previously fell outside the federal and state legislative frameworks. In addition, the government has introduced new fiscal instruments to support traditional populations in their economic activities. Yet a major future challenge facing policymakers is how to introduce new economic instruments — market and nonmarket, domestic and international — that will internalize the environmental benefits generated by traditional groups (Hall, n.d.).

It is important that such measures be expanded in order to provide a favorable policy context in which traditional knowledge can be put to maximum beneficial use for the purposes of promoting sustainable forms of development. Most productive conservation initiatives involve multi-institutional cooperation among a range of organizations like local communities, local and national governments, NGOs, the private sector, and international donors. The creation of synergy across public-private sectors (Evans 1996) is thus a fundamental part of this evolving process. The growth of such collaborative arrangements, in which each actor contributes in unique ways to development-conservation goals, is an encouraging sign that the beginnings of significant change may be under way.

References

Anderson, A. B., ed. 1990. *Alternatives to Deforestation: Steps Toward Sustainable Use of the Amazon Rain Forest*. New York: Columbia University Press.

Brazil. 1995. *Política Nacional Integrada para a Amazônia Legal*. Brasília: Ministry of the Environment.

Brazil. 1997. *Agenda 21 for Amazonia: Basis for Discussion*. Brasília: Ministry of the Environment.

Brazil. 2000. *Amazônia Sustentável*. Brasília: Ministry of the Environment.

Capobianco, J. P. 1996. Algumas questões relacionadas às unidades de conservão da Amazônia Legal Brasileira. In A. Ramos and J. P. Capobianco, eds., *Unidades de Conservão no Brasil: Aspectos Gerais, Experiências Inovadoras e a Nova Legislação (SNUC)*, pp. 17–27. São Paulo: Instituto Socioambiental (ISA).

Coleman, J. 1990. *Foundations of Social Theory*. Cambridge, Massachusetts: Harvard University Press.

Ehrlich, P. 1972. *The Population Bomb*. London: Ballantine.

Emperaire, L., ed. 1996. *La Forêt en Jeu*. Paris: Orstom/UNESCO.

Evans, P. 1996. Development strategies across the public-private divide. *World Development* (June) 24(6):1033–1037.

Foresta, R. 1991. *Amazon Conservation in the Age of Development: The Limits of Providence*. Gainesville: University of Florida Press.

Goulding, M., N. J. H. Smith, and D. J. Mahar. 1996. *Floods of Fortune: Ecology and Economy Along the Amazon*. New York: Columbia University Press.

Gross, A. 1989. *Fight for the Forest: Chico Mendes in His Own Words*. London: Latin America Bureau.

Hall, A. 1995. Towards new actions in social policies for sustainable development. Mimeo. Paper presented at UNDP meeting on Sustainable Human Development: Actions for New Generation Policies, Buenos Aires.

Hall, A. 1997a. *Sustaining Amazonia: Grassroots Action for Sustainable Development*. Manchester: Manchester University Press.

Hall, A. 1997b. Peopling the environment: A new agenda for research, policy and action in Brazilian Amazonia. *European Review of Latin American and Caribbean Studies* (June) 62:9–31.

Hall, A. 2000. Environment and development in Brazilian Amazonia: From protection to productive conservation. In A. Hall, ed., *Amazonia at the Crossroads: The Challenge of Sustainable Development*, pp. 99–114. London: Institute of Latin American Studies, University of London.

Hall, A. n.d. Extractive reserves: Building natural assets in the Brazilian Amazon. In J. K. Boyce, S. Narain, and E. A. Staunton, eds., "Reclaiming Nature: Worldwide Strategies for Building Natural Assets to Fight Poverty." Unpublished manuscript.

Hardin, G. 1968. The tragedy of the commons. *Science* 162:1243–1248.

McGrath, D. 2000. Avoiding a tragedy of the commons: Recent developments in the management of Amazonian fisheries. In A. Hall, ed., *Amazonia at the Crossroads: The Challenge of Sustainable Development*, pp. 171–187. London: Institute of Latin American Studies, University of London.

North, D. 1990. *Institutions, Institutional Change and Economic Performance*. Cambridge: Cambridge University Press.

Posey, D. 2000. Biodiversity, genetic resources and indigenous peoples in Amazonia. In A. Hall, ed., *Amazonia at the Crossroads: The Challenge of Sustainable Devel-*

opment, pp. 188–232. London: Institute of Latin American Studies, University of London.

Putnam, R. 1993. *Making Democracy Work: Civic Traditions in Modern Italy*. Princeton, New Jersey: Princeton University Press.

RAFI. 1994. *Conserving Indigenous Knowledge: Integrating Two Systems of Innovation*. New York: Rural Advancement Foundation International and UNDP.

Schmink, M. and C. Wood, 1992. *Contested Frontiers in Amazonia*. New York: Columbia University Press.

Smith, N. 2000. Agroforestry developments and prospects in the Brazilian Amazon. In A. Hall, ed., *Amazonia at the Crossroads: The Challenge of Sustainable Development*, pp. 150–170. London: Institute of Latin American Studies, University of London.

Smith, N., J. Dubois, D. Current, E. Lutz, and C. Clement. 1998. *Agroforestry Experiences in the Brazilian Amazon: Constraints and Opportunities*. Brasília: Pilot Program to Conserve the Brazilian Rain Forest.

Thompson, J. 1996. Moving the indigenous knowledge debate forward. *Development Policy Review* (March) 14(1):105–112.

WCED (World Commission on Environment and Development). 1987. *Our Common Future*. New York and Oxford: Oxford University Press.

Appendix

Findings and Recommendations

Findings	Recommendations
1. On the one hand, there is not enough interdisciplinary research and activity; on the other hand, there is too much uncoordinated and disconnected data from many disciplines and fields.	1. Interdisciplinary conferences and planning sessions should be encouraged among those whose aim is to compare, analyze, and integrate data and experiences from different disciplines and cultures. 2. The links between the biological and the social sciences should be encouraged and strengthened

* * * * *

Findings	Recommendations
2. Research and general understanding of local and traditional ecological knowledge (TEK) are inadequate and have not been related to ongoing international debates, e.g., the Convention for Biological Diversity (CBD) and the Food and Agricultural Organization of the U.N. (FAO).	1. Interdisciplinary field studies of TEK and local knowledge should be encouraged. 2. Databases of case studies and TEK should be developed. 3. Collaborative research led by indigenous and local communities should be given priority for funding. 4. Research initiatives should be linked to and directly address key issues on the wider use and application of "knowledge, innovations, and practices" of indigenous and local communities (Article 8.j of the CBD).

Findings	Recommendations
3. Integrated methodologies for interdisciplinary research and data evaluation are scant and insufficient to deal with complex economic, social, and ecological problems, especially in a region as large as Amazonia.	1. Emphasis should be given to developing field methodologies, as well as interdisciplinary approaches, for the collection and evaluation of data. 2. Priority should be given to implementing regional and global studies that compare TEK and local knowledge of different groups with knowledge based on Western science — especially knowledge relating to environmental health and monitoring.

* * * * *

4. Strategies seeking the wider use and application of local and traditional knowledge are weak and unsatisfactory	1. Priority should be given to applied research that employs local and traditional knowledge to the solving of regional, national, and global problems (e.g., forest management, agriculture, medicine, health) 2. There is an urgent need to develop adequate mechanisms for protection of TEK and traditional genetic resources, including encouraging its use, while guaranteeing equitable benefit sharing as required under the CBD.

* * * * *

5. Institutional and governmental priorities for funding, research, and policy are often contrary to well-established global initiatives (e.g., Agenda 21, CBD, U.N. Commission on Sustainable Development) and national policies (e.g., DFID White Paper) that guarantee support for local and indigenous peoples and the equitable use of traditional resources.	1. Funding priority should go to projects developed by local communities. 2. Emphasis should be put on projects that deal with restoration of degraded areas. 3. There should be increased research and emphasis on understanding local values, traditional assets, and community household structures in relation to economic exchange 4. Feasibility studies should be conducted to determine how to implement "environmental service payments" for communities that effectively conserve and manage natural resources and biodiversity

Contributors

Elizabeth Allen. Faculty of Social Sciences, Adam Smith Building, University of Glasgow, Glasgow G11 8QQ, Scotland, UK. gsia05@udcf.gla.ac.uk.

Michael J. Balick. Institute of Economic Botany, The New York Botanical Garden, Kazimiroff Boulevard & 200th Street, Bronx, NY 10458-5126, USA. mbalick@nybg.org.

Christopher Barrow. School of Environment, Environmental Biology & Society, University of Wales Swansea, Singleton Park, Swansea SA2 8PP, UK. c.j.barrow@swan.ac.uk.

Samuel Bridgewater. Royal Botanic Garden Edinburgh, 20A Inverleith Row, Edinburgh EH3 5LR, UK. lascuevas@btl.net.

Clóvis Cavalcanti. Institute for Social Research (INPSO), Fundação Joaquim Nabuco, R. Dois Irmãos, 92, 52071-440 Recife, PE, Brazil. clovati@fundaj.gov.br.

Charles R. Clement. Instituto Nacional de Pesquisas da Amazônia, Av. André Araújo, 2936–Aleixo, 69060-001 Manaus, Amazonas, Brasil. cclement@inpa.gov.br.

Philip M. Fearnside. National Institute for Research in Amazonia (INPA), C.P. 478, 69.011-970 Manaus, Amazonas, Brazil. pmfearn@inpa.gov.br.

Peter Furley. School of GeoSciences, University of Edinburgh, Edinburgh EH 8 9XP, Scotland. paf@geo.ed.ac.uk.

Anthony Hall. Department of Social Policy, London School of Economics, Houghton Street, London WC2A 2AE. a.l.hall@lse.ac.uk.

Mark Harris. Department of Social Anthropology, University of St Andrews, 71 North Street, St Andrews, KY16 9AL, Scotland. mh25@st-andrews.ac.uk.

John Hemming. Hemming Group Ltd., 32 Vauxhall Bridge Road, London SW1V 2SS, UK. j.hemming@hgluk.com.

William Milliken. Royal Botanic Gardens, Kew, Richmond, Surrey, TW9 3AB, UK. w.milliken@kew.org.

Stephen Nortcliff. Department of Soil Science, PO Box 233, The University of Reading, Reading, RG6 6DW, UK. s.nortcliff@reading.ac.uk.

Stephen Nugent. Department of Anthropology, Goldsmiths, University of London, New Cross, London SE14 6NW, UK. s.nugent@gold.ac.uk.

Joanna Overing. Department of Social Anthropology, University of St Andrews, St Andrews, Fife, KY16 9AL, Scotland. joanna.overing@st-andrews.ac.uk.

Christine Padoch. Institute of Economic Botany, The New York Botanical Garden, Kazimiroff Boulevard & 200th Street, Bronx, NY 10458-5126, USA. cpadoch@nybg.org.

Miguel Pinedo-Vásquez. Center for Environmental Research and Conservation, Columbia University, 1200 Amsterdam Avenue, New York, NY 10027, USA. map57@columbia.edu.

Claudio Urbano B. Pinheiro. Universidade Federal do Maranhão, Departamento de Oceanografia e Limnologia, Avenida dos Portugueses, s/n–Campus do Bacanga, CEP 65085-580 Sao Luis–Maranhão, Brasil. cpinheiro@elo.com.br.

Alcida Rita Ramos. Departamento de Antropologia, Instituto de Ciências Sociais, Universidade de Brasília, 70.910-900 Brasília, D.F., Brasil. arramos@unb.br.

James A. Ratter. Royal Botanic Garden Edinburgh, 20A Inverleith Row, Edinburgh EH3 5LR, UK. j.ratter@rbge.org.uk.

J. Felipe Ribeiro. Embrapa Cerrados, BR020 - km 18, CEP 73301, Planaltina, DF, Brazil. felipe@cpac.embrapa.br.

Michael Richards. Claywell Cottage, Aston Road, Ducklington, Witney OX29 7QZ, UK. emrichards@ntlworld.com.

Jan Salick. Missouri Botanical Garden, Box 299, St Louis, MO 63166-0299, USA. jan.salick@mobot.org.

Nancy Leys Stepan. Department of History, Fayerweather Hall 324, Columbia University, New York, NY 10027, USA. nls1@columbia.edu.

Index

Page numbers for boxes are indicated by b, figures by f, notes by *n*, and tables by t.